# ENGINEERING YOUR FUTURE

## A Project-Based Introduction to Engineering

**Alan G. Gomez, MS**
*University of Wisconsin & Madison West HS*

**William C. Oakes, PhD**
*Purdue University*

**Les L. Leone, PhD**
*Michigan State University*

### *Contributors*

Merle C. Potter, PhD
*Michigan State University*

Craig J. Gunn, MS
*Michigan State University*

Marybeth Lima, PhD
*Louisiana State University*

Heidi A. Diefes, PhD
*Purdue University*

### *Editor*

John L. Gruender
*Managing Partner & Editor*
*Great Lakes Press, Inc.*
*Education Resources Publisher*

**Great Lakes Press, Inc.**
Okemos, MI    St. Louis, MO
PO Box 550 / Wildwood, MO 63040
(636) 273-6016
custserv@glpbooks.com
www.glpbooks.com

International Standard Book Number: 1-881018-88-1 (hardcover)
International Standard Book Number: 1-881018-74-1 (softcover)

Copyright © 2004 by Great Lakes Press, Inc.

All comments and inquiries should be addressed to:

Great Lakes Press, Inc.
c/o John Gruender, Editor
PO Box 550
Wildwood, MO 63040-0550

jg@glpbooks.com
phone (636) 273-6016
fax (636) 273-6086

www.glpbooks.com

Library of Congress Control Number: 2004103274 (hardcover)
Library of Congress Control Number: 2004103269 (softcover)


Printed in the USA by Sheridan Books, Inc., Ann Arbor and Chelsea, Michigan.

10 9 8 7 6 5 4 3 2 1

## Important Information

**This book belongs to:** _____

_____

_____

_____

**phone:** _____

# Contents

# Preface

One of the most significant labor shortages confronting the United States pertains to technologically literate people. Every year the United States government accepts more and more people from foreign countries on work visas to place them in technology-related fields. The continued use of the H-1B visa program during one of the tech industry's most severe downturns has heightened renewed criticism of the program. It is a hot-button issue with many U.S. engineers who fear the country is giving away its tech jobs (Bjorhus, 2003).

Although we are doing more than we have in the past to give our students opportunities to become technologically literate, too often educators place students in front of computers and assume that technological literacy follows. When colleges use surveys to find out what kind of skills incoming freshman have, their skills, including computer skills, are much lower than expected. Baylor University reported that 24% of engineering students have some experience using CAD software, but that the expertise level, on a scale of one to five, was at one. Only 63% students had were familiar with PowerPoint, and then only at an expertise level of one out of five (DeJong, VanTreuren, Faris & Fry, 2001). While teachers need to teach content with computers, it can also be stated that educators can't teach with just computers. Students need to be given exposure to the creative nature of engineering through design projects, hands-on laboratories, and open ended problem solving (Sheppard & Jenison, 1996).

As budgets and, subsequently, programs are cut, parents, teachers, administrators, legislators, and taxpayers want to know that students are getting the best possible educational experience for their dollar. These "shareholders" are an integral part of education and their roles should not be slighted. "It's bound to hit K–12 education," says Jane Hannaway of the Washington-based Urban Institute, who is researching the ailing economy's impact on schools. "We're really beginning a significant, serious period of resource trouble" (Richard & Sack, 2003). Is it entirely subjective to determine what is a good course versus a bad one?

Parents are interested in an education for their children that focuses on more than just graduation requirements. They want to see their children enlightened to the variety of employment options available to them. Across the country, parents are pouring back into schools, questioning what goes on in their children's classrooms and pitching in to fill vacuums created by decreased funding (Walters, 1995). Teachers are also interested in offering academically viable courses and programs that provide highly valuable experiences for students, and not simply courses that "keep students busy for an hour."

Administrators are interested in effective curriculum and in involving their teachers in proactively educating students for the 21st century. During the course of a day a principal may engage in more than twenty-five meetings with parents, students, and teachers. Scheduled meetings coexist with the responsibility to deal with urgent problems that spring up at a moment's notice. While administrators are responsible for curriculum and programs, the great variety of demands on their schedule often prevents them from focusing on these matters until the deadline for determining allocation and scheduling.

Legislators are interested in the results of testing. Some of the courses and programs offered in schools fail to adequately cover the entirety of the content of the tests. One of the most significant questions in the minds of educators is whether courses are designed improperly (they do not directly contribute to better test scores, or fail to teach the precise material covered on the tests) or if the tests are designed improperly (whether they should include more reasoning, problem solving, or open ended creative questions).

Taxpayers are looking to get the best education for their tax money without wasting it on bureaucracy and programs. Presently, only about 25 percent of households have school-aged children—a historic low. Increasingly, therefore, taxpayers are looking at education as a financial investment that benefits other people's children (Labaree 1997, 62).

This book and its standards-based learning activities are organized around a set of concepts, skills, and attitudes which will help students become successful problem solvers in life, whether they pursue a career in engineering or not. Unfortunately, students in many schools can still matriculate having had no practical involvement with engineering-type problem solving concepts or case studies. Many students find the variety of choices in engineering schools daunting because they enter college with no real understanding of what engineering is. A major problem of secondary education is that schools teach science, technology, and mathematics only in the context of the specific disciplines.

This textbook addresses and solves that problem. It presents students with the major engineering concepts, and engages them in real-world case studies that resemble the problems that they would actually encounter in the field of engineering. The need for courses that stimulate interest in careers in engineering and technology has been apparent since the mid 1980's. The results of the Grinter Report of 1955 led to a change in the focus of curriculum from practical engineering-based to scientific-based, with more emphasis on theoretical approaches and less emphasis on the "machinery" of engineering (Sheppard & Jenison, 1996).

In engineering study, the transition from high school to college should become more seamless to significantly increase the likelihood of success and the retention of students in engineering. Many engineering schools are still dissatisfied with their ability to attract and retain engineering students. Many students continue to become discouraged during the first few semesters in engineering school due to a failure to adequately exposure them to actual engineering principles and design concepts. As a result, many transfer out of engineering.

Working with college students in the ME99 course at Stanford, a mechanical dissection course sponsored by the National Science foundation, it was realized that early exposure to engineering should ideally start well before the freshman year in college, if possible (Sheppard & Tsai, NDA). Criticisms leveled at US engineering schools include the offering of too few "practical" and "hands-on" courses (Sheppard & Jenison, 1996). These hands-on courses should be created and implemented with student success in mind. It is not logical to create large sections of courses in order to satisfy allocation issues in schools. This type of approach is taken at the expense of students. Laboratories employ active learning and a smaller class size to achieve two objectives: (1) to better inform students about the nature of engineering and its specific disciplines and (2) to improve these students retention in engineering (Hoyt & Oland, NDA). The first engineering schools in the United States used the laboratory as the primary mode of instruction (Durfee, 94).

Learning to navigate the road to the solution is just as important, if not more important, than finding the solution itself. Education needs to teach the learning process that students need to navigate that road. Typical classrooms have students sitting in rows, five across and six deep discouraging their talking with each other. Yet educators expect them to communicate and interact with people as a result. When the students graduate from high school, whether they go into the workforce, the military, two-year colleges or four-year institutions, they have to be able to work with others, in small groups and in teams. In a time of budget crisis or not, courses offered to high school students should not only be a contributing factor to their graduation, but to their lifelong success. There is an increasing need for students to not only to be successful in grade point averages, but to be creative thinkers and problem solvers.

# REFERENCES

Barker, Robert (8/19/2002–8/26/2002) The Art of Brainstorming. Business Week, Issue 3796, p168

Bjorhus, Jennifer (Sun, Jun 23, 2002) Layoffs at Sun prompt inquiry over work visas, Mercury News

DeJong, PhD., Nicole C., VanTreuren, PhD. Kenneth W., Faris, PhD., Donald R. and Fry, M.S., Cynthia C., Using Design To Teach Freshman Engineering Proceedings of the 2001 American Society for Engineering Education Annual Conference and Exposition

Daniel, Michelle (Jul/Aug 1995) When Creativity Counts. Women in Business, Vol. 47 Issue 4, p48

Durfee, W. K., (Feb/Mar 1994) "Design Education Gets Real," Technology Review, pp. 42–51

Gomez, Alan G. (Mar 2000) Engineering, but how? The Technology Teacher VOL 59 NO. 6 Pg 17–22. International Technology Education AssociationGomez, Alan G. (Mar 2003) Foundations of Technology International Technology Education Association and The Center to Advance The Teaching of Technology and Science (CATTS)

Grinter, L.E. (chair), 1955 Report on Evaluation of Engineering Education, ASEE, Washington, DC

Hedrick, J., The Freshman Engineering Course Balancing Act. Proceedings of the 2002 American Society for Engineering Education Annual Conference and Exposition

Hoyt, Marc, Department of Civil Engineering, and Oland, Matthew, Department of Chemical Engineering, University of Florida (No date available) The impact of a Discipline-Based Introduction to Engineering Course on Improving Retention

Kardos, Geza (Mar 1979) 'On Writing Engineering Cases', Proceedings of ASEE National Conference on Engineering Case Studies

Labaree, D. F., (1997) Public goods, private goods: The American struggle over educational goals. American Educational Research Journal 34:39–81

Lewis, Bob, Brainstorm no-brainer (Sep 30, 2002) InfoWorld, Vol. 24 Issue 39, p48, 1/2p

Moreno, Roxana, Mayer, Richard E., (1999) Gender Differences In Responding To Open-Ended Problem-Solving Questions. Learning & Individual Differences, 10416080, Vol. 11, Issue 4

Richard, Alan, Sack, Joettal, (Jan 8, 2003) States Brace For Tough New Year Education Week, 02774232, Vol. 22, Issue 16

Sheppard, Sheri & Jenison, Rollie, (Aug 1996) Freshman Engineering Design Experiences: an Organizational Framework. The International Journal of Engineering Education

Sheppard, Sheri D. Assistant Professor in the Department of Engineering at Stanford University & Tsai, June, Manufacturing Engineer at Hewlett Packard, (No date available) A Note on Mechanical Dissections with Pre-college Students

Verespej, Michael A., (May 7, 2001) VANISHING BREED Industry Week/IW, 00390895, Vol. 250, Issue 7

Walters, Laurel Shaper, (Feb 23, 1995) Christian Parents finding way back to schools Science Monitor, 08827729, Vol. 87, Issue 61

Wisconsin's Model Academic Standards. (1998). Wisconsin Department of Public Instruction. http://www.dpi.state.wi.us/standards/index.html

Wisconsin's Governor's Work-Based Learning Board WBL-10234 (R. 10/2000) Skills standard checklist

# Chapter 1

# Developmental Timeline for Engineering

## Introduction

The progress of engineering is a continuum, where every new innovation leverages previous innovation. The study of history is the best way to understand where we have been, where we are, and where we are going.

Engineering innovations were few and far between in the early years. However as time passed, innovations occurred more frequently. Today, engineering advances are made almost daily. The speed with which things now change indicates the urgency of understanding the process of innovation. For example, in 1965, Gordon Moore made an important observation regarding computer chips. When he started to graph data about the growth in memory chip performance, he realized there was a striking trend. Each new memory chip contained roughly twice as much capacity as its predecessor, and each chip was released within 18-24 months of the previous chip. If this trend continued, Moore reasoned, computing power would rise exponentially over relatively brief periods.

Moore's observation, now known as Moore's Law, described a trend that has continued and is still remarkably accurate. It is the basis for many computer performance forecasts. Over a 26-year period, the number of transistors on a chip has increased more than 3,200 times, from 2,300 transistors on the 4004 processor chip in 1971 to 7.5 million on the Pentium II processor. This is an excellent example of technology driving developments in science.

If a big-picture awareness regarding projects is not maintained, unintended consequences may occur. History provides a great opportunity to observe the context of before, during, and after of some of the greatest engineering problems ever faced, including those of today.

History is not about memorizing names and dates. History is about people and accomplishments. The insights that can be gained from stories about how engineers developed products can be very motivating, informative, and enlightening. These stories provide the foundation needed to learn and to design in the future.

Many types of professionals are required to study the history of their trade as part of their degree. Professional development is a significant wedge factor between those who advance in their careers, and those who stay in a rut, never advancing much beyond their entry position in their field. Successful professionals continue the educational process throughout their careers. Reading published papers, professional magazines, and journals builds on a foundational knowledge of the history of their field.

> "What news! How much more important to know what that is which was never old!"
> —Henry David Thoreau

The study of history not only helps us create new futures, but it also helps us understand what good qualities from the past are worth emulating. Craftsmanship, integrity, and dedication are clearly evident within engineering artifacts. History is also full of interesting, educational stories, characters, and clever developments.

## Definition of Engineering

Even if you already have a general knowledge of what engineering involves, a look at the definition of the profession may give you more insight. ABET (The Accreditation Board for Engineering and Technology) defines engineering as:

> "The profession in which knowledge of the mathematical and natural sciences, gained by study, experience, and practice, is applied with judgment to develop ways to use, economically, the materials and forces of nature for the benefit of mankind."

This definition places three responsibilities on an engineer: (1) to develop judgment so that you can (2) help mankind in (3) economical ways. A good way to develop the insight needed to fulfill these responsibilities is to look at case histories and historic overviews.

An engineering professor once said that an integral function of an engineer is to interpret the development and activity of mankind. Technical coursework teaches us skills, but history can teach us how to interpret scenarios, and how to sort out the pros and cons of various available options. History helps us forge bonds of fellowship that connect us to the past and inspire us to be our best for tomorrow.

## Getting Started

The engineer has always had a monumental impact on the human race at every stage of societal development. The few items mentioned in this chapter are only highlights of contributions that engineers have made to the progress of humanity.

## Prehistoric Culture

ABET's definition of engineering has within it an important statement: The profession in which knowledge of the mathematical and natural sciences . . . is applied. . . . Prehistoric people engaged in activities which we recognize today as engineering problem-solving, tool-making, etc., but they did not have the grasp of mathematical principles or the knowledge of natural science have today. These cultures and civilizations designed and built needed items by trial and error, using mere intuition. They built some spears that worked and some that failed, but in the end they perfected weapons that allowed them to kill game animals and feed their families. Because there was almost no written communication and only primitive transportation, little information or innovation was exchanged with other cultures. Each civilization inched ahead on its own.

However, many of these prehistoric innovators would have made fine engineers today. Even in light of their limited skill, their knowledge of their surroundings was more extensive than we, today, typically realize. Their skill in craftsmanship was often marvelous in its effectiveness, integrity, and intricacy. They passed on knowledge of all aspects of life with utmost seriousness to the next generation. This information was carefully memorized and kept accurate, evolving, and alive. Early man even tried to pass on vital information by way of coded cave paintings and etchings. Breakthroughs in transportation and exploration are being identified as having occurred ever earlier as we continue to make new discoveries of various peoples traveling great distances long before we thought possible, influencing others far away and bringing back novel ideas.

However, despite these positive traits, those in prehistoric times appear to have faced a harsh physical existence. The work of engineers can be seen in this light as a quest to expand outer capacity without sacrificing inner integrity. The physical limitations of prehistoric man can perhaps be highlighted as follows:

- They had no known written language.
- Their oral language was very limited.
- They had no significant means of transportation.
- They had no concept of formal education nor a specialized methodology for discovery.
- They survived by gathering food and by trying to kill game with primitive weapons.
- Improvement of the material aspects of life came about very slowly, by way of early, primitive engineering.

We will present a panoramic view of engineering by briefly stating some of the more interesting occurrences during specific time periods. Notice the kinds of innovations that were introduced. Take a careful look at the relationships that many inventions had with each other. Think about the present, and the connectivity between all areas of engineering and the critical importance of the computer. Innovations do not happen in a vacuum; they are interrelated with the needs and circumstances of the world at the time.

## The Speed of History

The rate of innovation involves some interesting points. As we contemplate the past 6000 years, you should realize that the rate at which we currently introduce innovations is far more rapid than in the past. It used to take years to accomplish tasks that today we perform in a very short time—tasks that we simply take for granted. In the past, decades could pass without noticeable technological progress. Think of the amount of time that it takes to construct a building today with the equipment that has been developed by engineers. It is not uncommon to see a complete home framed in a single day. However, there are churches that still stand today that took as long as 200 years to construct at the time they were built.

The proper connection to God, truth, justice, fate, reality, life, and ancestry was the goal of early science in many cultures. Not much was allowed to interfere with those quests. Even so, the ancients accomplished fantastic feats with only a rudimentary knowledge of the principles that we now learn as children. Some of this knowledge and the speed with which it was introduced or put into practice may have been purposely held back. Archimedes refused to release information that could have been used to make more effective weapons. He feared it would be used for evil and not the pursuit of wisdom. Only when his home city of Syracuse was no longer able to hold off Roman attackers did he release his inventions to the military. The study of history confronts us with dilemmas. Neither side of a true dilemma ever goes away. The story of history is never over.

Our technological roots can be traced back to the seed gatherer/hunter. Prehistoric people survived by collecting seeds and by killing whatever animals they could. They endured a very meager physical existence. As they gradually improved their security through innovation, humanity increased in numbers, and it became important to find ways to feed and control the growing populations. To support larger populations reliably, the methods and implements of farming and security were improved. Much later, after many smaller innovations, specialized industrial cultures came onto the scene, ready to introduce the world to increased productivity and greater material wealth.

## ENGINEERING'S BEGINNINGS

### The Earliest Days

The foundations of engineering were laid with our ancestors' efforts to survive and to improve their quality of life. From the beginning they examined their environment and saw areas where life could be made easier and more stable. They discovered improved ways to hunt and fish. They developed better methods for providing shelter for their families. Their main physical concern was day-to-day survival. As life became more complicated and small collections of families became larger communities, new concerns arose: power struggles, the acquisition of neighboring tribes' lands, more complex religious observances. All of these involved some work with tools. Engineering innovations were needed to further these interests. Of course, in those days projects weren't thought of as separate from the rest of life. In fact, individuals weren't generally thought of as being separate from their community. They didn't look at life from the point of view of specialties and individual interests. Every person was an engineer, to an extent.

Modern aborigines still live today much as their ancestors did in prehistoric times. However, frequently, even they take advantage of modern engineering in the form of tools, motors, and medicine. The Amish can be included in this category of roots-oriented culture. Such cultures tend to use tools only for physical necessities so that the significance of objects doesn't pollute their way of life.

## Egypt and Mesopotamia

As cities grew and the need for addressing the demands of the new fledgling societies increased, a significant change took place. People who showed special aptitude in certain areas were identified and assigned to ever more specialized tasks. This labeling and grouping was a scientific breakthrough. It gave toolmakers the time and resources to dedicate themselves to building and to innovation. This new social function created the first real engineers, and for the first time innovation flourished rapidly.

Between 4000 and 2000 B.C., Egypt and Mesopotamia were the focal points for engineering activity. Stone tools were developed to help in the quest for food. Copper and bronze axes were perfected through smelting. The development of the plow was allowing man to become a farmer so that he could leave the nomadic life and reside in one place. Mesopotamia also made its mark on engineering by giving birth to the wheel, the sailing boat, and methods of writing. Engineering skills that were applied to the development of everyday items immediately improved life as they knew it.

We will never fully appreciate the vast importance of the Greeks, Romans, and Egyptians in the evolution of engineering.

During the design and construction of the Pyramids (c. 2700–500 B.C.) the number of engineers required was immense. They had to make sure that that stones were properly transported long distances, that stones could be effectively lifted into place at great heights, that everything fit correctly, and that the tombs would be secure against robbery. Imhotep (chief engineer to King Zoser) was building the stepped pyramid at Sakkara (pictured in Fig. 1.1) in Egypt about 2700 B.C. The more elaborate Great Pyramid of Khufu (pictured in Fig. 1.2) would come about 200 years later. By investigating the construction of the Pyramids, we learn about the need for designing, building, and testing, as with any engineering project. These early engineers, using simple tools, performed with great acuity, insight, and technical rigor, tasks that even today give us a sense of pride in their achievements.

The Great Pyramid of Khufu (pictured in Fig. 1.2) is the largest masonry structure ever built. Its base measures 756 feet on each side. The 480-foot-tall structure was constructed of over 2.3 million limestone blocks with a total weight of over 58 million tons. Casing blocks of fine limestone were attached to all four sides. These casing stones, some weighing as much as 15 tons, have been removed over the centuries for a wide variety of other uses. It is hard for us to imagine the engineering expertise needed to quarry and move these base and casing stones, and then piece them together so that they would form the pyramid and its covering.

**Figure 1.1 The stepped Pyramid of Sakkara.**

**Figure 1.2 The Great Pyramid of Khufu.**

Here are additional details about this pyramid given by Roland Turner and Steven Goulden in Great Engineers and Pioneers in Technology, Volume 1: From Antiquity through the Industrial Revolution:

> *"Buried within the pyramid are passageways leading to a number of funeral chambers, only one of which was actually used to house Khufu's remains. The granite-lined King's Chamber, measuring 17 by 34 feet, is roofed with nine slabs of granite which weigh 50 tons each.*

*To relieve the weight on this roof, located 300 feet below the apex of the pyramid, the builder stacked five hollow chambers at short intervals above it. Four of the "relieving chambers" are roofed with granite lintels, while the topmost has a corbelled roof. Although somewhat rough and ready in design and execution, the system effectively distributes the massive overlying weight to the sturdy walls of the King's Chamber.*

*Sheer precision marks every other aspect of the pyramid's construction. The four sides of the base are practically identical in length—the error is a matter of inches—and the angles are equally accurate. Direct measurement from corner to corner must have been difficult, since the pyramid was built on the site of a rocky knoll (now completely enclosed in the structure). Moreover, it is an open question how the builder managed to align the pyramid almost exactly north-south. Still, many of the techniques used for raising the pyramid can be deduced.*

*After the base and every successive course was in place, it was leveled by flooding the surface with Nile water, no doubt retained by mud banks, and then marking reference points of equal depth to guide the final dressing. Complications were caused by the use of blocks of different heights in the same course."*

The above excerpt mentions a few of the fascinating details of the monumental job undertaken to construct a pyramid with primitive tools and only human labor.

## Outline of Engineering History

As we enter the modern age, the scope of invention appears to narrow, with much of the activity relating to computers. What is the role of the inventor in history? Sometimes the name is important, other times not. At times during certain eras, innovations were being made simultaneously by a number of people. So no individual really stands out. Perhaps the developments were more collective during such times. Perhaps the players involved were often racing each other to the patent office. At other times, with certain inventions, a single person made a significant breakthrough on their own.

Below is an outline containing many of the most significant engineering innovations. Be encouraged to research on your own any which are of particular interest to you.

### 1200 B.C.–A.D. 1

- The quality of wrought iron is improved
- Swords are mass produced

- Siege towers are perfected
- The Greeks develop manufacturing
- Archimedes introduces mathematics in Greece
- Concrete is used for the arched bridges, roads, and aqueducts in Rome

## A.D. 1–1000

- The Chinese further develop the study of mathematics
- Gunpowder is perfected
- Cotton and silk are manufactured

## 1000–1400

- There is growth in the silk and glass industries
- Leonardo Fibonacci (1170–1240), medieval mathematician, writes the first Western text on algebra

## 1400–1700

- Georgius Agricola's De Re Metallica, a treatise on mining and metallurgy, is published posthumously
- Federigo Giambelli constructs the first time bomb for use against Spanish forces besieging Antwerp, Belgium
- The first water closet (toilet) is invented in England
- Galileo begins constructing a series of telescopes, with which he observes the rotation of the sun and other phenomena supporting the Copernican heliocentric theory
- Using dikes and windmills, Jan Adriaasz Leeghwater completes drainage of the Beemster-meer body of water, the largest project of its kind in Holland (17,000 acres)
- Otto von Guericke, mayor of Magdeburg, first demonstrates the existence of a vacuum
- Christian Huygens begins work on the design of a pendulum-driven clock
- Robert Hooke develops the balance spring used to power watches
- Charles II charters the Royal Society, England's first organization devoted to experimental science
- Isaac Newton constructs the first reflecting telescope
- Work is completed on the Languedoc Canal, the largest engineering project of its kind in Europe at that time
- Thomas Savery patents his Miners Friend, the first practical steam pump
- The agriculture, mining, textile, and glassmaking industries are expanded
- The concept of the scientific method of invention and inquiry is originated
- The humanities and science are first thought to be two distinctly separate entities

- Robert Boyle finds that gas pressure varies inversely with volume (Boyle's Law)
- Leibniz makes a calculating machine to multiply and divide

## 1700–1800

- The Leyden jar stores a large charge of electricity
- The Industrial Revolution begins
- James Watt makes the first rotary engine
- The instrument-maker Benjamin Huntsman develops the crucible process for manufacturing steel, improving quality, and sharply reducing cost
- Louis XV of France establishes the Ecole des Ponts et Chausses, the world's first civil engineering school
- John Smeaton completes construction of the Eddystone lighthouse
- James Brindley completes construction of the Bridgewater Canal, beginning a canal boom in Britain
- James Watt patents his first steam engine
- The spinning jenny and water frame, the first successful spinning machines, are patented by James Hargreaves and Richard Ark Wright, respectively
- Jesse Ramsden invents the first screw-cutting lathe, permitting the mass production of standardized screws
- The Society of Engineers, Britain's first professional engineering association, is formed in London
- David Bushnell designs the first human-carrying submarine
- John Wilkinson installs a steam engine to power machinery at his foundry in Shropshire, the first factory use of the steam engine
- Abraham Darby III constructs the world's first cast iron bridge over the Severn River near Coalbrookdale
- Claude Jouffroy d'Abbans powers a steamboat upstream for the first time
- Joseph-Michel and Jacques-Etienne Montgolfier construct the first passenger-carrying hot air balloon
- Henry Cort patents the puddling furnace for the production of wrought iron
- Joseph Bramah designs his patent lock which remains unpicked for 67 years
- British civil engineer John Rennie completes the first building made entirely of cast iron

## 1800–1825

- Automation is first used in France
- The first railroad locomotive is unveiled
- Jean Fourier, French mathematician, states that a complex wave is the sum of several simple waves
- Robert Fulton begins the first regular steamboat service with the Clermont on the Hudson River in the U.S.

- Chemical symbols as they are used today are developed
- The safety lamp for protecting miners from explosions is first used
- The single wire telegraph line is developed
- Photography is born
- Electromagnetism is studied
- The thermocouple is invented
- Aluminum is prepared
- Andre Ampere shows the effect of electric current in motors
- Sadi Carnot finds that only a fraction of the heat produced by burning fuel in an engine is converted into motion. This forms the basis of modern thermodynamics

## 1825–1875

- Rubber is vulcanized by Charles Goodyear in the United States
- The first iron-hulled steamer powered by a screw propeller crosses the Atlantic
- The rotary printing press comes into service
- Reinforced concrete is used
- Isaac Singer invents the sewing machine
- George Boole develops symbolic logic
- The first synthetic plastic material celluloid is created by Alexander Parkes
- Henry Bessemer originates the process to mass-produce steel cheaply
- The first oil well is drilled near Titusville, Pennsylvania
- The typewriter is perfected
- The Challenge Expedition (1871–1876) forms the basis for future oceanographic study

## 1875–1900

- The telephone is patented in the United States by Alexander Graham Bell
- The phonograph is invented by Thomas Edison
- The incandescent light bulb also is invented by Edison
- The steam turbine appears
- The gasoline engine is invented by Gottlieb Daimler
- The automobile is introduced by Karl Benz (see photo below)

## 1900–1925

- The Wright brothers complete the first sustained flight
- Detroit becomes the center of the auto industry
- Stainless steel is introduced in Germany
- Tractors with diesel engines are produced by Ford Motor Company
- The first commercial airplane service between London and Paris commences
- Diesel locomotives appear

## 1925–1950

- Modern sound recordings are introduced
- John Logie Baird invents a high-speed mechanical scanning system, which leads to the development of television
- The Volkswagen Beetle goes into production
- The first nuclear bombs are used
- The transistor is invented

## 1950–1975

- Computers first enter the commercial market
- Computers are in common use by 1960
- The first artificial satellite, Sputnik 1 (USSR), goes into space
- Explorer I, the first U.S. satellite, follows
- The laser is introduced
- Manned space flight begins
- The first communication satellite, Telstar, goes into space
- Integrated circuits are introduced
- The first manned moon landing occurs

## 1975–1990

- Supersonic transport from U.S. to Europe begins
- Cosmonauts orbit the earth for a record 180 days
- The Columbia space shuttle is launched and reused for space travel
- The first artificial human heart is implanted

## 1990–Today

- The Hubble Space Telescope (HST) is carried into orbit in 1990 in the space shuttle Discovery. Built from1978–1990, the HST cost $1.5 billion
- The Internet Society is chartered, and 1,000,000 host computers are connected in a network
- Computer processor speed is dramatically improved
- The Channel Tunnel (the "Chunnel") between England and France is completed
- The first rendezvous of a NASA spacecraft with the Russian Mir space station occurs
- MP3 audio format traded widely though computer servers
- World's new tallest building (1,483 feet) opens in Kuala Lumpur, Malaysia
- Global Positioning Satellite (GPS) technology is declassified, resulting in hundreds of safety, weather, and consumer applications
- High Definition Television signals / products become available
- The genetic code of a human chromosome is mapped
- Robots walk on Mars

## LEARNING ACTIVITIES

### Activity 1.1    Moore's Law

Time:    3–5 hours

In this activity, students will research the development of the memory chip as well as forecast the future of processor technology. (A good start would be to look into molecular electronics.) Have students prepare detailed information based on the timeline for processor chip development and advancements in new technologies. Students should identify other significant technological developments that have a direct association with memory chips. Students can present findings orally and in display format to the entire class.

In 1965, Gordon Moore made an important observation regarding computer chips. When he started to graph data about the growth in memory chip performance, he realized there was a striking trend. Each new memory chip contained roughly twice as much capacity as its predecessor, and each chip was released within 18-24 months of the previous chip. If this trend continued, Moore reasoned, computing power would rise exponentially over relatively brief periods.

Moore's observation, now known as Moore's Law, described a trend that has continued and is still remarkably accurate. It is the basis for many computer performance forecasts. Over a 26-year period, the number of transistors on a chip has increased more than 3,200 times, from 2,300 on the 4004 processor chip in 1971, to 7.5 million on the Pentium II processor. This is an excellent example of technology driving developments in science.

## RESOURCES

http://www.intel.com/intel/museum/25anniv/hof/moore.htm. This site presents Intel's Processor Hall of Fame, and presents an overview of the development of their processors.

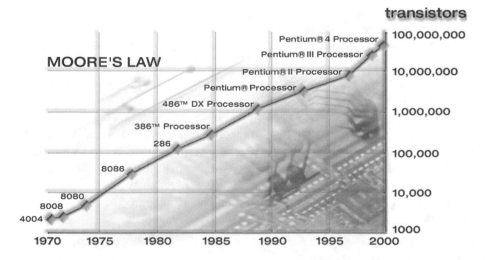

PRISM Magazine, a publication of the American Society of Engineering Educators (ASEE), November, 2000, *Quantum Leap.* By Corinna Wu. As silicon chips are pushed closer to their technological limits, engineers are moving to the molecular level to deliver the next step in computing speed.

## Activity 1.2   VHS vs. BETA

Time:      1–2 hours

In this activity, students will trace the development of the videotape. Students should evaluate each product and present their findings in a brief report. Students will trace the origin of a product's development and develop a rating scale for evaluating the product against its competition. The teacher may help students develop a rating scale, possibly discussing criteria with the entire class. Students should answer the following questions: 1. What were the pros and cons of each format? 2. Why did one product prevail over the other? A written or oral presentation may also be required.

Shortly after SONY introduced the format Betamax (BETA), Victor Company of Japan (Japan Victor or JVC) and its parent company Matsushita Electric introduced their own format called VHS (Video Home System). JVC's VHS machine also used U-Matic technology, although the recording format, tape-handling mechanisms, and cassette sizes differed. By 1977, these two formats were battling for supremacy in a market that was so large that neither format could serve all of it.

Although VHS was growing in popularity, BETA enjoyed steady sales until 1985. SONY introduced new features like the wireless remote control, half-speed and one-third speed machines, multi-function machines (scan, slow, still), high-fidelity (hi-fi) sound, and camcorders. However, JVC and other VHS manufacturers would quickly follow SONY's lead. SONY failed to get important company allies on its side when it first introduced BETA; companies were initially interested but were waiting until the technology was proven and accepted by consumers. The length of the videotape was another problem. BETA's one-hour recording length was half that of the VHS format, and consumers preferred the two-hour tapes.

## RESOURCES

http://www.digitalcentury.com/encyclo/update/sony.html. This site overviews the digital century and gives a good perspective on where VHS and BETA had their battles.

Cusumano, Michael A.; Mylonadis, Yiorgos; and Rosenbloom, Richard S.; "Strategic Maneuvering and Mass-market Dynamics: The Triumph of VHS over BETA"; *Business History Review;* Spring 1992.

http://cybercollege.com/tvp048.htm. This site has a brief history of the BETA format and where it is today.

http://www.urbanlegends.com/products/beta_vs_vhs.html. A good timeline of the video format battle is presented on this website. There are many additional resources such as magazines and articles that the instructor may use in the classroom.

## REFERENCES

A History of Technology, Volume II, Ed. Charles Singer, Oxford University Press, New York, 1956.

Burghardt, M. David, Introduction to Engineering, 2nd Ed., New York, Harper-Collins, 1995.

Burstall, A., A History of Mechanical Engineering, London: Faber, 1963.

De Camp, L. Sprague, The Ancient Engineers, Cambridge, The MIT Press, 1970.

Great Engineers and Pioneers in Technology, Volume 1: From Antiquity through the Industrial Revolution, Eds. Roland Turner and Steven L. Goulden, New York, St. Martin's Press, 1981.

Kirby, R., et al., Engineering in History, New York, McGraw-Hill, 1956.

Miller, J. A., Master Builders of Sixty Centuries, Freeport, New York, Books for Libraries Press, 1972.

Red, W. Edward, Engineering The Career and The Profession, Monterey, California, Brooks/Cole Engineering Division, 1982.

*Chapter 2*

# The History of the Early Engineering Disciplines

As a prelude to the in-depth look at the engineering fields, which we provide in the Engineering Disciplines chapter later in the book, this section takes a brief look at the historic backgrounds of Civil, Industrial, and Mechanical Engineering.

## The History of Civil Engineering

The need for boundaries and surveys precipitated the evolution of the field of civil engineering. Noting and marking the foundations of monumental structures and accurately dividing land into parcels were the job of a surveyor. Because conventional landmark noting was not accurate enough, the ancient Egyptians used surveying to predict and mark the location of the Nile River flood waters. This led to the development of special instruments for measuring. The Romans learned from the Egyptians and the Greeks that it was important to document and measure land. They also laid out aqueducts and designed roads though accurate surveying.

After the fall of the Roman Empire, the Arabic people developed proficiency with the astrolabe, a fixed surveying method linked to the stars. This culture was also responsible for the development of trigonometry and the practice of using triangulation in order to achieve accuracy. Though significant scientific and mathematical knowledge was maintained in Arabic civilizations, the advancements of this non-Western culture have not always received clear attributions in the West. Michael Chamberlain, Associate Professor of History at the University of Wisconsin-Madison writes, "Islamic science merged three relatively independent scientific, mathematical, and engineering cultures: the Greek, the Near Eastern, and the Indian. This had obvious consequences in the domains of mathematics and astronomy, where the introduction of the zero enabled Muslim mathematicians to develop algebra and to describe mathematically astronomical phenomena that had been observed for two millennia in the Near East."

The civil engineering profession came about because of the need to recognize the difference between military engineers and other engineers. In the 18th century, Europe and the United states recognized those who could complete large-scale projects as civil engineers. In 1782, John Smeaton of England molded himself as a civil engineer, the first to do so. In 1793, a society of civil engineers was created in England. Their name was later changed in 1818 to the "Institution of Civil Engineers." After 1824, in the United States, many engineers became part of the Franklin Institute in Philadelphia. This was merely an informal society of engineers. The American Society of Civil Engineers (ASCE) formed November 5, 1852, and is still present and active today.

## BRIDGES

Bridges can be classified in three basic classes. The *beam bridge* was often made of wood in early times and its more modern variation is the cantilevered beam. A *suspension bridge* was originally constructed from rope in the early days, and currently exists as the cable stayed model. An *arched bridge* was typically made from stone or brick.

The first known written description of a bridge is from 2000 B.C. The bridge was made of wooden timber beams and built on stone pillars in order to cross the Euphrates River. The design was commissioned for Babylon by Queen Semiramis. She gathered workers from the entire known world to work on this project, suggesting that bridge building was a widely known skill and practice at that time. The beginning of the Golden Era of Roman stone-arch bridge and aqueduct construction began in 300 B.C. Hellenistic (Greek) "invented" the wooden truss in 250 B.C. for construction purposes. A brief history of some of the greatest achievements in bridge building follows.

The Zhaozhou (Anji) Bridge was constructed in 610 A.D. and is the world's oldest known open-spandrel stone-arch bridge. Its courageous flat arch reduced the overall height of the structure. In 1697 the Frankford Avenue Bridge that spans the Pennypack Creek in Philadelphia, Pennsylvania, was constructed. It was the first known stone arch bridge built in the United States, and remains in use today.

The first known pile supported highway bridge was Sewall's Bridge, built over the York River in York, Maine. The piles were driven into the river bottom by standing them in place while heavy oak logs were dropped on them. It was approached technically, built with engineering plans and site surveys. The bridge was replaced with a similar structure in 1934 and remains in use.

The world's first all-metal bridge was fittingly named the "Iron Bridge," built of cast iron. This structure was designed by Abraham Darby III, and spans the Severn River near Coalbrookdale, England. The main span is 30.5 m (100 feet), its total length is 60 m (196.85 feet), and the weight of the iron used is 378.5 tons.

The world's first modern suspension bridge was the Jacob's Creek Bridge in Uniontown, Pennsylvania. James Finley designed and built this bridge which cost just $600 in 1801. Built using iron chains and a stiffened floor system, this bridge barely outlasted the life span of its creator. James Finley died in Uniontown in 1828 and the bridge was demolished in 1833.

## DAMS

Dams must take into account many factors when they are designed. First, dams must be strong enough to resist the reservoir water that is backed up behind it.

When a dam is constructed on a river, typically a pool of water builds up behind it and forms a man-made lake. An example of this type is Lake Meade behind Hoover Dam. A dam must also be impervious to water, resisting leaks and erosion. The early dams, like the Sadd el Kafara, an ancient Egyptian structure, were built with earth and rock. This oldest-known dam stood 62 feet high and had to take into account what water would do to the earthen materials that it was made of. Dams must be constructed so that water cannot find its way into the structure. The greatest historical example of such a failure was when the 43-foot-high Mill River Dam, nine years old at the time, failed due to uncontrollable seepage. This failure drew public outrage, and from then on it was determined that there was a need for professional engineering in dam construction.

A dam must also accommodate overflow. If a river begins to supply more water than the dam can hold, an active or passive system must prevent that water from building up too much pressure on the structure, or rising over the structure—either of which could lead to catastrophic failure. The level of the water behind a dam and the water controlled through a dam has to be regulated by a discharge system with gates. The Assyrian King Sennacherib built dams in 700 B.C. that diverted the river Khost in order to supply Nineveh with water. Masonry arch dams built 800 years later by the Romans in Greece and North Africa probably employed these same systems, as the dams' primary functions were to divert water and store it.

## ROADS

Early on, markings were used to designate where travel paths were so that people knew which path and direction to take to get to their desired destinations. With the invention of the wheel, roadways began to appear, and evolved into paved surfaces with drainage systems to divert water off of them. In 3000 B.C., the Herappa and Mohenjo-Daro civilizations in the Indus valley developed paved streets with a drainage system underneath the pavement. The Lake Moeris Quarry Road, the world's oldest paved road (2500 B.C.), was eight miles long, but only four miles of it remain today. The Romans built their first paved road in 312 B.C. and it extended from Rome to Capula—some 130 miles. A century and a half later, in 144 B.C., the Romans built the first high-level aqueduct and introduced hydraulic cement into its design. The Romans constructed over 372 major roads running a combined distance of 53,000 miles. When the Romans retreated from Britain, the roadways suffered. 600 years later, with the Norman invasion, it was realized that a good road system was of considerable value. The church had the greatest influence and responsibility for the maintenance and construction of inns and other accommodations where travelers could rest. This remained the case until Henry VIII adopted the position of Supreme Head of the Church in England in 1534 and dissolved the monasteries.

The 14th century saw the creation of the Royal Road in the Inca Empire. It was 300 miles long, but not intended for use by wheeled vehicles. In the 1500's, Spaniards and Colonials developed the El Camino Real. This trail was the first European inland transportation route in the continental United States. Its original use was for military and political use only. This may be why it is considered the beginning of the interstate highway system, as the El Camino Real symbolizes the importance of regional and national networks of roadways for political, social, defense, and economical benefit.

In 1662, Paris saw omnibuses become a part of its transportation system. These omnibuses constituted the world's first horse-drawn public transportation. It was also operated on a regular schedule so people could plan their travel. It wasn't until the later 1700's that the United States saw the rapid development of large-scale road projects. In 1775, the Kings Highway was established as a military route linking St. Augustine, Florida, with Barrington, Georgia. The first toll road in the US was established in Virginia in 1785. In 1794, the private Lancaster turnpike (Philadelphia) became the first paved (gravel and stone) rural road in the US. The first road in the world used only by motorized vehicles was not built until 1914. The 40.5-mile single-lane stretch was built on Long Island, New York. It wasn't until 1921 that a double lane carriageway, with no access along the way, was experimented with just west of Berlin, Germany.

The early roadways had much in common with today's roadways. However, the sub-surface materials and the surface materials continue to evolve to suit the needs of travelers.

## TUNNELS

In 600 B.C., the Samos Aqueduct Tunnel routed the water supply though a hill on a Greek island. Tunnels dug by the Persians and Armenians in Iran in the eighth century B.C. brought water to towns. By the 17th century tunnels were widely used. At first they were used to route canals though hills instead of around them, similar to early practices with water. Railroads and roadways soon followed suit, tunneling though hills and mountains instead of laying tracks or pavement for miles to go around them.

In 1820, Marc Isambard Brunel developed a shield for boring under the Thames River. The dual tunnels that are still in use today were completed in 1841, and are 1200 yards long. This was the first time in the world that a tunnel was cut underneath a body of water.

While machine drills, boring machines, and the use of compressed air to force water out of tunnels were all great advancements in tunnel development, the most significant advancement was the use of Alfred Nobel's dynamite.

Alfred Nobel was born in 1833 in Stockholm, Sweden to a family of engineers. His family was descended from Olof Rudbeck, the best-known technical genius of Sweden's 17th century. His father, Immanuel Nobel, was an engineer and inventor who built bridges and buildings in Stockholm. Immanuel was not happy about Alfred's love of poetry, and considered him to be somewhat overly introverted. At the age of 17, his father tried to expand his horizons by sending him abroad to learn more about chemical engineering.

Alfred returned to Sweden and concentrated his efforts on developing nitroglycerine as an explosive. After creating several explosions, including one that killed his brother and several others, Alfred convinced authorities that the production of this substance was very dangerous. Within the Stockholm city limits, experimentation with nitroglycerin was no longer legal. Alfred Nobel was now forced to move his laboratory and experiments onto a barge anchored on Lake Mälaren. 1864 found Alfred Nobel mass-producing nitroglycerine using additives that made handling it safer. While mixing the nitroglycerine with an additive of silica, he found that the combination would form a paste-like substance. This paste was then shaped into rods for insertion into drilled holes, and patented as dynamite in 1867. This invention, along with the detonators or blasting caps that Nobel invented, significantly reduced the costs associated with blasting rock, drilling tunnels, and construction in general.

Nobel built up companies and laboratories in more than 20 countries; he holds more than 350 patents; he amassed an immense fortune; and, he managed to enjoy and write poetry and drama. On November 27, 1895 in Paris, Nobel signed his last will and testament and set aside his estate to establish the Nobel Prize to be awarded annually after his death.

## WATER SUPPLY AND CONTROL

The ancient Sumerians and Egyptians were the first to build and maintain several different types of water systems, including dams, aqueducts, and canals. *Dams* are barriers constructed across a waterway to control the flow or raise the level of water. *Aqueducts* are pipes or channels designed to transport water from a remote source, usually by taking advantage of gravity and bridge-like structures supporting a conduit or canal passing over a river or low ground. *Canals* are artificial waterways or artificially improved rivers used for travel, shipping, or irrigation. The ancient Sumerians and Egyptians also built bridges and distribution systems for water.

R.A. Buchanan writes "The complexity of modern life and, in particular, the organization of large towns and cities on which it is based, can only be sustained by an intricate network of public services. The magnificent Roman system of roads and aqueducts should remind us that even ancient cities required basic services to function efficiently." Some of the historical highlights in water distribution are as follows:

- In the sixth century B.C. underground sewers were installed throughout Rome to drain surface water away from the city.
- In 690 B.C. Sennacherib's Canal supplied water for Nineveh and featured a concrete lining for the canal and an arched aqueduct.
- 500 B.C. found the Persians and Syrians building qanats, which are underground irrigation systems. These were different than aqueducts because the water was already available and was merely being tapped and brought out horizontally to the surface.
- In 250 B.C. the Guan Xing flood control and irrigation project was established in China.
- From 300 B.C. to 100 A.D. Roman engineers developed extensive municipal water supply systems. These systems moved pure water long distances to

Rome and other cities. Water was then distributed by gravity in public fountains and households. Unfortunately the piping created by the engineers had large amounts of lead and silver content within it and some believe that lead poising was one of the contributing factors of the fall of the Roman Empire.

- In 110 A.D. the Acquedotto Traiano-Paolo was ordered to be built by the Roman emperor Trajan. This 25-mile aqueduct gathered water from five different springs, and still supplies water to fountains in Rome today.
- In the 4th century, Rome had over 1,350 Fountains and 850 public baths, and water was also used to flush the sewer system.
- In 600 A.D. the Hohokam irrigation system was built by ancient people in what is now the state of Arizona.
- England passed the Urban Sanitary Act in 1388 that forbid throwing garbage and sewage into ditches and rivers.
- In the 16th century, water powered pumps were installed in Germany and England to pump water from rivers and reservoirs for municipal use.
- The 17th century saw the authorization of private companies to supply London and other cities in Britain with fresh water.
- In 1718 The Acequias of San Antonio was one of the first engineered water supplies in what is now the United States. The eight canals were 15 miles in length and were built by Spanish settlers who learned the techniques from the Moors. Sections are still in use today.
- The Bernoulli brothers were working on experiments in 1738 that were originally begun by Galileo, da Vinci, Castelli, Torricelli, and Huygens in which they generated the equation that describes water flow in a pipe. This launched the modern study of hydraulics.
- Gravity-fed water systems were commonly used in colonial America. However, the first pumped public waterworks in the United States was the water-powered Bethlehem Waterworks in Bethlehem, Pennsylvania. George Washington and other early Americans considered making Bethlehem the nation's capitol, partly because of the city's technological leadership in water systems.
- In the 1790's it was determined that the water supply in Philadelphia was to blame for several disastrous yellow fever epidemics. In 1801, the Philadelphia Municipal Water Supply was completed and was one of the first major metropolitan waterworks to use steam-powered pumping stations.

## The History of Industrial Engineering

This section will provide a snapshot of some of the major accomplishments of what we now know as industrial engineering.

As international commerce increased, so did competition amongst suppliers. This competition spawned the development of industrial engineering. Though an industrial engineer's role can change from company to company, their main role is to combine workers, machines, and materials in order to increase productivity and reduce waste. The application of this philosophy can be traced back to early tribal cultures which first began to create more efficient tools and make the best use of people's specific skills.

Early machines helped drive the industrial revolution and were built by inventors that could be classified either as industrial engineers or mechanical engineers.

The first mechanically-assisted cutting device was actually used for drilling. It was a rocking drill that was cord driven and required that an assistant manipulate the cord in order to give it an alternating rotary movement. This type of a system was probably used in a horizontal application as a lathe to create the first turned product. An Etruscan wooden bowl found in the Tomb of the Warrior at Corneto from 700 B.C. was likely produced in this fashion. The earliest illustration of a lathe was found on a wall in the Egyptian tomb of Petosiris from the third century B.C. The drawings include illustrated bearings and tool rests, and demonstrated load bearing and tool rigidity as related to a produced item.

The pole lathe was developed in the 12th century as the size and complexity of work to be done increased. This lathe was designed with heavier wooden construction to be more rigid and powerful than previous designs. This wood lathe was inadequate for turning metal. In order to turn metal, a continuous drive machine was created and used a large wheel cranked by an assistant. The first illustration of this type of a machine was from 1475 A.D. An illustration also appeared in Jost Amman's *Panoplia* in 1565. A similar method of continuous driving employed a treadle and crankshaft designed by Leonardo da Vinci in 1500. This was later developed by Spaichel in 1561 using human power. This paved the way for ornamental turners and foot pedal sewing machines. Continuous drive machines using large wheels also made use of alternate power supplies (other than human-generated power). Examples of alternate power sources were: horse gins, water wheels, steam engines, and electric motors.

Several other machines created between the late 1700's and mid 1800's are listed below:

John Wilkinson's cylinder boring mill of 1776 was the first truly industrial machine tool. The machine used James Watt's steam engine for power. Wilkinson is considered the "father of the industrial revolution."

Henry Maudslay's workshops produced general machine tools, lathes, and special purpose machines. Mark Isambard Brunel designed block making machinery in 1800, and in 1810 completed the machine that supplied rigging blocks to the British Royal Navy. Maudslay's workshops were the training ground for many great engineers such as Joshua Field (his partner), Richard Roberts, Joeseph Clement, James Seaward, William Muir, Joeseph Whitworth, and James Naysmyth.

In 1806, David Wilkinsons made a screw cutting industrial lathe that was used extensively. This earned him a reputation as the "founder of the American machine tool industry."

Richard Roberts built many machines after he left Henry Maudslay's workshop. He produced a planing lathe in 1817 and a large industrial lathe with a back gear that allowed spindle speed changes (which is still in use today). Of great significance were his automatic spinning mule and differential gear from 1825. In

1855 he designed a power loom, the first effective weaving machine. Roberts was considered the most creative and innovative engineer of his time.

James Naysmyths was appointed Maudslay's personal assistant in 1829 and left to start his own workshop when Maudslay died in 1831. Naysmyth wrote about his mentor and maintained the techniques and practices he learned there. Naysmyth and Gaskell started a firm which was later renamed Naysmyth Wilson and Company. They manufactured locomotives and machine tools and devices to improve the operation of various machines. Naysmyth's most notable invention was the steam hammer of 1839 that was used to forge the 30 inch diameter paddle wheels for Isambard Kingdom Brunel and his *Great Britain* steamship.

## History of Mechanical Engineering

This section briefly describes some of the achievements within mechanical engineering. It is not a complete history of every product that could be classified under mechanical engineering though it does provide some insight into the trials and tribulations of early mechanical engineers.

As coke replaced charcoal in the blast furnaces of England in the early 1700s, the beginnings of the modern mechanical engineering profession dawned. The introduction of coke allowed for larger blast furnaces and an ability to produce higher quality wrought iron. These improvements laid the foundation for the Industrial Revolution, during which the production of great quantities of steel was made possible. As these materials became more readily available, mechanical engineers began to design improved lathes and milling and boring machines. They realized that a wide variety of devices were needed to work with the quantities of iron and steel being produced. As the pace of manufacturing increased, the number of mechanical engineers also showed a marked increase. As a new major classification of engineering pertaining to tools and machines, mechanical engineering received formal recognition in 1847 in England. In the United States, the corresponding recognition did not take place until after 1850.

## Boats

James Watt, a man with little formal education, invented the new model steam engine in 1778. This engine cooled the used steam in a condenser separate from the main cylinder. Watt's successful engine spurred the application of steam power to water, land, and air transportation. The engineering problems related to adapting the steam engine to ship use were the least difficult, and by 1786 several inventors had developed steam-powered boats. Robert Fulton was an interested observer at several demonstrations of these boats, and after careful analysis, came up with the winning combination of a dependable Watt engine and an improved hull design.

While in Pennsylvania, Fulton made acquaintance with Benjamin Franklin, who not only allowed him to paint his portrait but also provided him introductions to people of significance. Franklin wrote a letter of introduction to the West family. While staying at Torquay in 1793, Fulton read about a canal project proposed by the Earl of Stanhope that was to link the English and Bristol channels. Fulton wrote Lord Stanhope a letter and enclosed a sketch of his design which involved no use of locks (watertight basins with gates and valves at both ends that allow vessels to move from one water level to another).

In this same communication, he informed Lord Stanhope that he had a design for moving ships by steam. Fulton went on to work on steamboats, torpedo experiments, and other engineering ventures in the United States. From its first run on the Hudson River, August 17, 1807, Fulton's *Clermont* steamboat was a financial success.

## Trains

Steam-powered vehicles had been devised for land operation, but the weight and size limitations were finally overcome with the invention of high pressure boilers and iron rails. Now, the steam locomotive became practical. The first railways were built for use in mines and ironworks. A Cornish engineer named Richard Trevithick built the first steam-powered locomotive. It first ran in South Wales in 1804. George Stephenson, a mining engineer, built the first passenger railway from Stockton to Darlington in north-eastern England. It was opened in 1825.

In 1829, a competition was held to test for the design of the proposed rail line between Liverpool and Manchester. An important condition of the competition was that each locomotive had to consume its own smoke and haul a load equal to three times its own weight at an average speed of not less than 10 miles per hour. Crowds gathered to view the "Rainhill Trials."

The "Perseverance" was entered by Timothy Burstall. It had a vertical boiler with a furnace beside it. Fuel was fed to the fire from a hopper on top. The "Perseverance" was too slow and was pulled from the competition along with the horse-propelled "Cycloped." Both only attained speeds of 6 miles per hour.

The "Sans Pareil" was built by Timothy Hackworth. The two-cylinder engine ran for 27 miles at an average speed of 14 miles per hour and hauled a load of 14.3 tons. The maximum speed for one trip was 17 miles per hour, but the failure of the boiler feed pump ended the trial. It was a good solution but broke down too often.

The $500 prize was collected by George Stephenson, whose "Rocket" traveled 70 miles at an average speed of 15 miles per hour. Its highest speed, 29 miles per hour, set the bar for all future locomotives.

The Liverpool to Manchester Railway was engineered by George Stephenson. George's son, Robert Stephenson, studied the models of earlier builders and then investigated locomotive traction. He discovered that even a 1 percent grade tripled the traction force required. Keeping grades low, he and his father built several successful English railroads including the Liverpool to Manchester Railway that was opened by the Duke of Wellington in September 1830.

## Early Road transportation

Early road transportation methods had been pretty much limited to chariots used in warfare by Middle Eastern nations. These designs were handed down to the Greeks and Romans and had either two or four wheels. The Greeks and Romans used them predominantly for the transportation of goods, however chariot racing became popular just prior to 770 B.C.

The Romans, under direction of Julius Cesar, tried and failed to invade Britain. They attempted once in 55 B.C. and again in 54 B.C. Both times the combination of British soldiers and horrible weather forced them to retreat. The Romans finally invaded with success almost one hundred years later in 43 A.D. under the direction of Emperor Claudius. The Romans introduced many of their transportation methods and laid the groundwork for improved transportation in Britain. After the Roman Empire in Britain finally collapsed in 410 A.D., the infrastructure of roadways deteriorated and led to the decline of wheeled vehicle use. Most chose to travel by horseback in the times to come.

Between 1550 and 1600 A.D. carriages and coaches began to be imported into Britain. These were confined initially to baggage travel between towns. Most people in this time were poor, and if they traveled at all, they traveled by foot. British manufactures spent much of their time and money copying these European imports to sell themselves. In 1555 Walter Rippon built the first British coach for the Earl of Rutland.

The Post Office Act of 1765 decreed that the mail had to be delivered at a rate of at least 6 miles per hour. Because of this legal proclamation, mail coaches began to be regularly used starting in 1784. While the concept of the vehicles didn't change that much over the next few decades, evolution of the individual parts improved the vehicle as a whole. Problem solving and re-engineering lead to overall improvement in speed, punctuality, and service to customers. New parts were engineered to keep wheels from falling off, and new axel bearings were designed with face plates that were lubricated from an oil cup and a groove that was machined into the axel. Springs began to replace leather straps in suspensions providing a smoother ride. Dished wheels were also used. One-piece iron tyres (tires) replaced short strips of iron nailed onto the wooden wheel. In the early 19th century, lever brakes operated a friction pad on the iron tyre replacing the old wedge systems that rather endangered the driver.

Powered road transportation began around the mid 18th century in a serious attempt to replace the horse as a source of power. Steam engines were the only viable solution at this time as the internal combustion engine was not yet available.

## Early Automobiles

In 1769, 44-year-old French engineer Nicolas Joseph Cugnot (1725–1804) invented a military gun-carriage tractor that was used to haul artillery for the French army. Cugnot's steam-powered tractor had only three wheels and a top speed of 2.5 miles per hour. This first automobile had a significant shortcoming. It had to make frequent stops in order for the boiler to build up enough pressure to power the drive wheels. The French military was not impressed with this slow machine that had to make so many pressure-rebuilding stops and it ended up as a museum piece. However, one year later Cugnot invented a tricycle that carried four passengers successfully.

James Watt was born in Greenock, Scotland, in 1736. At the age of nineteen Watt traveled to Glasgow to learn how to make mathematical instruments. A while later, after a year in London, Watt returned to Glasgow where he began his own instrument-making business. Watt soon developed a reputation as a high-quality engineer. In 1763 he was sent a Newcome steam engine that needed to be fixed. This was his first introduction to steam engines, and while he was reassembling it, he figured out how to make the engine more efficient. Watt took his idea to a businessman named John Roebuck who provided the funding to back Watt's project. Roebuck subsequently went bankrupt in 1773 and Watt took his idea to another investor, Matthew Boulton. Boulton's factory built and sold Watts's engines for the next eleven years. Watts's new engine was four times as powerful as the original Newcome steam engine he repaired.

In 1785, William Murdock, who was James Watt's staff engineer, designed a three-wheeled steam-driven vehicle that was much lighter than Cugnot's design. However, James Watt felt that Murdock spent too much work time designing this personal project and, ultimately, Murdock lost his job.

In 1796, Wiliam Symington, a Scottish engineer, designed and built a model of a steam road carriage and demonstrated it for a group of financial investors. This, too, would prove to be a disappointment as the investors had little interest. Richard Trevithick built and tested several models before driving his full-size steam carriage in Cornwall in 1801. The vehicle tested successfully until it flipped over on its side. Trevithick and his cousin Andrew Vivian righted the machine and stopped at an inn to celebrate their efforts. Trevithick and Vivian were quite surprised afterwards to find that they had left the engine hot and that all the water had boiled off, heating the iron red hot. The steam carriage and the garage caught fire and burned to the ground.

## The Bicycle

Human energy was not seriously developed as a means of vehicular propulsion until the 19th century in a series of prototypes beginning with the "Dandy Horse" era in 1817. It was most likely called a "Dandy Horse" due to the transition from riding an actual horse. Baron Karl Drais von Sauerbronn of Manheim created this first prototype and was thus coined "the father of the bicycle." He first demonstrated his "Draisine" in the Luxemburg gardens in Paris. His two-wheeled prototype was made of a wooden frame with a front wheel that could be controlled and turned by the operator. An arm rest sat behind the handlebars of the vehicle so the rider could rest their arms while the vehicle was propelled by thrusting one's feet against the ground to propel the vehicle in a forward motion. This machine gained popularity not for practical purposes, but merely as a novelty.

Kirkpatrick MacMillian, a blacksmith of Courthill, Dumfriesshire, invented and built the first pedal-powered bicycle. These pedals powered the bicycle much like a child's pedal-powered toy car. A back and forth motion sent power to the rear wheel while the front was used for steering. Both wheels were mounted on brass bearings, one of the first to use this type of a system. This contraption used a saddle seat and ran on iron tyred wooden wheels. This design did not catch on either, but many others began to copy the idea of pedal-powered bicycles. (However the experimentation in the following twenty or so years was more limited to tricycles and quadra-cycles.)

The "Boneshaker" of 1860 was a variance of a velocipede built by Pierre Michaux of Paris. It had a frame made of wrought iron, and the pedals were mounted in line with the front wheel and axel. This bicycle had a friction shoe brake that could be engaged on the rear tire which slowed the machine. In 1865 Michaeux built 400 and they sold for about 13 US dollars. It was the first two-wheeled bicycle that actually caught on for practical use.

# LEARNING ACTIVITIES

## Activity 2.1    Bridge Designer

Time:    3–5 hours

In this activity, students will design a steel truss bridge to carry a two-lane highway over a river. Use the WestPoint Bridge Designer software available on the Marine Academy web site listed below. There are several different design projects to choose from. Each offers a unique set of site conditions to consider in the bridge design.

Once the bridge design can successfully carry the highway load without collapsing, students should continue to refine their design. Students should minimize bridge design costs but still ensure that it is strong enough to carry the specified loads. This activity introduces concepts such as optimization and resource management through a realistic design and simulation experience.

Vehicular travel is one of the greatest and most liberating conveniences we enjoy today. To have the freedom to go where you like and complete tasks in a timely manner is made much easier with an automobile. However, the costs as-

**Explosives detonate a 130-foot, 520 ton section of the northbound lane of the Hoan Bridge, dropping a damaged span precisely underneath to avoid damage to the Milwaukee Metropolitan Sewerage District's gallery building below. December 28, 2000. Photo Courtesy of the Milwaukee Journal Sentinel.**

**A safety manager with United Water Services walks on a pile of gravel that was used to absorb the shock when the damaged section of the Hoan Bridge was precisely dropped. Photo Courtesy of the Milwaukee Journal Sentinel.**

sociated with driving include insurance, regular maintenance, and fuel. Roads and bridges are a necessary element of car travel. Figure 2 shows how part of the Hoan Bridge in Milwaukee, Wisconsin, was demolished when a section of the northbound lanes dropped more than three feet. The closing of the bridge for repairs caused major traffic congestion on surrounding city streets, creating a real headache for motorists. On a much larger scale, when the Loma Prieta Earthquake of 1989 hit the San Francisco Bay area in California, many roadways and bridges were impassable for weeks and months to follow.

## RESOURCES

http://www.dean.usma.edu/cme. This site is the home of the Department of Civil and Mechanical Engineering of the United States Military Academy. Visit here to download the WestPoint Bridge Designer software for free.

## Activity 2.2    How Can We Save Water?

Time:    1–2 class periods, 4 days at home collecting data.

In this activity, you will examine and compare ways that your classmates use water as they brush their teeth. You will determine a more efficient way to brush your teeth and conserve water at the same time.

Have you ever wondered how much water you use around the house? Rarely do people consider how much water they use in a typical day. Some water is wasted during activities like showering, washing dishes, or brushing teeth. Water is a precious resource and its use should be carefully monitored. In this way, more water can be conserved.

## REFERENCES

Walch (1998) Hands On Science Series: Water. Pp. 47–50

## Activity 2.3     Guidelines for a Waste Treatment Facility

Time:      1–4 class periods

In this activity, students will design a waste treatment facility for their region. Present the students with this scenario: "Water supplies have become polluted in your community from both industrial and residential waste. You have the responsibility of designing and locating a new waste treatment facility for your region." Discuss with students the general considerations for the waste treatment plant and its ideal location. Encourage students to debate issues such as safety, costs to the public, regulations for water quality, and industrial pollution controls.

Each team of three or more students will develop a list of social, technical, and environmental guidelines to be followed. Next, students will develop conceptual drawings and written descriptions for their design solutions. An extension of this activity is to have each team develop a mock-up of their facility showing the various phases of water treatment and the associated technological systems. This extended activity will require an additional 5-7 class periods.

Water supply and treatment facilities exist near every populated area, yet we often take for granted the quality of the water we drink and use, and the elimination of water after use. Removing pollutants and wastes from used water before releasing the water into the environment involves treatment systems.

## REFERENCES

Horton, Komacek, Thompson, & Wright. (1991). *Exploring Construction Systems.* Davis Publishing, Pg. 292

## CHAPTER REFERENCES

A History of Technology, Volume II, Ed. Charles Singer, Oxford University Press, New York, 1956.

American Society of Civil Engineers, The Civil Engineer: His Origins, New York: ASCE, 1970

American Society of Civil Engineers historical timeline and project listing website. www.asce.org

Burghardt, M. David, Introduction to Engineering, 2nd Ed., New York, Harper Collins, 1995.

Burstall, A., A History of Mechanical Engineering, London: Faber, 1963.

De Camp, L. Sprague, The Ancient Engineers, Cambridge, The MIT Press, 1970.

Dickinson, H.W., Robert Fulton: Engineer and Artist, his life and works. London, 1913. 327 pp

Great Engineers and Pioneers in Technology, Volume 1: From Antiquity through the Industrial Revolution, Eds. Roland Turner and Steven L. Goulden, New York, St. Martin's Press, 1981.

Greaves, W. F., and J. H. Carpenter, A Short History of Electrical Engineering, London: Longmans, Green and Co Ltd, 1969.

Kirby, R., et al., Engineering in History, New York, McGraw-Hill, 1956.

Miller, J. A., Master Builders of Sixty Centuries, Freeport, New York, Books for Libraries Press, 1972.

Red, W. Edward, Engineering The Career and The Profession, Monterey, California, Brooks/Cole Engineering Division, 1982.

The Engineering Specific Career Advisory Problem-Solving Environment (ESCAPE) Purdue University Department of Freshman Engineering. West Lafayette, IN 47907

The School of Mechanical Engineering 2003 at the University of Adelaide SA 5005 AUSTRALIA

http://www.railfan.net/. This site hosts forums, photo galleries and links to Rail history and information. A very large source for anything associated with rail travel and construction

*Chapter 3*

# Profiles of Historical Engineers

Over the course of history, many engineers have had a huge impact on society. Engineers design the bridges we cross, the roads we travel, the automobiles we drive, and the computers we operate. They are the backbone of society, improving the present and inventing the future. Without engineers, society would slowly progress as people randomly thought of new solutions, whereas engineers dedicate entire careers to solving problems. But, history is about people.

To understand the story of an invention, one should study and understand the people involved. An invention and its inventor are integrally woven together. When studying the history of an invention or innovation, people find themselves becoming more interested as they get to know more about the people involved and the challenges they faced. The field of engineering has been around for centuries. Ever since human beings first walked the earth, we have prided ourselves in making objects that will accomplish a task more efficiently, allowing people to do things they have never done or letting them go places they have never been before. Who were these great people who paved the way for the rest? A brief list of history's greatest engineers follows.

## Nolan K. Bushnell (1943–    )

Nolan K. Bushnell is considered the father of the video game industry. He founded Atari in 1972 and launched the video game revolution that same year with Pong. Selling Atari in 1976 for 28 million dollars, he went on to found Chuck E. Cheese's restaurant, selling that as well after a few years.

Bushnell has started over 20 companies since he began with Atari. Now he provides service to many corporations, including IBM, Cisco System, and Commodore International, while he also sits on the board of several leading companies.

He traced the spark that started his career, in an interview with Joyce Gemperlien of the San Jose Mercury News, to Mrs. Cook's third grade class when he had an assignment on electricity. Bushnell put together his contraption, which he showed to the entire class. He then went home, found his flashlight, wires, and odds and ends around the house and began to tinker with them and has been tinkering ever since!

Nolan K. Bushnell holds patents on some of the basic technologies for many of the early video games he made, as well as some he is currently working on. He is also the inventor/co-inventor for many worldwide patents in various other fields and industries.

He received his B.S. in electrical engineering from the University of Utah, where he was a Distinguished Fellow. While attending Stanford University's Graduate School, Bushnell gave lectures at major universities and corporations. He inspired others with his views on entrepreneurship and innovation.

## Thomas Alva Edison (1847–1931)

Thomas Alva Edison is known to have said, "If we did all the things we are capable of doing, we would literally astound ourselves." Edison was a man of dreams and determination. Applying his great curiosity to the world around him, Edison managed to lay the groundwork for many technological advances still in use today.

In 1862, at the age of fifteen, Edison left his home in Port Huron, Michigan, to work as a telegrapher. He roamed the United States and Canada delivering telegraph messages. In 1869, he moved to New York where he worked on inventions related to the telegraph. He came out with the Universal Stock Printer, a device that automatically printed stock telegraphs. For this novel invention, Edison received $40,000 in compensation, an enormous sum at time, which enabled Edison to become a full-time inventor.

He completed one of his greatest inventions in 1877. His cylinder phonograph could record and play back using tin foil and needles. Edison took out a patent on his new machine on February 19, 1878. In order to promote the phonograph, he sold the manufacturing and sales rights to a new company being formed, The Edison Speaking Phonograph Company. Thomas Edison received $10,000 plus 20% of the company's profits. Edison would return to the phonograph later in his career, but for many years he concentrated instead on his most famous work, the incandescent light bulb.

In 1879, no one had invented a product practical for home use. Not only was Edison challenged to create an electrical light source, he also had to make the bulbs durable, long lasting, safe, and economical for consumers. Edison did produce a light bulb that met all of these requirements, and it only took him one year.

Thomas Edison could be considered history's greatest engineer due to his great vision and his many technological contributions. Over his 84-year life span, Thomas Edison took out 1,093 patents and started ten different companies. One of these companies became General Electric, a leader in the electronics industry and one of the largest, most productive and powerful corporations in the world. Edison was a man of innovation, who summed up his life's work by saying "Genius is one percent inspiration, ninety-nine percent perspiration."

## Leonardo da Vinci (1452–1519)

Leonardo da Vinci, shown in Fig. 3.1, had an uncanny ability to envision mechanized innovations that would one day see common usage, especially in the field of

**Figure 3.1 Leonardo da Vinci (born April 15, 1452, Vinci, Italy, died May 2, 1519).**

weaponry, e.g., helicopters, tanks, artillery (see Figs. 3.2–3.6). He seemed to know almost everything that was knowable in his time. He was a true Renaissance man who harnessed his immense genius to make improvements to almost every aspect of the lives of everyday people and had a great impact on future generations. Da Vinci's handicap, however, was his inability to read Latin. This prevented him from reading and learning from the common scientific writings of the day, which focused on the works of Aristotle and other Greeks and their relation to the Bible. Instead, da Vinci was forced to make his assessments solely from his observations of the world around him. He was not interested in the thoughts of the ancients; he simply wanted to use his engineering skills to improve his environment.

Da Vinci did indeed bring his genius to bear on an enormously wide range of subjects, but little of his work had any relationship to that of his contemporaries. However, essential aspects of development were always lacking from his innovations. He lacked a community of engineering peers with whom he could integrate

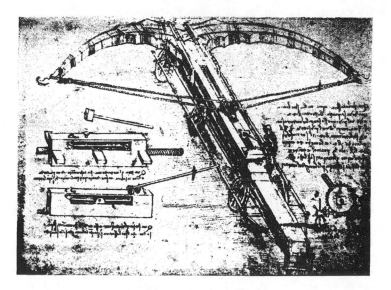

**Figure 3.2 Weapon design.**

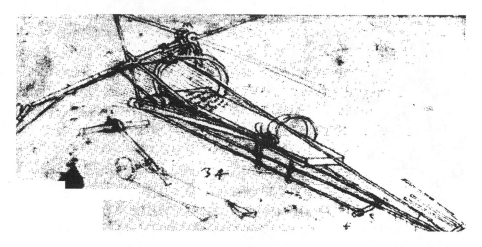

**Figure 3.3 A flying machine.**

and refine his efforts. Because he was so far ahead of his time, many of his innovations did not come to fruition for centuries.

Da Vinci's life as an engineer was one of total immersion in the intellectual activities of his time. He created frescoes, he tried sculpting he was an architect, and he engineered hundreds of useful and useless devices. He was a respected member of the community and was consistently in the employ of the aristocracy.

In order to obtain his first commission as an engineer he wrote a letter claiming that he could: construct movable bridges, remove water from the moats of fortresses under siege, destroy any fortification not built of stone, make mortars and dart throwers, make flame throwers and cannon capable of firing stones and making smoke, design ships and weapons for war at sea, dig tunnels without making any noise, make armored wagons to break up the enemy in advance of the infantry, design buildings, sculpt in any medium, and paint.

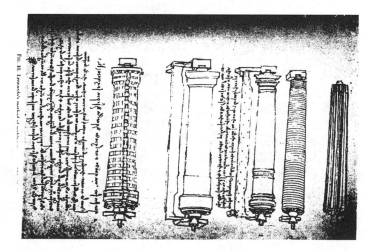

**Figure 3.4 Cannon design.**

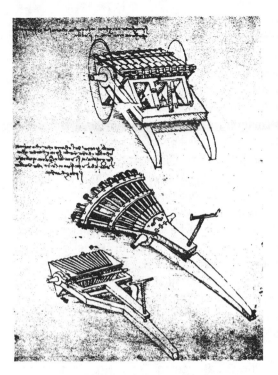

**Figure 3.5 Machine guns.**

With that résumé, he was hired with the title of Painter and Engineer to the Duke. Did he do all the things that he said he could? We do not know. But we do know that, among other things, he worked diligently perfecting the canal system around Milan.

Da Vinci's chief function, during the period he that was employed by the Duke, was to produce spectacular shows for the entertainment of the aristocracy who came to court. He produced musical events, designed floats, and delighted the court with processions that included spectacular flying devices, dazzling the eyes. The interesting thing is that during this time he wrote over 5,000 pages of notes, detailing every conceivable kind of invention. He truly was a great engineer at work.

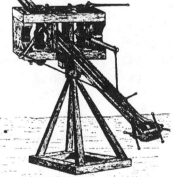

**Figure 3.6 Stone throwers.**

# Isambard Kingdom Brunel (1806–1859)

Isambard Kingdom Brunel was quite possibly one of the most diverse and most capable engineers in history. Brunel's projects showed his diversity and excellence. Isambard designed and built a broad variety of projects ranging from railways to bridges to ships. Because of his flexibility as an engineer, Isambard Kingdom Brunel makes it into the top ten historical engineers.

Born the son of an engineer, he was sent off to college at the age of 14, after which he went to work in his father's office at the age of sixteen. His first position was as an apprentice on the early stages of the Thames Tunnel, which his father had designed. He advanced to resident engineer in charge of the project in 1826, at the age of 20, when William Armstrong resigned from that position due to health reasons. On May 18, 1827, the Thames River broke through the tunnel for the first time. Brunel dove to the breach in a diving bell to inspect the damage that would end up being repaired with bags of clay and gravel. Eight months later on January 12, 1828, the river broke through a second time and Isambard was seriously injured during rescue operations but managed to save the lives of several men. All work in the tunnel stopped for seven years because of financial problems.

While recuperating from his injuries in the tunnel accident, Brunel entered a design competition and submitted designs for a suspension bridge that would pass over the Avon Gorge. The judge of the contest was Scottish Engineer, Thomas Telford, who was also a competitor of Brunel's. All of Brunel's designs were rejected and Telford's designs were accepted. After a great fight to win acceptance of his ideas, a second contest was opened and one of Brunel's designs was accepted in 1830.

Brunel was the lead designer for the Great Western Railway project. He designed the track and its layout for the project. As part of this project, he designed the ocean liner that would take the passengers from the end of the rail to America. His designs even extended to the lamp posts in the train stations. The Great Western Railway was, at the time, the longest railway ever conceived.

Throughout his career, Brunel made an effort to seek out new technologies and anticipate developing markets. He used fundamental logic and analysis to reshape the mechanical and structural engineering of his time. In doing so, he helped reshape the art and technology of architecture.

A great irony, as happened with many other early engineers and their projects, was that the Clifton Suspension Bridge was not finished until after Brunel's death, who died at the age of 53 in 1859. Brunel's father had died ten years earlier at the age of eighty.

## Charles Proteus Steinmetz (1865–1923)

"No man really becomes a fool until he stops asking questions," said Charles Steinmetz. Steinmetz was a pioneer in the field of electrical engineering. Without his passion for knowledge and understanding, the use of electricity would not exist in the way it does today.

One of his first research jobs at General Electric was on hysteresis, or when power is lost because of magnetic resistance. Webster defines hysteresis as "a retardation of the effect when the forces acting upon a body are changed (as if from viscosity or internal friction); *esp.:* a lagging in the values of resulting magnetization in a magnetic material (as iron) due to a changing magnetizing force". His research led him directly to alternating current, which solved the power loss issue but did not solve the problem completely. No theories existed at the time for alternating current. Over the next twenty years, he successfully developed these theories. Without his theories many inventions using electric power would have never existed or would have been created at a much later date.

An incident that occurred after Charles retired illustrates just how valuable he was as an engineer. One of the generators he had worked on was broken and none of the technicians had any idea about how to fix it. They asked him to come back to fix the problem. With chalk, he marked the malfunctioning part with an "X." He gave them a bill for one thousand dollars, and they asked him for an itemized invoice. He sent back an invoice, which said: "Marking chalk X on side of generator: $1. Knowing where to mark chalk X: $999." When no one else knew how to fix it, he did.

Steinmetz took out 200 electrical patents in his lifetime and he considered his work in the field of electromagnetism to be his greatest accomplishment. He invented the metallic electrode arc lamp and worked to develop devices to protect power lines from lightning bolt strikes. Steinmetz even developed designs for electric vehicles, but the production work on these new cars and trucks stopped when he died.

## Henry Ford (1863–1947)

Henry Ford was born on July 30, 1863, in Wayne County, Michigan. As a young man, Henry disliked school and farm life, so he walked to Detroit in search of a job. He found just what he was looking for in a small machine shop working as an apprentice. It was in this shop that Henry learned of engines and how they work. He worked for several years learning everything he could about engines, and then opened his own shop back in his hometown of Dearborn.

Ford worked his way up to chief engineer at the Detroit Edison Company and this is where he designed his first vehicle, the Quadricycle, with a buggy frame with four bicycle tires as wheels. Ford continually revised his designs for vehicles, and in 1903, he founded the Ford Motor Company. He introduced the Model T or "everyman's car" in 1908 and it stayed on the market for 19 years. It was due to Henry Ford's advancements in production technology that he was able to market the Model T so well.

> "His general policy of resistance to labor is as out of step with the times as the oxcart."
> —The Washington News, in a 1937 editorial against Ford

In 1913, the creation of the assembly line allowed the production of automobiles in greater quantities than previously possible and lowered costs. These reduced costs were passed on to the consumer and made the automobile affordable for the average American. Along with the assembly line, Ford also instituted a minimum wage at his plant of $5.00 a day, claiming that it increased productivity. One little known fact about Henry Ford was his contribution to society of the five-day workweek. Up until Ford created his company, most businesses operated on a six-day workweek. Henry believed that the extra day off would allow people to have more leisure time, and therefore, spend more money. Society has operated on the five-day workweek ever since.

By the 1920s, the Ford Motor Company was almost entirely self-sufficient. However, in 1926, Ford began to lose sales to General Motors because the Model T was becoming outdated. Ford Closed its plants for five months while plans were implemented to build a Model A and a V8 model.

Lee Iacocca later wrote of Henry Ford in Time Magazine "The only time I ever met Henry Ford, he looked at me and probably wondered, 'Who is this kid?' He wasn't real big on college graduates, and I was one of 50 in the Ford training course in September 1946, working in a huge drafting room at the enormous River Rouge plant near Detroit. One day there was a big commotion at one end of the floor and in walked Henry Ford with Charles Lindbergh. They walked down my aisle asking men what they were doing. I was working on a mechanical drawing of a clutch spring and I was worried that they'd ask me a question because I didn't know what I was doing—I'd been there only 30 days. I was just awestruck by the fact that there was Colonel Lindbergh with my new boss, coming to shake my hand. Almost half a century before Ray Kroc sold a single McDonald's hamburger, Ford invented the dealer-franchise system to sell and service cars. In the same way that all politics is local, he knew that business had to be local. Ford's "road men" became a familiar part of the American landscape. By 1912, there were 7,000 Ford dealers across the country."

## Dr. Robert Goddard (1882–1945)

Dr. Robert Goddard grew up in New England and he invented the process used for liquid fueled rockets. It was his design that first took a rocket off the ground that was not fueled by solid fuels like black powder. Designs similar to his are used in the rockets that launch astronauts into space, deliver warheads to distant countries, and carry the communications and espionage satellites into orbit.

Dr. Goddard's design of the bi-fuel system, which combined the fuel and the oxidizer right before ignition, allowed for the safe transportation of the highly volatile liquid fuels such as liquid oxygen and liquid hydrogen. Although he did not use liquid hydrogen or oxygen, his fuel system design paved the way for man to travel to and walk on the moon, and gave the military the ability to launch a multi-ton warhead from one side of the world to the other.

With the knowledge of Newton and Galileo, he investigated the physics necessary to launch a rocket. If you examine the equation, $\Sigma F = ma$, you must have a net force to overcome the acceleration of gravity on the mass of the rocket. Using Newton's second law (every action has an equal and opposite reaction), he produced a rocket engine that gave him a force large enough to push his rocket off the ground for a short while. Although he did not have enough fuel to reach escape velocity (11 km/s), his rocket did lift off from earth using liquid fuel, giving the Germans the knowledge to produce the V series weapons and the Americans the ability to land men on the moon.

Goddard was the first scientist who not only realized the potentials of missiles and space flight but also contributed directly in bringing them to practical realization. This rare talent in creative science and practical engineering places Goddard well above the other European rocket pioneers. The dedicated labors of this modest man went largely unrecognized in the United States until the dawn of what is now called the space age. High honors and wide acclaim, belated but richly deserved, now come to the name of Robert H. Goddard.

On September 16, 1959, the 86th Congress authorized the issuance of a gold medal in the honor of Professor Robert H. Goddard.

In memory of the brilliant scientist, a major space science laboratory, NASA's Goddard Space Flight Center, Greenbelt, Maryland, was established on May 1, 1959.

## Wilbur (1867–1912) and Orville Wright (1871–1948)

Growing up in a rather small area of the country at the time, Orville and his brother Wilbur began their lives tinkering around in a workshop making pointless little mechanical objects. The brothers were still young when they decided that they wanted to become members of the new field of aeronautical engineering. Both brothers taught themselves how to read and write and began to keep up with the world's engineering issues by reading the newspaper and studying books. They were most curious, however, with the work of Otto Lilienthal (a German scientist who tested gliders and kites). Wilbur and Orville were determined to design the first powered flying model. While they began to work on their newest project of testing curved wing design, Orville built the first machine that would decrease the testing time of designs, what we call today a wind tunnel. Using the wind tunnel, the Wright brothers now could build scale model wing designs and test the levels (quantitative) of lift and drag on any shape they wanted. This major step in aeronautical history set the Wrights ahead of every other leading scientist in the world. Here is an excerpt from the Journal of the Western Society of Engineers, December 1901:

> "My own active interest in aeronautical problems dates back to the death of Lilienthal in 1896. The brief notice of his death which appeared in the telegraphic news at that time aroused a passive interest which had existed from my childhood and led me to take down from the shelves of our home library a book on *Animal Mechanism,* by Professor Marey, which I had already read several times. From this I was led to read more modern works, and as my brother soon became equally interested, we soon passed from the reading to the thinking, and finally to the working stage. It seemed to us that the main reason why the problem had remained so long unsolved was that no one had been able to obtain any adequate practice. We figured that Lilienthal in five years of time had spent only about five hours in actual gliding through the air. The wonder was not that he had done so little, but that he had accomplished so much . . . . yet Lilienthal with this brief practice was remarkably successful in meeting the fluctuations and eddies of wind gusts. We thought that if some method could be found by which it would be possible to practice by the hour instead of by the second there would be hope of advancing the solution. . . . It seemed feasible to do this by building a machine which would be sustained at a speed of 18 miles per hour, and then finding a locality where winds of this ve-

locity were common. With these conditions, a rope attached to the machine to keep it from floating backward would answer nearly the same purpose as a propeller driven by a motor, and it would be possible to practice by the hour, and without any serious danger, as it would not be necessary to rise far from the ground, and the machine would not have any forward motion at all. We found, according to the accepted tables of air pressures on curved surfaces that a machine spreading 200 square feet of wing surface would be sufficient for our purpose, and that places could easily be found along the Atlantic coast where winds of 16 to 25 miles were common. When the winds were low, it was our plan to glide from the tops of sand hills, and when they were sufficiently strong we would use a rope for our motor and fly over one spot. Our next work was to draw up the plans for a suitable machine. After much study, we concluded that tails were a source of trouble rather than of assistance; and we decided to dispense with them altogether. It seemed reasonable that if the body of the operator could be placed in a horizontal position instead of the upright, as in the machines of Lilienthal, Pilcher, and Chanute, the wind resistance could be materially reduced since only one square foot instead of five would be exposed. As a full half horsepower could be saved by this change, we tried the horizontal position. Then the method of control used by Lilienthal, which consisted in shifting the body, did not seem quite as quick or effective as the case required; after long study, we contrived a system consisting of two large surfaces on the Chanute double-deck plan, and a smaller surface placed a short distance in front of the main surfaces in such a position that the action of the wind upon it would counterbalance the effect of the travel of the center pressure on the main surfaces. Thus, changes in the direction and velocity of the wind would attend only to the steering of the machine, which was to be affected by curving the forward surface up or down. The lateral equilibrium and the steering to right or left was to be attained by a peculiar torsion of the main surfaces, which was equivalent to presenting one end of the wings at a greater angle than the other. In the main frame a few changes were also made in the details of construction and trussing employed by Mr. Chanute. The most important of these were (1) the moving of the forward main crosspiece of the frame to the extreme front edge; (2) the encasing in the cloth of all crosspieces and ribs of the surfaces; (3) a rearrangement of the wires used in trussing the two surfaces together, which rendered it possible to tighten all the wires by simply shortening two of them."

By far the task of actually piloting a successful flight is what the Wright brothers are credited for. However, the manner in which they went about the experiments that led up to this can and should be considered their greatest achievement. Without the practices they employed and the extensive testing and formulating they did, the flight would have never happened. Read more about the Wright brothers in the full version of the text online.

## Grace Murray Hopper (1906–1992)

Grace Murray Hopper, a computer engineer and Rear Admiral in the U.S. Navy, developed the first computer compiler in 1952 and the computer program language COBOL. Upon discovering that a moth had jammed the works of an early computer, Hopper popularized the term "bug." In 1983, by special presidential appointment, Hopper was promoted to the rank of Commodore. Two years later, she became one of the first women to be elevated to the rank of Rear Admiral. In 1986, after forty-three years of service, she retired. At 80 years old, she was then the oldest active duty officer. She spent the remainder of her life as a senior consultant to Digital Equipment Corporation.

Hopper received numerous honors over the course of her lifetime. In 1969, the Data Processing Management Association awarded her the first Computer Science Man-of-the-Year Award. She became the first person from the United States, and the first woman, to be made a Distinguished Fellow of the British Computer Society in 1973. She also received multiple honorary doctorates from universities across the nation. The Navy even christened a ship in her honor. In September 1991, she was awarded the National Medal of Technology, the nation's highest honor in engineering and technology.

## Joseph B. Strauss (1827–1870)

The idea of connecting each side of the Golden Gate Strait was proposed as early as 1872 by Charles Croker. The idea was given life again in 1916 by a newspaper editor of the San Francisco Call Bulletin. Croker began an editorial campaign for a bridge that eventually inspired San Francisco city engineer Michael M. O'Shaughnessy to promote a national inquiry about the feasibility of such a bridge. Most engineers said that it would be impossible. Speculation costs of those who said it could be done ran a high as $100 million. Joseph Baermann Strauss said it could be done and for between $25–$30 million. He did have experience in such matters as the designer of nearly 400 spans.

Strauss submitted preliminary designs with a cost estimate of $27 million on June 28, 1921. This effort did not have the financial support of the federal government as the San Francisco-Oakland Bay Bridge had already received all funds available. Strauss dedicated himself to the bridge, informing civic leaders that the bridge could pay for itself with tolls alone. A special district was formed to oversee the bridge's financing, design, and construction. This led to the Golden Gate Bridge and Highway District Act, which allowed counties the right to organize as a district in order to borrow money, issue bonds, construct the bridge and collect tolls.

The U.S. War Department had final say in the authorization of construction because of the impact of the structure on shipping and military logistics, and it owned land on both sides of the Golden Gate Strait. A provisional permit was granted on December 20, 1924. Strauss's work, however, was not complete. This bridge design met with considerable opposition from ferry companies and other well-financed special interest groups. However, the bridge prevailed and construction began on January 5, 1933.

Strauss created the Golden Gate, which had a span of 1,280 meters, and connected San Francisco to Marin County. When the bridge was opened, it held the world record for its extensive length. Since the bridge was completed, millions of cars have passed over it.

The Golden Gate is linked from one end to the other with long cables that are suspended through the air. The total length of these cables is long enough to circle the entire world three times. It took two unbelievably long years to hang these cables back and forth creating the design.

The total weight of the bridge, including bridge, anchorages and approaches after the re-decking project of 1986 is 887,000 tons or 804,700,000 kilograms.

## Leo Szilard (1898–1964)

After studying engineering in Budapest and Berlin, Szilard moved to London in order to escape Nazi persecution. In 1933, he obtained a patent of the nuclear chain reactor. He first tried to use indium and beryllium to create a chain reaction but soon found that these elements would not work. In 1939, two years after moving to New York City, he learned about fission and subsequently found out that uranium would work for such a reaction. Worried about the Germans developing an atomic bomb, Leo recommended that the U.S. government immediately start work on the atomic bomb. As a result of his letter, the U.S. government was persuaded to launch the Manhattan Project. Finally, on December 2, 1942, Szilard and his partner Enrico Fermi successfully generated the first controlled nuclear chain reaction.

Even though he is credited with building the first atomic bomb, he advocated against its use. He also organized an opposition to the May Johnson Bill, which would have given the responsibility for all nuclear energy to the U.S. military. In 1955, Leo Szilard and Enrico Fermi were given a patent for the nuclear fission reactor. Leo Szilard not only helped the field of nuclear engineering, he also made contributions to political science, nuclear physics, statistical mechanics, genetics, and molecular biology.

Szilard was a collaborator with Albert Einstein on 45 patent applications over seven years in six countries, mostly on refrigeration technology.

## LEARNING ACTIVITIES

## Activity 3.1   The Great Pizza Factory

Time:      3–5 hours

In 1907, Henry Ford announced his goal for the Ford Motor Company: to create "a motor car for the great multitude." At that time, automobiles were expensive, custom-made machines.

Ford's engineers took the first step towards this goal by designing the Model T, a simple, sturdy car with no factory options—not even a choice of color. The Model T, first produced in 1908, kept the same design until the last one—number 15,000,000—rolled off the assembly line in 1927. From the start, the Model T was

**Moving assembly line at Ford Motor Company's Michigan plant**

less expensive than most other cars, but it was still not affordable for the "multitude." Ford realized he needed a more efficient way to produce cars in order to lower their cost. He and his team looked at other industries and found four principles that would further their goal: interchangeable parts, continuous flow, division of labor, and reducing wasted effort.

In this activity, students work in teams of three to design an assembly line to make pizzas. This assembly line should be able to be modified to handle any type of pizza style (e.g., deluxe or just cheese and sausage). Students will need to locate local pizza supply companies and makers. They may want to visit food production companies to learn more about aspects of production, how resources are managed, and what steps are taken to ensure quality. Students will need to analyze costs and consider the best ways to customize the pizzas. Time and resources permitting, students may be able to carry out the tasks for a fundraiser or a team lunch.

## Activity 3.2   Aviation History

Time:     2–5 hours

America's commercial aviation industry was born in 1927 when Congress forced the Post Office to pay private companies to fly mail between distant cities. While McDonnell-Douglas capitalized on the opportunity by selling M Series mail planes to airlines, Boeing went one step further. It won the contract to carry mail between Chicago and San Francisco. Occasional paying passengers made flights even more profitable.

Boeing developed the Model 40 specifically for this market. As many as four passengers, and 500 pounds of mail, stayed dry in an enclosed cabin while the pilot navigated from an open cockpit. Passenger travel proved so successful that, in 1928, Boeing introduced the Model 80—a passenger plane designed with luxury in mind. Leather upholstered seats, hot and cold running water, reading lamps and in-flight stewardesses ensured the utmost in passenger comfort during 23-hour flights from San Francisco to Chicago.[1]

In this activity, students will analyze the effects of aviation on the individual, society, and the environment. Historical events, global trends, economic factors, and ethical issues are areas to be covered. The students should develop an electronic

**"For many years I have been afflicted with the belief that flight is possible to man. My disease has increased in severity and I feel that it will soon cost me an increased amount of money if not my life." A May 1900 letter from Wilbur Wright to Octave Chanute, an engineer and bridge builder in Chicago.**

presentation including timelines and examples for the web or display for others outside the classroom.

## REFERENCES

*Great Engineers and Pioneers in Technology, Volume 1: From Antiquity through the Industrial Revolution.* Eds. Roland Turner and Steven L. Goulden. New York: St. Martin's Press, 1981.

Kirby, R., et al. *Engineering in History.* New York: McGraw-Hill, 1956.

Miller, J. A. Master Builders of Sixty Centuries. Freeport, NY: Books for Libraries Press, 1972.

Placzek, Adolf K. *Macmillan Encyclopedia of Architects. Vol. 1-4.* London: The Free Press, 1982.

Wright, Wilbur, December 1901. "Some Aeronautical Experiments". *Journal of the Western Society of Engineers.* (presented) September 18, 1901.

http://www.pbs.org/wgbh/aso/databank/entries/dt13as.html This site has an overview of Henry Ford's creation of the assembly line and the four main objectives.

http://www.faa.gov/ The Federal Aviation Administration's website. Many links and resources are here for students and educators.

http://www.faa.gov/education/resource.htm#curic Curriculum links for educators off the FAA website.

http://www.nara.gov/ This is the site for the U.S. National Archives.

[1]  http://www.boeing.com/companyoffices/history/index2.htm Boeing's website. Follow the link to their homepage where the history of Boeing is listed. The above text and picture are directly from Boeing's website.

*Chapter 4*

# What Is Engineering?

## Introduction

Engineers produce things that have an impact on us everyday. They invent, design, develop, manufacture, test, sell, and service products and services that improve people's lives. The Accreditation Board for Engineering and Technology (ABET), which is the national board establishing accreditation standards for all engineering programs, defines engineering as follows [Landis]:

> "Engineering is the profession in which a knowledge of the mathematical and natural sciences, gained by study, experience, and practice, is applied with judgment to develop ways to utilize, economically, the materials and forces of nature for the benefit of mankind."

Frequently, students early in their educational careers find it difficult to understand exactly what engineers do, and more to the point, where they fit best in the vast array of career opportunities available to engineers.

Common reasons for a student to be interested in engineering include the following:

1. Proficiency in math and science
2. High school counselor suggestion
3. Has a relative who is an engineer
4. Heard it's a field with tremendous job opportunity
5. Read that it has high starting salaries
6. Always took things apart when a child
7. Likes building things

Though these may be valid reasons, they do not imply a firm understanding of engineering. What is important is that a student who is interested in engineering understands the career and its options. We all have our own aptitudes and talents; finding places to use them, however, is the key to a rewarding career.

The purpose of this chapter is to provide information about what engineers do.

## THE ENGINEER AND THE SCIENTIST

To understand what engineers do, a contrast must be made between the roles of engineers and of the scientist.

The main difference between the engineer and the scientist is in the object of their respective work. The scientist searches for answers to questions to understand why a phenomenon occurs. The engineer searches for answers to similar questions but always with an application in mind.

Theodore Von Karman, one of the pioneers of America's aerospace industry, said, "Scientists explore what is; engineers create what has not been." [Paul Wright, Introduction to Engineering].

In general, science is about discovering things or acquiring knowledge. Scientists are always asking why. They are interested in increasing their knowledge base. The answers they seek may be abstract in nature, such as understanding the universe's beginning, or more practical, such as the reaction of a virus to a new drug.

The engineer asks the same question but asks because a problem is preventing a product or service from being produced. The engineer is always thinking about the application when asking why. The engineer becomes concerned with issues such as the demand for a product, the cost of producing the product, and the impact on society and the environment by the product.

Scientists and engineers work in many of the same fields and industries but have different roles. Here are some examples:

- Scientists study the planets in our solar system to understand them; engineers study the planets so they can design a spacecraft to operate in the environment of that planet.
- Scientists study atomic structure to understand the nature of matter; engineers study the atomic structure in order to build smaller and faster microprocessors.
- Scientists study the human neurological system to understand the progression of neurological diseases; engineers study the human neurological system to design artificial limbs.
- Scientists create new chemical compounds in a laboratory; engineers create processes to mass produce new chemical compounds for consumers.
- Scientists study the movement of tectonic plates to understand and predict earthquakes; engineers study the movement of tectonic plates to design safer buildings.

## THE ENGINEER AND THE ENGINEERING TECHNOLOGIST

Engineering technology is another profession closely related to engineering though both have different career opportunities. ABET, which accredits engineering technology programs, as well as engineering programs, defines engineering technology as follows:

"Engineering technology is that part of the technological field which requires the application of scientific and engineering knowledge and methods combined with technical skills in support of engineering activities; it lies in the occupational spectrum between the craftsman and engineering at the end of the spectrum closest to the engineer."

Technologists work with existing technology to produce goods for society. Engineering technology students spend time in their curricula working with the actual machines and equipment they will eventually use in their careers. By doing this, technologists are equipped to be productive in their occupation from the first day of work.

Engineers and technologists apply technology for society's betterment. The main difference between the two fields is that the engineer is able to create new technology through research, design, and development. Rather than being trained to use specific machines or processes, engineering students study additional mathematics and engineering science subjects. This equips engineers to use these tools to advance the state of the art in their field and move technology forward.

Engineers and engineering technologists often perform similar jobs. For example, in manufacturing settings engineers and technologists work supervisors of assembly line workers. In the technical service fields, both work as technical support personnel, supporting equipment purchased by customers. Most opportunities, however, are different for engineering and engineering technology graduates.

- The technologist identifies the computer networking equipment necessary for a business to meet its needs and oversees the installation of that equipment; the engineer designs new computer boards to transmit data faster.
- The technologist develops a procedure to manufacture a shaft for an aircraft engine using a newly developed welding technique; the engineer develops the new welding machine.
- The technologist analyzes a production line and identifies new robotic equipment to improve production; the engineer develops a computer simulation of the process to analyze the impact of the proposed equipment.
- The technologist identifies the equipment necessary to assemble a new CD player; the engineer designs the new CD player.
- The technologist identifies the proper building materials and oversees the construction of a new building; the engineer determines the proper support structures, taking into account the local soil, proposed usage, earthquake risks, and other design requirements.

## WHAT DO ENGINEERS DO?

Within engineering, basic job classifications have commonalities across the engineering disciplines. What follows are brief descriptions of these different engineering job functions. A few examples are provided for each function. It is important to realize that all the fields of engineering have roles in each of the main functions described here.

## Research Engineers

The role of the engineering researcher is the one closest to that of a scientist. Research engineers explore fundamental principles of chemistry, physics, biology, and mathematics in order to overcome barriers preventing advancement in their field. Engineering researchers differ from scientists in that they are interested in the application of a breakthrough, whereas scientists are concerned with the knowledge that accompanies a breakthrough.

Research engineers conduct investigations to extend knowledge using various means. One of the means used is conducting experiments. Research engineers may be involved in designing and implementing experiments and interpreting the results. Typically, the research engineer does not perform the actual experiment. Technicians usually do the actual testing. Large-scale experiments may involve the coordination of additional support personnel including other engineers, scientists, technologists, technicians, and craftspeople.

Research is also conducted using the computer. Computational techniques are developed to calculate solutions to complex problems without doing costly and time-consuming experiments. Computational research requires the creation of mathematical models to simulate naturally occurring phenomena under study. Research engineers might develop the computational techniques to perform the complex calculations in a timely and cost-effective manner.

Most research engineers work for some type of research center. A research center might be a university, a government laboratory such as NASA, or an industrial research center. In most research positions, an advanced degree is required. Often, a Ph.D. is needed. If research appeals to you, a great way to explore it is by doing an undergraduate research project with an engineering professor. This will allow you to observe the operation of a laboratory and find out how well you enjoy being part of a research team.

## Development Engineers

Development engineers take the knowledge acquired by the researchers and apply it to a specific product or application. Development engineers bridge the gap between laboratory research and full-scale production. The development function is often coupled with research in research and development (R&D) divisions. The researcher may prove something is possible in a laboratory setting; the development engineer shows that it will work on a large, production-size scale and under actual field conditions. This is done in pilot manufacturing plants or by using prototypes.

Development engineers are continuously looking for ways to incorporate researchers' findings into prototypes in order to test their feasibility for use in tomorrow's products. Often, an idea proven in a laboratory needs to be altered before it can be introduced on a mass production scale. The development engineers must identify these areas and work with the design engineers to correct them before full-scale production begins. An example of a development process is the building of concept cars within the automotive industry. These cars incorporate advanced design concepts and technology. The cars are then used as a test case to see if the design ideas and technology actually perform as predicted. The concept cars are put through exhaustive tests to determine how well the new ideas enhance a vehicle's performance. Each year, new technology is introduced into production automobiles that was first proven in development groups using concept vehicles.

## Testing Engineers

Test engineers are responsible for designing and implementing tests to verify the integrity, reliability, and quality of products before they are introduced to the public. The test engineer devises ways to simulate the conditions a product will encounter during its life. Test engineers work closely with development engineers in evaluating prototypes and pilot facilities. Data from these initial development tests are used to decide whether full production versions will be made or if significant changes are needed before a full-scale release. Test engineers work with design engineers to identify the changes in the product to ensure its integrity.

A challenge which engineers face is simulating the product's conditions during its life span, and doing so in a timely, cost-effective manner. Often the conditions the product will face are difficult to simulate in a laboratory. A constant problem for the test engineer is simulating the product's aging. An example of such a challenge is testing a pacemaker for regulating a patient's heart. This device is designed to last several decades, and an effective test cannot take 20 years or the product will be obsolete before introduced. The test engineer must simulate conditions within the human body to avoid exposing people to unnecessary risks.

Another challenge facing test engineers involves acquiring accurate and reliable data. The test engineer must produce data that show that the product is functioning properly or that identify areas of concern. Test engineers develop data acquisition and instrumentation methods to achieve this. Techniques such as radio telemetry can transmit data from the inside of the tested device. The measurement techniques must not interfere with the operation of the device, presenting a tremendous challenge for small, compact products. Test engineers must cooperate with design engineers to determine how the tested device can be fitted with instrumentation yet still meet its design intent.

Test engineers must have a wide range of technical and problem-solving skills. They also must be able to work in teams involving a wide range of people. They work with design and development engineers, technicians, and craftspeople, as well as management personnel.

## EXAMPLE 1

Test engineers must understand the important parameters of their tests. The development of a certain European high-speed train provides an example of the potential consequences, which may result when test engineers fail to understand these parameters. A test was needed to show that the windshield on the locomotive could withstand the high-velocity impacts of birds or other objects it might encounter, a common design constraint encountered in airplane design. The train test engineers borrowed a "chicken gun" from an aerospace firm for this test. A chicken gun is a mechanism used to propel birds at a target, simulating in-flight impact. With modern laws governing cruelty to animals, the birds are humanely killed and frozen until the test.

On the day of the test, the test engineers aimed the gun at the locomotive windshield, inserted the bird, and fired. The bird not only shattered the windshield but put a hole through the engineer's seat.

The design engineers could not understand what went wrong. They checked their calculations and determined that the windshield should have held. The problem became clear after the test engineers reviewed their procedure with the engineers from the aerospace firm from whom they had borrowed the equipment. The aerospace test engineers asked how long they had let the bird thaw.

The response was, "Thaw the bird?"

There is a significant difference in the impact force between a frozen eight-pound bird and that of a thawed bird. The test was successfully completed later with a properly thawed bird.

## Design Engineers

The design function, where the largest number of engineers are employed, is what many people think of when they think of engineering. The design engineer is responsible for providing the detailed specifications of the products society uses.

Rather than being responsible for an entire product, most design engineers are responsible for a component or part of the product. The individual parts are then assembled into a product such as a computer, automobile, or airplane. Design engineers produce detailed part dimensions and specifications to ensure that the components fit properly with adjoining pieces. They use modern computer design tools and are often supported by technicians trained in computer drafting software.

The part's form is also a consideration for the design engineer. Design engineers use their knowledge of scientific and mathematical laws, coupled with experience, to generate a shape to meet the part specifications. Often, a wide range of possibilities and considerations exist. In some fields, these considerations can be calculated. In others, such as in consumer products, the reaction of a potential customer to a shape may be as important as how well the product works.

The design engineer must verify that the part meets the reliability and safety standards established for the product. The design engineer verifies the product's integrity. This often requires coordination with analysis engineers to simulate complex products and field conditions, and with test engineers to gather data on the product's integrity. The design engineer is responsible for making design corrections based on the  test results.

In today's world of increasing competition, the design engineer must also involve manufacturing engineers in the design process. Cost is a critical factor in the design process and may make the difference between a product that fails or suc-ceeds. Communication with manufacturing engineers is, therefore, critical. Often, simple design changes can radically change a part's cost and affect the ease of the part's construction.

Design engineers also work with existing products. Their role includes re-designing parts to reduce manufacturing costs and time. They work on redesign-ing products that have not lived up to their expected life span or have suffered fail-ure. They also modify products for new uses. This usually requires additional analysis, minor redesigns, and significant communication between departments.

## Analysis Engineers

Analysis is an engineering function performed in conjunction with design, devel-opment and research. Analysis engineers use mathematical models and compu-tational tools to provide the information for design, development, or research en-gineers to help them perform their function.

Analysis engineers typically are specialists in a technology area important to the products or services. Technical areas might include heat transfer, fluid flow, vi-bration, dynamics, system modeling, and acoustics. They work with computer models of products to make these assessments. Experienced analysis engineers often possess an advanced degree.

In order to produce the required information, these engineers must validate their computer programs or mathematical models. To do so may require compar-ing test data with their predictions. This requires coordination with test engineers to design an appropriate test and to record the data. An example of analysis' role is temperature prediction in an aircraft engine. Material selection, component life estimates, and design decisions are based in large part on the temperature the parts attain and the duration of those temperatures. Heat transfer analyses are used to determine these temperatures. Engine test results validate the tempera-ture predictions. The permissible time between engine overhauls can depend on these temperatures. The design engineers use these results to ensure reliable aircraft propulsion systems.

## Systems Engineers

Systems engineers work with the overall design, development, manufacture, and operation of a complete system or product. Design engineers are involved in the design of individual components, but systems engineers are responsible for the integration of the components and systems into a functioning product.

Systems engineers are responsible for ensuring the components interface properly and work as a complete unit. Systems engineers are also responsible for identifying the overall design requirements. This may involve working with cus-tomers or marketing personnel to determine market needs. From a technical standpoint, systems engineers are responsible for meeting the overall design requirements.

Most engineers enter systems engineering after becoming proficient in an area important to the systems, such as component design or development. Some graduate work often is required prior to taking on these assignments. However, some schools offer an undergraduate degree in systems engineering.

## Manufacturing and Construction Engineers

Manufacturing engineers turn the specifications of the design engineer into reality. They develop processes to make the products we use every day. They work with diverse teams of individuals, from technicians on the assembly lines to management, in order to maintain the integrity and efficiency of the manufacturing process.

It is the responsibility of manufacturing engineers to develop the processes for taking raw materials and changing them into the finished pieces that the design engineers detailed. They utilize advanced machines and processes to accomplish this. As technology advances, new processes must be developed for manufacturing products.

The repeatability or quality of manufacturing processes is an area of increasing concern to modern manufacturing engineers. These engineers use statistical methods to determine the process precision. A lack of precision in the manufacturing process may result in inferior parts that cannot be used or that may not meet the customer's needs. Manufacturing engineers are concerned about product quality. High-quality manufacturing means lower costs since more parts are usable, resulting in less waste. Ensuring quality means having the right processes in place, understanding the processes and working with the people involved to make sure the processes are being maintained at peak efficiency.

Manufacturing engineers keep track of plant equipment. They schedule and track required maintenance to keep the production line moving. They must track the raw material inventories, partially finished parts, and completely finished parts. Excessive inventories tie up substantial amounts of cash used in other parts of a company. The manufacturing engineer is also responsible for maintaining a safe and reliable workplace, including the safety of facility workers and the environmental impact of the processes.

Manufacturing engineers must work with diverse teams including design engineers, tradesmen, and management. Just in time manufacturing practices reduce inventories needed in factories, but require manufacturing engineers at one facility to coordinate their operation with their counterparts at other facilities. They must coordinate the work of the line workers who operate the manufacturing equipment. Manufacturing engineers must maintain a constructive relationship with their company's trade workers.

Manufacturing engineers play a critical role in modern design practices. Since manufacturing costs are such an important component for the product success, the design process must take into account manufacturing concerns. Manufacturing engineers identify high-cost or high-risk operations in the product's design phase. When problems are identified, they work with the design engineers to generate alternatives.

In large item production, such as buildings, dams, and roads, the production engineer is called a construction engineer rather than a manufacturing engineer. The main difference is that the construction engineer's production facility is typically outdoors while the manufacturing engineer's facility is indoors in a factory. Construction engineer functions are the same as mentioned above when "assembly line" and "factory" are replaced with terms like "job site," reflecting the construction of a building, dam, or other large-scale project.

## Operations and Maintenance Engineers

After a new production facility is brought on-line, it must be maintained. The operations engineer oversees the ongoing facility performance. Operations engineers must have much expertise dealing with mechanical and electrical issues to maintain a production line. They must be able to interact with manufacturing engineers, line workers, and technicians who service the equipment. They must coordinate the technician service schedule to ensure efficient service of the machinery, thus minimizing its downtime impact on production.

Maintenance and operations engineers work in non-manufacturing roles. Airlines have staffs of maintenance engineers who schedule and oversee safety inspections and repairs. These engineers must have expertise in sophisticated inspection techniques to identify all possible problems.

Large medical facilities, as well as other service sector businesses, require operations or maintenance engineers to oversee their equipment operation. Emergency medical equipment must be maintained in peak working order.

## Technical Support Engineers

A technical support engineer serves as the link between customer and product, assisting with installation and setup. For large industrial purchases, the purchase price may include technical support. The engineer may visit the installation site and oversee a successful startup. For example, a power station for irrigation might require the technical support engineer to supervise the installation and assist the customer with any startup problems getting the product operational. In order to be effective, the engineer must have good interpersonal and problem-solving skills, as well as solid technical training. The technical support engineer may also troubleshoot problems with a product, e.g., serving on a computer company's help line. Diagnosing design flaws found in the field once the product is in use is another example. Technical support engineers do not have to have in-depth knowledge of each facet of the product. They must, however, know how to tap into such knowledge bases.

Modern technical support is being used as an added service feature. Technical support engineers work with customers to operate and manage their own company's equipment, as well as that of other companies. For example, a medical equipment manufacturer might sell its services to a hospital and offer to manage and operate its highly sophisticated equipment. The manufacturer's engineers would not only maintain the equipment but would also help the hospital use its facilities efficiently.

## Customer Support

Customer support functions are similar to those of technical support as a link be-tween the manufacturer and the customer. However, customer support personnel also are involved in the business aspect of the customer relationship. Engineers are often used for this function because of their technical knowledge and prob-lem-solving ability. These positions require experience with products and cus-tomers and require some business training as well.

The customer support person works with technical support engineers to en-sure a high degree of customer satisfaction. Customer support is also concerned with actual or perceived value by the customer. Are they getting what they paid for? Is it cost effective to continue current practices? Customer support person-nel are involved in warranty issues, contractual agreements, trade-in value, or credits for existing equipment. They work closely with technical support engineers and management personnel.

## Sales

Engineers are valuable members of the sales force in numerous companies. These engineers must have interpersonal skills conducive to effective selling. Sales engineers bring many assets to their positions.

Engineers have the technical background to answer customer questions and concerns. They are trained to identify which products, services, and solutions are right for the customer and how they can be applied. Sales engineers can identify other applications or products that might benefit the customer once they become familiar with their customer's needs.

In some sales forces, engineers are utilized because the customers are engi-neers themselves and have engineering-related questions. When airplane manu-facturers market their aircraft to airlines, they send engineers. The airlines have engineers whose technical concerns are overseeing the operation and mainte-nance of their aircraft. Sales engineers have the technical background to respond to these questions.

## Consulting Engineers

Consulting engineers are either self-employed or they work for a firm that does not provide goods and services directly to consumers. Such firms provide techni-cal expertise to the organizations that do. Many large companies do not have technical experts for all areas of operation. Instead, they use consultants to han-dle technical issues as they arise.

For example, a manufacturing facility in which a cooling tower is used in part of the operation might have engineers who are well versed in the manufacturing processes but not, however, in cooling tower design. The manufacturer would hire a consulting firm to design the cooling tower and related systems. The consultant might oversee the installation of such a system or simply provide a report with rec-ommendations. After the system is in place or the report is delivered, the con-sultant would move on to another project with another company.

Consulting engineers also might be asked to evaluate the effectiveness of an organization. In such a situation, a consultant team might work with a customer and provide suggestions and guidelines for improving the company's internal processes. These might be design methods, manufacturing operations, or business practices. While some consulting firms provide only engineering-related expertise, other firms provide engineering and business support, requiring the consulting engineers to work on business-related issues as well as technical issues.

Consulting engineers interact with a wide range of companies on many projects from all engineering disciplines. Often a consultant needs to be registered as a professional engineer in the state where he or she does business.

## Management

In many instances engineers work themselves into project management positions and, eventually, into full-time management. National surveys show that more than one half of all engineers will be involved in some management responsibilities, supervisory or administrative, before their career is over. Engineers in management are chosen for their technical ability, problem-solving ability, and leadership skills.

Engineers may manage other engineers and support personnel, or they may oversee the corporation's business aspects. Often, prior to being promoted to this management level, engineers acquire some business or management training. Some companies provide this training for their employees or offer incentives to take management courses in the evening on their own time.

## Other Fields

Some engineering graduates enter fields other than engineering, such as law, education, medicine, and business. Patent law is one area in which an engineering or science degree is almost essential. In patent law, lawyers research, write, and file patent applications. Patent lawyers must have the technical background to understand what an invention does so that they can describe the invention and, thus, protect it for the inventor.

Another area of law that has become popular for engineering graduates is corporate liability law. Corporations are faced with decisions every day over whether to introduce a new product and must weigh potential liability risks. Lawyers with technical backgrounds have the ability to weigh technical as well as legal risks. Such lawyers are also used in litigation over liability issues. Understanding what an expert witness is saying in a suit can be a tremendous advantage in a courtroom and can enable a lawyer to cross-examine the expert witness. Often, when a corporation is sued for product liability, the lawyers defending the corporation have technical and engineering backgrounds.

Engineers are involved in several aspects of education. The one with which students are most familiar is that of an engineering professor. College professors usually have their Ph.D.s. Engineers with Master's degrees can teach at community colleges and in some engineering technology programs. Engineering graduates also teach in high schools, middle schools, and elementary schools. To do this full-time usually requires additional training in educational methods, but the engineering background is a great start. Thousands of engineers are involved in part-time educational projects, visiting classes as guest speakers to show students what real engineers do (See example two below). The American Society for Engineering Education (ASEE) is a great resource if you are interested in engineering education.

Engineers also find careers in medicine, business, on Wall Street, and in other professions. In modern society, with its rapid expansion of technology, the combination of problem solving ability and technical knowledge makes engineers valuable and versatile.

## EXAMPLE 2

Listed below are things that a mechanical engineer (ME) might perform in a design function. This illustrates how a specific job in one industrial sector can vary from the same job in another sector. Everything listed below is what a mechanical engineer might do in the specified field.

- **Aerospace.** Design of an aircraft engine fan blade: Detailed computer analyses and on-engine testing are required for certification by the FAA. Reliability, efficiency, cost, and weight are all design constraints. The design engineer must push current design limits, optimizing design constraints, make tradeoffs between efficiency, cost, and weight.
- **Biomedical.** Design of an artificial leg and foot prosthesis giving additional patient mobility and control: Computer modeling is used to model the structure of the prosthesis and the natural movement of a leg and foot. Collaboration with medical personnel to understand the needs of the patient is critical to the design process. Reliability, durability, and functionality are the key design constraints.
- **Power.** Design of a heat recovery system in a power plant, increasing plant productivity: Computer analyses are performed as part of the design process. Cost, efficiency, and reliability are the main design constraints. Key mechanical components can be designed with large factors of safety since weight is not a design concern.
- **Consumer products.** Design of a pump for toothpaste for easier dispensing: Much of the development work might be done with prototypes. Cost is a main design consideration. Consumer appeal is another consideration and necessitates extensive consumer testing as part of the development process.
- **Computer.** Design of a new ink-jet nozzle with resolution approaching laser printer quality: Computer analyses are performed to ensure that the ink application is properly modeled. Functionality, reliability, and cost are key design concerns.

# LEARNING ACTIVITIES

## Activity 4.1    Stress Analysis

Time:      3–4 hours

## CASE STUDY

An exercise stress test is a special type of electrocardiogram (EKG) that compares the heart's electrical activity at rest and under exertion. The test is noninvasive, generally safe, and painless. A physician may recommend an exercise stress test for a number of reasons:

- To diagnose conditions such as coronary artery disease (a chronic disease in which there is hardening or atherosclerosis of the arteries). Coronary artery disease can be diagnosed through an exercise stress test if it is causing cardiac ischemia (in which the heart is not getting enough oxygen-rich blood) and/or arrhythmia (irregular heart rhythm).
- To diagnose a possible heart-related cause of symptoms such as chest pain, shortness of breath, or lightheadedness.
- To determine a safe level of exercise.

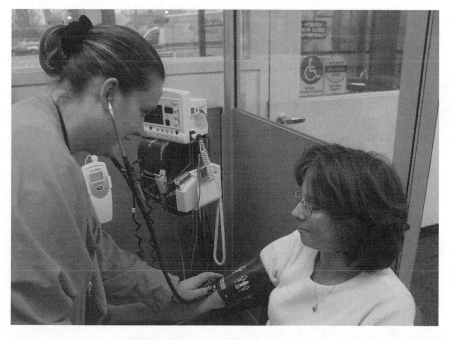

**A nurse used many different pieces of equipment to measure blood pressure. Illustration copyright © medmovie.com 2003. All rights reserved.**

- To check the effectiveness of a balloon angioplasty (in which plaque in the arteries is pushed back against the artery walls to make more room for blood flow), or other procedures that have been done.
- To predict a future risk of dangerous heart-related conditions, such as heart attack.

Depending on the results of the exercise stress test, the physician may recommend additional testing, e.g., a nuclear stress test or a cardiac catheterization).[1]

Students will complete a test using their own backpacks and classroom books. This case study should be used when conditions are acceptable for being outside.

1. Students should take their pulse at rest in the classroom after they have been sitting for at least five minutes. Log this rate in the data table.
2. Next, students should walk around the school block at a regular pace. The students should take their pulse when a COMPLETE lap is completed. Log these data and the lap time into the data chart.
3. A backpack with books is now needed for the next step. Make sure to rest for about five minutes before going on the next walk. Students should now put their backpacks on and walk around the block, keeping track of the time it takes to complete one lap. At the completion of one lap, they should take their pulse and note the time it took to get around on this lap. Log these data in the chart.
4. The final stage is to use the same backpack to complete the lap with an additional load of books that are carried by hand. Again, at the completion of one lap, the students should take their pulse and note the time it took to get around on this lap. Log the data in the chart.

| Test number | Time to complete lap | Pulse rate |
|---|---|---|
| 1 (at rest) | N/A | |
| 2 (no pack) | | |
| 3 (backpack) | | |
| 4 (backpack and books) | | |

## RESOURCES

www.heartcenteronline.com HeartCenterOnline provides a valuable service to patients with heart-related conditions and individuals seeking to minimize their risk of developing heart conditions. Their foremost objective is demonstrating their commitment to an individual's cardiovascular health and wellness.

The text in part is used with permission of HeartCenterOnline

# Chapter 5

# Engineering Disciplines

## Introduction

Engineering careers continually evolve over time. Some disciplines have even become extinct because of society's changing needs. As we progress through this new century, it is inevitable that new engineering disciplines will develop and more will fall to the wayside.

The salary information provided is courtesy of the National Society of Professional Engineers. This information is based on an average salary of an engineer with a BA or BS.

## ACOUSTICAL ENGINEERING

What factors determine a comfortable environment? Do the people and noises around us help or hinder our productivity? How can we better reproduce the subtle nuances of recorded music? Acoustical engineers face these and more questions everyday.

Acoustical engineering is a fairly confusing field, as it is part of both architectural engineering and engineering mechanics. Most commonly, however, acoustical engineers are drawn upon to plan, perfect, or improve the sound of an architectural space, hence the connection to architectural engineering. When architects design the plans for a building, they carefully consider the building's usage, whether it is office, commercial or residential. In any of these cases, placement of essential equipment and skeletal structures of the building is dictated by function, the amount and type of required space. Once these items are incorporated into the design, the acoustical engineer is usually brought in to set the mood of the environment.

From birth, much human function is dependent upon the workings of the ear and its related substructures. The tiny hairs within the ear are responsible for balance, and the eardrum is responsible for auditory perception. Seasickness and airsickness are related to the ear's inability to perceive a horizon or constant level. Acoustical engineers not only assist architects, they also investigate how different noises and background sounds affect productivity. For instance, acoustical engineers can help design and treat a building so that it exhibits acoustically dead properties. This means that all echo and background noise at any frequency in the human range is eliminated. However, such dead spaces are not always the easiest to work in. Imagine going to a museum, and hearing none of the echo, or grandeur implied by such a space. Imagine going to an office building with the reverb, or echo, of a church or concert hall. Acoustical engineers find the proper balance of background noise and foreground conversation.

When working on large, high-productivity office spaces, the hardest sounds to eliminate are those of the HVAC or air conditioning systems. Mechanical engineers could conceivably eliminate the source noises, but that is not usually the case. HVAC resonates between 50 and 70 Hertz. This is approximately the range of human hearing where sound becomes omni-directional, i.e., we cannot locate the sound's source. This is also the range of human hearing where we can feel a small amount of vibration, which, after prolonged exposure, could cause nausea. Nausea is not a productivity enhancing feeling, so acoustical engineers try to isolate HVAC systems completely from working spaces.

Acoustical engineers deal with two basic properties of sound: reflection and absorption. When used in conjunction, these properties can create a room with a comfortable background noise level, as well as an easy transfer of primary sounds (person-to-person conversation). In the late 1980s, George Lucas began the THX sound certification system. The system dealt with the problem of non-uniform sound between theaters. Lucas envisioned every theater having the same quality of sound, equaling that of the studio where the movie was mixed. To make the sounds as similar as possible, the stringent certifications not only dealt with what materials and shapes could be used to construct the space, but also the brands and styles of speakers. In the THX certification process, the horizontal and vertical dispersions of sounds from each speaker are carefully measured, and the speakers are placed accordingly. Extensive sound deadening materials are used in THX theaters to deaden the room. However, some spaces are left to be semi-reflective, such as the area next to the rear surround speakers. Each speaker is designed to play part of its signal directly to the listener, and the remaining part off the walls, to be reflected to the listener. How were these direct/reflected speakers designed? Again, acoustical engineers were responsible for the development.

Acoustical engineering, as a broad definition, deals with the properties of sound in a space. This definition implies that it deals not only with large spaces, like auditoriums, but also small spaces, like speaker boxes and car doors. In almost every high-end speaker system, the boxes that house the speaker drivers are specially engineered to accentuate certain characteristics of the driver. Acoustical engineers examine the properties of the speaker, such as linearity (ability of a speaker to reproduce the same ratio of frequencies at all volume levels), excursion (movement of speaker cone), and power handling, in order to design a perfectly matched enclosure. B&W loudspeakers are some of the finest conventional technology speakers in the world. Their model line starts at $120 per pair and extends to Nautilus speakers, at $110,000 per pair. Their flagship line, the Nautilus line, represents the culmination of over seven years of research. Acoustical engineers worked with mechanical, electrical, aerodynamic, and materials science engineers to produce strong, efficient, linear drivers. Then, acoustical engineers at B&W worked endlessly to create an enclosure that would allow the driver to produce all frequencies in the human range effortlessly and without coloration. Eventually, they succeeded, by creating special cones for each driver. Each cone is shaped in an exponentially decreasing size. This design minimizes standing waves and unwanted reflections within the enclosures, as well as allowing the maximum amount of sound to be sent forward. This is just one example of some of the amazing accomplishments of acoustical engineers.

Whether examining the innumerable surfaces of a church, or drawing CAD plans for a subwoofer enclosure, acoustical engineers are always making the sound quality in our lives better. While they are in high demand, there are few acoustical engineers.

## AUTOMOTIVE ENGINEERING

The automobile is the most important means of transportation for millions of people around the globe. Worldwide, there are more than 400 million passenger cars plus more than 100 million light trucks, such as vans and pickups. People depend on their cars and trucks to travel to and from work, run errands, visit friends and relatives, and take vacations.[i]

The development of automobiles has had an enormous effect on people's way of life throughout much of the world. Probably no other invention, discovery, or technological advance has created greater or more rapid changes in society.[ii]

This invention has also had a huge economic impact. Developed nations such as the United States, Japan, Germany, and Italy depend on automotive production to provide jobs for millions of workers. Gas stations, motels, restaurants, and other businesses that serve automobile travelers are of major importance to the economic well-being of all developed countries and increasingly to developing ones.[iii]

The people behind this life-changing invention are automotive engineers. Automotive engineering encompasses a broad field. Automotive engineers plan, coordinate, and implement the specifications for the new car, engineering every needed part. Such parts entail everything from steering wheels and road wheels to headlights and taillights. Engineers rely on computer-aided engineering programs to design and draw parts, combine the parts into components, and integrate the components into the car's systems.[iv] As engineers design the body, computers show how well the car will protect occupants in a crash. But actual cars must be crashed into walls to be sure they meet government standards for impact protection. Prototypes cost hundreds of thousands of dollars. If the prototypes fail the tests, they must be redesigned and reengineered until they are able to pass.

Automotive engineers are responsible for taking the aesthetic design developed by the automotive designer and making the mechanical aspects of the car fit into it. Once the car body has been developed, automotive engineers, usually trained with Bachelors degrees in mechanical engineering, design the appropriate suspension, drive train, and other vehicle components. By doing this job, automotive engineers earn between 36,000 and 55,000 dollars per year. In addition,

automotive engineers may be employed by an automotive components manufacturer, developing car parts ranging from seats to door handles to shifter knobs.
Automotive engineers will be needed as long as there is an automotive industry
and as long as cars prove the major form of transport.

Helping the world move from point A to point B presents many challenges.
Some of these challenges are: emissions laws, cost, shared platforms, performance, and consumer demands.

Emissions Laws—As states and the federal government get tougher on pollution,
the engineers responsible for engine design must make their engines burn fuel
more cleanly without sacrificing performance. In addition, those responsible for
transmissions and drive trains must develop more efficient gear ratios.

Cost—Automotive engineers must develop a product that is financially viable and
appealing to consumers. This may mean making trade offs, sacrificing certain features for others, or trimming the cost of component production by making it easier to manufacture.

Shared Platforms—One of the many ways auto manufacturers work toward lower
costs lies in sharing the platform of the car. This translates into cars of different
brands, but of the same parent company, sharing sixty to eighty percent of their
components and having similar overall chassis dimensions. Automotive engineers must then rise to the challenge of tuning the engine and suspension to give
two similar cars different characteristics to appeal to different markets. A prime example of this is the VW-Audi vehicles that have as many as five different cars on
the same platform across their multiple European divisions.

Performance—With consumers constantly wanting their new car to be better than
their last, automotive engineers must constantly improve their company's vehicles. From making engines last longer to adding horsepower without skyrocketing
the price, the engineers must find ways to improve their vehicles.

Consumer Demands—Automotive engineers cannot design a car they think
would be cool; instead, they must design what consumers want to buy. In recent
years, the demand has been for trucks. Auto companies have called on their engineers to develop more trucks, placing more focus on them, making their truck
more desirable to the buying public than that of their competitors.

Automotive engineers develop and test self-propelled vehicles. Their job provides many challenges but it is combined with job security due to the size of their
employers and the necessity of the automotive engineers work.

An automotive engineer does many jobs every day. They may design a prototype of a new SUV, limit the amount of automobile pollution, or design a better,
more fuel efficient car.

Automotive engineers can go into many different professions. Such professions include mechanical engineering, applications engineering, electronic engi-

neering, test engineering, laboratory technology/plastics engineering, and program engineering.

Many companies, such as Ford, GM, and Chrysler, hire automotive engineers. Without an engineering degree, though, getting such a job is difficult. No certification exists to work for a company. The average salary, depending on the engineer's demand, ranges from 50,000 to 70,000 dollars and larger signing bonuses are possible.

Advancement possibilities are endless. This is a profession that will be in demand for years to come. The invention of the automobile was one that has made our world what it is. Engineers are constantly designing new models that consume less fuel and are better for the environment. In the coming years, our supply of petroleum may need to be rationed. Engineers are designing and building solar powered, battery powered, and hydrogen fuel cell-powered cars.

To be an automotive engineer, one must have good interpersonal skills with customers, and excellent written and oral communication skills.[v] Also one should have strong multitasking abilities, technical knowledge, and design experience.

## AEROSPACE ENGINEERING (AERO OR ASTRONAUTICAL)

Aerospace engineers are responsible for developing planes and spacecraft that weigh up to half a million pounds or travel at up to 17,000 miles per hour. Aerospace engineers design, develop, test, and help manufacture aircraft, missiles, and spacecraft. They supervise the manufacture of these products. Aerospace engineers develop new technologies for military or commercial uses. Aerospace engineers specialize in many fields ranging from propulsion to thermodynamics to spacecraft.

Aerospace engineering can be divided into two fields. The first field is aeronautical engineering, which works with aircraft like private jets, helicopters and commercial airplanes. The second field is astronautical engineering, which works with spacecraft like the Apollo Project, the Mars Lander, and the International Space Station.

Aeronautical engineers develop technology for use in aviation, defense systems, and attack projectile weaponry. They specialize in fields such as structural design, guidance, navigation and control systems, instrumentation and communication, and production systems. They may also specialize in commercial transport, military fighter aircraft, military and commercial helicopters, missiles, and rockets. Aeronautical engineers may also choose to develop expertise in aerodynamics, thermodynamics, celestial mechanics, propulsion, and acoustics.

Astronautical engineers develop technology for use in space exploration. They develop planes that can operate in space and on earth. They develop booster rockets for spacecraft and they may specialize in propulsion, guidance and control systems, structural design, and defense systems.

Many responsibilities come along with the job of an aerospace engineer. The engineers are responsible for the conception, design, development, and production of many products. For example, engineers at National Aeronautics and Space Administration (NASA) are responsible for building spacecraft that are safe and able to achieve flight in space. The engineers test the spacecraft in many ways to assure its successful performance. They design the propulsion and guidance systems. These engineers have to be responsible for the lives of others.

Most successful engineers go to a four-year college or university and major in a field that applies to engineering. But engineers can be trained at a technical college. After this first step, some engineers go on to formal or informal training with an employer. A formal training program would be like an apprenticeship. Informal on-the-job training is useful. Some people also acquire engineering skills in the armed forces or through a home study program. Licensing or an examination is required to practice independently in most fields.

Advancement opportunities would be through promotions to project or research supervisor. Some engineers continue their education in order to acquire skills for management and administrative positions in high tech industry. Some of them however remain as teachers and project managers.

Statistics indicate that the federal government, specifically the Department of Defense (DoD) and NASA, hire one out of seven of aerospace engineers. California, Washington, Texas, and Florida are the states in which most aerospace engineers are hired.

The American Institute of Aeronautics and Astronautics (AIAA) is one of the most prominent aerospace professional societies. It is composed of the following seven technical groups, which include 66 technical committees.

- Engineering and Technology Management
- Aircraft Technology Integration and Operations
- Propulsion and Energy
- Space and Missile Systems
- Aerospace Sciences
- Information and Logistics Systems
- Structures, Design and Testing

Aeronautical/aerospace/astronautical engineers' salaries according to the National Society of Professional Engineers:

| Average in 2003 | 25%tile | 75%tile | Average in 2000 | 25%tile | 75%tile |
|---|---|---|---|---|---|
| $75,956 | $50,508 | $90,000 | $56,639 | No Data | No Data |

## AGRICULTURAL ENGINEERING

Agricultural engineering traces its roots back thousands of years to the time when people began to examine ways to produce food and food products more efficiently. Today, the role of the agricultural engineer is critical to our ability to feed the ever-expanding world population. The production and processing of agricultural products is the primary concern of agricultural engineers. Specializations within agricultural engineering include power machinery, food processing, soils and water, structures, electrical technologies, and bioengineering.

The mechanical equipment used on modern farms is highly sophisticated and specialized. Harvesting equipment not only removes the crop from the field but also begins to process it and provides information to the farmers on the quality of the harvest. This requires complicated mechanical and electrical systems that are designed and developed by the agricultural engineer.

Once the food is harvested, it must be processed before it reaches the marketplace. Many different technologies are involved in the efficient and safe processing and delivery of food products. Food process engineers are concerned with providing healthier products to consumers who increasingly rely on processed food products. This often requires the agricultural engineer to develop new processes. Another modern concern is increased food safety. Agricultural engineers design and develop means by which food is produced free from contamination, such as the irradiation techniques used to kill harmful bacteria.

Effective soil and water resource management is a key aspect of productive agriculture. Agricultural engineers design and develop means to address effective land use, proper drainage, and erosion control. This includes such activities as designing and implementing an irrigation system for crop production and designing a terracing system to prevent erosion in a hilly region.

Agricultural structures are used to house harvested crops and livestock and their feed. Agricultural engineers design structures including barns, silos, dryers, and processing centers. They look for ways to minimize waste or losses, optimize yields and protect the environment.

Electrical and information technology development is another area important to the agriculture community since farms can be located in isolated regions. Agricultural engineers design systems that meet the needs of the rural communities.

Bioengineering has rapidly evolved into a field with wide application in health products as well as agricultural products. Agricultural engineers are working to harness these rapidly developing technologies to further improve the quality and quantity of the agricultural products necessary to continue to feed the world's population.

The American Society of Agricultural Engineers (ASAE) is one of the most prominent professional societies for agricultural engineers. It has eight technical divisions that address areas of agricultural engineering.

- Food Processing
- Information and Electrical Technologies
- Power and Machinery
- Structures and Environmental
- Soil and Water
- Forest
- Bioengineering
- Aquaculture

Agricultural engineers' salaries according to the National Society of Professional Engineers:

| Average in 2003 | 25%tile | 75%tile | Average in 2000 | 25%tile | 75%tile |
| --- | --- | --- | --- | --- | --- |
| $65,898 | $51,750 | $73,500 | $62,614 | $45,345 | $72,750 |

## BIOENGINEERING (BIOMEDICAL, BIOMECHANICAL, BIOCHEMICAL)

Bioengineering is the application of engineering principles to biological systems. This can be seen in the production of food or in genetic manipulation to produce a disease-resistant plant or animal strain. Bioengineers work with geneticists to produce bioengineered products. New medical treatments are produced using genetically altered bacteria to produce human proteins needed to cure diseases. Bioengineers also work with geneticists in producing these new products in the massive quantities needed by consumers.

Biomedical engineering encompasses many different fields of study, including biology, chemistry, physics, technology science, and medicine. Biomedical engineers use skills gained from the studies of these fields to solve health care problems.

Biomedical engineering is one of the newer fields of engineering, first recognized as a discipline in the 1940s. Its origins, however, date back to the first artificial limbs made of wood or other materials. Biomedical engineering encompasses the three basic categories of medical, clinical, and bioengineering.

Biomedical engineering is a broad field that overlaps with several other engineering disciplines. In some institutions, it may be a specialization within another discipline. Biomedical engineering applies the fundamentals of engineering to meet the needs of the medical community. Because of the wide range of skills needed in the fields of engineering and medicine, biomedical engineering often requires graduate study.

The medical aspect of biomedical engineering involves the design and development of devices to solve medical challenges. This includes designing mechanical devices such as prostheses, which give individuals more mobility. It also involves the development of the chemical processes necessary to make an artificial kidney function and the electrical challenges in designing a new pacemaker.

Medical engineers develop instrumentation for medical uses including non-intrusive surgical instruments. Much of the trauma of surgery results from the incisions made to gain access to the area of concern. Procedures allowing a surgeon to make small incisions, as done in arthroscopic surgery, have reduced the risk and recovery time for patients.

Biomedical engineering is one of the fastest growing fields in engineering. As people live longer and as doctors are able to prolong life by extraordinary means, a greater need exists for medical monitoring equipment and integrated medical devices to replace failing organs, limbs, etc. The advances in technology, such as smaller microchips, longer lasting batteries, and stronger materials, have allowed much smaller, less cumbersome devices to be developed by biomedical engineers. Improved pacemakers, ocular implants, and artificial heart valves are just some of the newest advances in biomedical engineering. Researchers soon hope to be able to engineer and grow new tissue to replace diseased, damaged, or missing tissue in victims of diseases such as cancer or in victims of injuries caused by gunshot wounds or fire.

To enter the field of biomedical engineering, one needs to pursue an undergraduate, and often a graduate, degree in biomedical engineering, usually at an ABET accredited school. Once obtaining a Bachelor's and a Master's degree, one could also pursue a Ph.D. to enter a college-level teaching profession. If one pursues a practicing career, a biomedical engineer can be promoted through salary increases or through promotion to management positions.

Biomechanical/biomedical engineers' salaries according to the National Society of Professional Engineers: (Data were not available for BA or BS and an average was computed among all that returned information for this category.)

| Average in 2003 | 25%tile | 75%tile | Average in 2000 | 25%tile | 75%tile |
|---|---|---|---|---|---|
| $94,137 | $80,000 | $120,000 | ————————No Data———————— | | |

## CHEMICAL ENGINEERING

Chemical engineering differs from most of the other fields of engineering in its emphasis on chemistry and the chemical nature of products and processes. Chemical engineers take what chemists do in a laboratory and, by applying fundamental engineering, chemistry and physics principles, design and develop processes to produce products for use in our society. These products include detergents, paints, plastics, fertilizers, petroleum products, food products, pharmaceuticals, and electronic circuit boards.

The most common employment of chemical engineers is in the design, development, and operation of large-scale chemical production facilities. In this area, the design functions involve the design of the processes needed to safely and reliably produce the final product. This may involve controlling and using chemical reactions, separation processes, or heat and mass transfer. While the chemist might develop a new compound in the laboratory, the chemical engineer would develop a process to make the compound in a pilot plant, which is a small-scale version of a full-size production facility. An example would be the design of a process to produce a lower-saturated fat product with the same nutritional value yet still affordable. Another example would be the development of a process to produce a stronger plastic used for automobile bumpers.

The nature of the work for chemical engineers, like all other engineers, is solving problems. Chemical engineers, using chemical principles, specialize in solving problems that involve the production and use of chemicals. They deal with a wide variety of materials some of which are medicines, bleach, polymers, fuels, cosmetics, and synthetic fibers. Chemical engineers work to design and improve equipment that produces the chemicals created in labs. An ongoing mission of environmentally friendly chemical engineers is to find a way to reduce a factory's pollution output. They do this by designing less wasteful methods of chemical creation or more efficient methods. Chemical engineers even play a part in eliminating world hunger by developing processes to produce fertilizers.

Chemical engineers can hold many jobs. Two out of every three chemical engineers, in 1996, worked for the manufacturing industry. Pharmaceutical, paper, and petroleum plants hire chemical engineers to design more efficient production facilities that produce less pollution. They also hire chemical engineers as workers and supervisors. Private firms hire some chemical engineers for consulting. Even government agencies, like the Food and Drug Administration (FDA), hire chemical engineers for research positions.

Many of the chemicals and their byproducts used in industry can be dangerous to people and the environment. Chemical engineers must develop processes that minimize harmful waste. They work with new and traditional processes to treat hazardous byproducts and reduce harmful emissions.

Chemical engineers are  active in the bioproducts arena. This includes the pharmaceutical industry, where chemical engineers design processes to manufacture affordable drugs. Geneticists have developed the means to produce human proteins artificially that can be used successfully to treat many diseases. They use genetically altered bacteria to produce these proteins. It is the job of the chemical engineer to take the process from the laboratory and apply it to a larger-scale production process to produce the quantities needed by society.

Educational requirements to be a chemical engineer include a strong grounding in math and the sciences to earn a Bachelor's degree. Chemical engineers often have to apply different ideas from chemistry, biology, physics, and earth science to produce an end product. They also need to be able to understand advanced mathematics like calculus. Chemical engineers need to use a computer in their work, so they should have at least a working knowledge of computer science.

The American Institute of Chemical Engineering (AIChE) is one of the most prominent professional organizations for chemical engineers. It is organized into 13 technical divisions that represent the diversity of the chemical engineering field:

- Catalysis and Reaction Engineering
- Computing and Systems Technology
- Engineering and Construction Contracting
- Environmental
- Food, Pharmaceutical and Bioengineering
- Forest Products
- Fuels and Petrochemicals
- Heat Transfer and Energy Conversion
- Management
- Materials Engineering and Sciences
- Nuclear Engineering
- Safety and Health
- Separations

Chemical engineers salaries' according to the National Society of Professional Engineers:

| Average in 2003 | 25%tile | 75%tile | Average in 2000 | 25%tile | 75%tile |
|---|---|---|---|---|---|
| $94,904 | $64,000 | $108,123 | $94,762 | $64,875 | $106,278 |

## CIVIL ENGINEERING

Civil engineering is the oldest branch of engineering. It is responsible for designing and supervising the construction of roads, buildings, airports, tunnels, bridges, and water and sewage systems. The main objective of each engineer is to design systems that are functional, efficient, and durable. While they engineer systems, they are also concerned for the environment's well-being. They record all environmental impact data and then from these data, they plan projects that are not environmentally harmful. While being environmentally aware, civil engineers must keep in mind other factors that might affect their projects, e.g., population shifts, urban planning/renewal efforts, zoning laws, and building codes. The broad field of civil engineering includes these categories: structural, environmental, transportation, water resources, surveying, urban planning, and construction engineering.

Structural engineers are the most common type of civil engineer. They are primarily concerned with the integrity of the structure of buildings, bridges, dams and highways. Structural Engineers evaluate the loads and forces to which a structure will be subjected. They analyze the structural design with regard to how it will withstand earthquakes, wind forces, snow loads, or wave impacts, depending on the area in which the building will be constructed.

A related field of structural engineering is architectural engineering, which is concerned with the form, function, and appearance of a structure. The architectural engineer works alongside the architect to ensure the structural integrity of a building. The architectural engineer combines structural and functional concerns with the aesthetic concerns of the architect.

Civil engineers in the environmental area are concerned with the proper disposal of residential and industrial wastes. They may design and adapt landfills and waste treatment facilities to meet community needs and standards. Industrial waste presents a greater challenge because it may contain heavy metals or other toxins that require specialized disposal procedures. Environmental engineering encompasses more and is detailed in a later section of this chapter.

Transportation engineers are concerned with the design and construction of highways, railroads, and mass transit systems. They are involved in the optimization and operation of the systems, e.g., traffic engineering. Civil engineers develop the tools to measure the need for traffic control devices such as signal lights and to optimize these devices to allow proper traffic flow. This can become extremely complex in cities where the road systems were designed and built long ago but the areas around those roads have changed significantly.

Civil engineers work with water resources as they construct and maintain dams, aqueducts, canals, and reservoirs. Water resource engineers are charged with providing safe and reliable supplies of water for communities. This includes the design and operation of purification systems and testing procedures. As communities continue to grow, so do the challenges for civil engineers to produce safe and reliable water supplies.

Surveyors locate property lines and establish alignment and proper placement of engineering projects. Surveyors use satellite technology as well as aerial and terrestrial photogrammetry and rely on computer processing of photographic data.

Urban planning engineers are involved in this process. They incorporate the components of a city (e.g., buildings, roads, schools, airports) to meet the population's needs. These needs include adequate housing, efficient transportation, and open spaces. The urban planning engineer always looks to the future to fix problems before they reach a critical stage.

Construction engineers are concerned with the management and operation of construction projects and are interested in the improvement of construction methods and materials. Construction engineers design and develop building techniques and building materials that are safer, more reliable, cost effective and environmentally friendly.

The American Society of Civil Engineers (ASCE) is one of the most prominent professional organizations for civil engineers. It is organized into the following 16 technical divisions covering the breadth of civil engineering:

- Aerospace
- Air Transport
- Architectural Engineering
- Construction Division
- Energy
- Engineering Mechanics
- Environmental Engineering
- Geomatics
- Highway
- Materials Engineering
- Pipeline
- Urban Planning and Development
- Urban Transportation
- Water Resources Engineering
- Water Resources Planning and Management
- Waterways, Ports, Coastal and Ocean Engineering

Civil engineers' salaries according to the National Society of Professional Engineers:

| Average in 2003 | 25%tile | 75%tile | Average in 2000 | 25%tile | 75%tile |
|---|---|---|---|---|---|
| $85,341 | $57,242 | $95,000 | $74,113 | $50,500 | $83,000 |

## COMPUTER ENGINEERING

Computer engineers focus on a balanced mix of hardware and software. They design and build computer-related hardware products for many applications. Those applications include personal computers, cell phones, automobiles, aircraft, washing machines, and anything that has a computer or program in it.

Computer engineers usually apply the theories and principles of science and mathematics to design hardware, software, networks, and processors. Computer engineers also build computer chips. They often work as part of a team that designs new computing devices, computer-related equipment, systems, and software.

Much of what computer and electrical engineers do overlaps. Many computer engineering programs are part of electrical engineering or computer science programs. However, computer technology and development have progressed so rapidly that this field requires specialization in this field. Given the wide range of computer applications and their continued expansion, computer engineering has an exciting future.

Computer engineering is similar to yet distinct from computer science. Both fields are extensively involved with the software design and development. The main difference is that the computer scientist focuses primarily on the software and its optimization. The computer engineer focuses primarily on computer hardware or the machine itself. Software written by the computer engineer is often designed to control or to interface more efficiently with the hardware of the computer and its components.

Computer engineers are involved in the design and development of operating systems, compilers, and other software that requires efficient interfacing with the computer's components. They work to improve computer performance by optimizing the software and hardware in applications such as computer graphics.

Computer engineers develop and design electronics to interface with computers. These include modems, Ethernet connections, and other means of data transmission. Computer engineers created the devices that made the Internet possible.

In addition to having computers communicate with each other, the computer engineer is interested in having computers work together. This may involve increasing the network communication speed or using tightly coupled multiple processors to improve the computing speed.

Security is becoming a bigger concern as more information is transferred using computers. Computer engineers are developing new means of commercial and personal security to protect the integrity of electronic communications.

Computer engineers work on the design of computer architecture. Designing faster and more efficient computing systems is a tremendous challenge. As faster computers are developed, applications arise for even faster machines. The continual quest for faster and smaller microprocessors involves overcoming barriers introduced by the speed of light and by circuitry so small that the molecular properties of components become important.

Everyday computer engineers work in teams to design, implement, and test computer systems. They meet with the team members daily to discuss design problems. They design circuit boards, chips, and peripherals, (disks, memory, etc.). Each day, computer engineers simulate computer systems to check the design. They test computer systems in manufactured products and research and develop future computers.

Computer engineers' salaries according to the National Society of Professional Engineers:

| Average in 2003 | 25%tile | 75%tile | Average in 2000 | 25%tile | 75%tile |
|---|---|---|---|---|---|
| $102,876 | $61,423 | $126,000 | $77,109 | No Data | No Data |

## CONSTRUCTION ENGINEERING

Construction engineers are concerned with the management and operation of construction projects. They are interested in the improvement of construction methods and materials. Construction engineers design and develop building techniques and building materials that are safer, more reliable, cost effective, and environmentally friendly.

Construction engineers and managers incorporate the technical requirements, with the given financial and legal requirements, in their planning to meet project deadlines. The architectural engineer is responsible for implementing the design to insure the quality of construction and meets the project's cost and schedule requirements. Modern computer tools and project management skills are implemented to manage such complex projects.

## ELECTRICAL ENGINEERING

Electrical engineers are responsible for the design, development, testing and supervision of the manufacture of electrical equipment. Electrical equipment includes household appliances and guidance systems for rockets and satellites. They work with all products and systems that use electricity. Examples of electrical equipment are transmission and power generation equipment used by electric utilities, electric motors, machinery controls, and lighting and wiring used in buildings, automobiles, and aircraft. Radar, computer hardware, communications and video equipment are examples of electrical equipment. Several major areas are specialties of the electrical engineer: power generation, transmission and distribution; communications; computer electronics; and electrical equipment manufacturing.

Electrical engineers design new products. They write performance specifications and develop maintenance schedules. They want their designs to be efficient, long lasting, and safe. Electrical engineers test equipment, estimate the time and cost of engineering projects, and solve operating problems. They need to make sure their designs are not too expensive to manufacture. Electrical engineers submit their ideas to management for approval, and then, after approval, direct the installation and testing of the equipment they have designed.

Considering the wide range of electronic devices people use, electrical engineering has become the most populated of the engineering disciplines. Electrical engineers have many career opportunities in almost all of the industrial sectors. To provide a brief discussion of such a broad field, we will divide electrical engineering into eight areas: computers, communications, circuits and solid state devices, control, instrumentation, signal processing, bioengineering, and power.

Engineers specializing in computer technology are in such high demand that numerous institutions offer a separate major for computer engineering. (See the computer engineering section above.)

Electrical engineers are responsible for the explosion in communication technologies. Satellites provide nearly instantaneous global communication. Global Positioning Systems (GPS) allow anyone with the required handheld unit to pinpoint precisely where they are anywhere in the world. Fiber optics and lasers are rapidly improving the reliability and speed with which information can be exchanged. Wireless communication allows people to communicate almost anywhere and with anyone.

Electrical engineers design and develop electronic circuits. Circuit design has changed rapidly because of microelectronics. As circuits continue to shrink, new barriers such as molecular size emerge. As we reach the limits of current technology, incentives to develop new and faster ways to accomplish the same tasks will appear.

Almost all modern machines and systems are digitally controlled. Digital controls allow for safer and more efficient operation. Electronic systems monitor processes and make corrections faster and more effective than human operators. This improves reliability, efficiency, and safety. The electrical engineer is involved in the design, development, and operation of these control systems. The engineer must determine the kind of control required, what parameters to monitor, the correction speed, and many other factors. Control systems are used in chemical plants, power plants, automotive engines, and airplanes.

For a control system to operate correctly, it must be able to measure the important parameters of what it is controlling. Doing so requires accurate instrumentation, which electrical engineers specialize in developing. Electrical engineers who work in this area develop electrical devices to measure quantities such as pressure, temperature, flow rate, speed, heart rate, and blood pressure. Often, the electrical devices convert the measured quantity to an electrical signal that can be read by a control system or a computer. Instrumentation engineers design systems to transmit the measured information to a recording device using telemetry. Such systems are needed for transmitting a satellite's measurements to the recording computers back on earth while it orbits a distant planet.

Signal processing is another area where electrical engineers are needed. The electrical signals coming from instrumentation or other sources must be conditioned before the information can be used. Signals may need to be electronically filtered, amplified, or modified, such as the active noise control system on a stethoscope, which allows a paramedic to listen to a patient's heart while in a helicopter. The active noise control system can block out the sound of the helicopter so the paramedic can make a quick, accurate assessment of the patient. Signal processing also comes into play in areas such as voice recognition for computers.

Electrical engineers work in biomedical or bioengineering applications as described earlier. Electrical engineers work with medical personnel to design and develop devices used in patient diagnosis and treatment. Examples include non-intrusive techniques for detecting tumors through Magnetic resonance Imaging (MRI) or through Computer Automated Tomography (CAT) scans. Other examples include pacemakers, cardiac monitors, and controllable prosthetic devices.

The generation, transmission, and distribution of electric power are perhaps the most traditional aspect of electrical engineering. Electrical engineers work closely with mechanical engineers in the production of electrical power. They oversee the distribution of power through electrical networks and must ensure reliable supplies of electricity to our communities. With modern society's dependence on electricity, interruptions in the flow of electricity can be catastrophic. Today's power-related challenges revolve around reliability and cost of delivery. Electrical engineers also work with materials engineers to incorporate superconductivity and other technology in more efficient power transmission.

The Institute of Electrical and Electronics Engineers (IEEE) is the largest and most prominent professional organization for electrical engineers. It is organized into 37 technical divisions, which indicates the breadth of the field of electrical engineering.

- Aerospace and Electronic Systems
- Antennas and Propagation
- Broadcast Technology
- Circuits and Systems
- Communications
- Components Packaging and Manufacturing Technology
- Computer
- Consumer Electronics
- Control Systems
- Dielectrics and Electrical Insulation
- Education
- Electromagnetic Compatibility
- Electron Devices
- Engineering in Medicine and Biology
- Engineering Management
- Instrumentation and Measurement
- Lasers and Electro-Optics
- Magnetics
- Microwave Theory and Techniques
- Neural Networks
- Nuclear and Plasma Sciences
- Oceanic Engineering
- Power Electronics
- Power Engineering
- Professional Communication
- Reliability
- Robotics and Automation
- Signal Processing
- Social Implications of Technology

- Solid-State Circuits
- Geoscience and Remote Sensing
- Industrial Electronics
- Industrial Applications
- Information Theory
- Systems, Man and Cybernetics
- Ultrasonics, Ferroelectrics and Frequency Control
- Vehicular Technology

Electrical engineers' salaries according to the National Society of Professional Engineers:

| Average in 2003 | 25%tile | 75%tile | Average in 2000 | 25%tile | 75%tile |
|---|---|---|---|---|---|
| $99,019 | $70,000 | $105,000 | $83,107 | $63,100 | $93,350 |

Electronics engineers' salaries according to the National Society of Professional Engineers:

| Average in 2003 | 25%tile | 75%tile | Average in 2000 | 25%tile | 75%tile |
|---|---|---|---|---|---|
| $81,266 | $50,662 | $97,137 | $72,355 | $58,397 | $89,120 |

## ENVIRONMENTAL ENGINEERING

The professional discipline of environmental engineering is defined as "the application of engineering principles to improve and maintain the environment for the protection of human health, for the protection of nature's beneficial ecosystems, and for environment-related enhancement of the quality of human life." In other words, environmental engineering is a career that applies science in the pursuit of making the world a safer place for humans and animals. Environmental engineering has been around for more than one hundred years. For most of those years it was called sanitary engineering, but now it has grown to include areas such as air pollution control, hazardous waste management, industrial hygiene, public health engineering, radiation protection, solid waste management, water supply engineering, and wastewater control.

Environmental engineering is a field that has evolved to improve and protect the environment while keeping pace with the growth of industrial activity. This challenging task has three parts to it: disposal, remediation, and prevention.

Disposal is similar to that topic as covered under civil engineering. Environmental engineers are concerned with disposal and processing of industrial and residential waste. Landfills and waste treatment facilities are designed for residential waste concerns. The heavy metals and other toxins found in industrial wastes require special disposal procedures. The environmental engineer develops the techniques to dispose of such waste properly.

Remediation involves the cleaning up of a contaminated site. Such a site may contain waste that was improperly disposed and requires the ground and/or water to be removed or decontaminated. The environmental engineer develops the means to remove the contamination and return the area to a usable state.

Prevention is an area in which environmental engineers are becoming more involved. Environmental engineers work with manufacturing engineers to design processes that reduce or eliminate harmful waste. One example is in the cleaning of machined parts. Coolant is sprayed on metal parts as they are cut to extend the life of the cutting tools. The oily fluid clings to the parts and has to be removed before the parts are assembled. An extremely toxic substance had been used in the past to clean the parts because no suitable alternative existed.

Environmental engineers discovered that oil from orange peels works just as well and is perfectly safe (even edible). Since this new degreasing fluid did not need to be disposed of in any special way, manufacturing costs were reduced, and the workers' environment improved.

Environmental engineers must be well grounded in engineering fundamentals and environmental regulations. Within their companies, they are the experts on compliance with the changing environmental laws. An environmental engineer must understand the regulations and know how to apply them to the various processes the company encounters.

Environmental engineering is a broad field that encompasses many different responsibilities. Possible job titles are researcher, designer, planner, pollution control facilities operator, teacher, government regulatory agency official, and programs manager. Employers could be private consulting engineering firms, universities, private research firms, testing laboratories, government agencies (federal, state, and local), and all types of industry and private business. Jobs are available around the world. Work can be inside or outside, and sites can be rural villages or booming metropolises. Environmental engineering is an especially diverse career that offers options for everyone.

Environmental engineers' salaries according to the National Society of Professional Engineers:

| Average in 2003 | 25%tile | 75%tile | Average in 2000 | 25%tile | 75%tile |
|---|---|---|---|---|---|
| $87,793 | $55,000 | $100,500 | $66,319 | $47,608 | $75,500 |

## FIRE PROTECTION ENGINEERING

Fire protection engineers design fire sprinkler and alarm systems as well as exit and control systems. These professionals work in the public and private sector for consulting firms, petrochemical societies, federal agencies, insurance companies, and health care industries. They compile risk analyses of major facilities as well as consult with architects on large projects ranging from skyscrapers to hotels to stadiums.

A fire protection engineer also plays a role in the investigation of fires and explosions. Local fire departments have their fire inspectors on staff but sometimes hire fire protections engineers as consultants to help solve a crime. These engineers also assure safety throughout the NASA space program and do fire experimentation and research in order to prevent and fight fires.

The Society of Fire Protection Engineers was established in 1950 and incorporated in 1970. This society has over 3,500 members in the United States and overseas, with 41 regional chapters within the country and ten internationally.

Fire protection engineers salaries according to the National Society of Professional Engineers:

| Average in 2003 | 25%tile | 75%tile | Average in 2000 | 25%tile | 75%tile |
|---|---|---|---|---|---|
| $122,243 | $73,548 | $166,150 | —————No Data————— | | |

# FOOD PROCESS ENGINEERING

Many different technologies are involved in the efficient and safe processing and delivery of food products. Food process engineers are concerned with providing healthier products to consumers who increasingly rely on processed food products.

Food process engineers design processing, handling, and packaging equipment for the food industry. They can work as project engineers, plant engineers and in management, sales, consulting, or marketing. They work in food, chemical, biochemical, and pharmaceutical industries and can work for the public or private sector.

# GENETIC ENGINEERING

In recent times the demand for genetic engineers, in agriculture and fighting against certain diseases, has grown. Higher yielding crop seeds and superior prescription drugs owe their existence to the research carried out by geneticists and genetic engineers.

The best description of a genetic engineer's work is: "Geneticists use scientific methods to research genes found in the cells of plants and animals to develop better products medicines and services." Genetic engineering is essentially the same as genetics. Genetic engineering is the more specific application of genetic manipulation techniques to certain engineering projects. Some present types of genetic engineering deal with cloning, xeno-technology and eugenics, clinical integration of genetic technologies (including genetic testing/counseling), risk neutralization, threats and hazards as they appear, facilitating public participation in decision-making, citizens' juries, consensus conferences, science courts, genetic literacy, and the public understanding of the new genetics.[vii]

Genetic engineering and the mapping and sequencing of the human genome have deeply personal effects on everyone. It has developed into a complicated political, economic, and moral conflict involving key social institutions. While it can be thought to pose a threat to the boundaries that conventionally define each person and distinguish humans from other animals, it paves the way for new approaches to health, medicine, family life, industry, and the law. It is likely to generate significant risks and dangers, with possibly unforeseen and irreversible outcomes, while calling into question our understanding of morality and ethics.

*The new genetic technologies are rapidly developing as the key to understanding social change in the next century.*

The variety of geneticists is vast due to the hundreds of different research focuses into which they can be categorized. A common type of geneticist is one that specializes in the disease study and its effect on humans. The daily duties of such a person would include identifying genes in humans that have a natural defense against disease and developing pharmaceuticals that resemble these in order to slow the disease's spread or even cure it. They must follow prescribed safety measures and work with dangerous chemicals, electron microscopes, inflated spectroscopes, chromatographs, or even gene guns to carry out their research.

Many people associate genetic engineers with DNA recombination. However, genetic engineers are sometimes called biotechnologists. Mankind began using biotechnology about 10,000 years ago when farmers first began to pick the seeds of wild plants for cultivation or breeding wild animals selectively. Today, genetic engineers remove DNA from an organism and change it with radiation or chemicals and place it into the same or a different organism. For example, some scientists have created a banana containing an edible vaccine by splicing a harmless genetic portion from a bacterium into a banana (Schoofs).

Genetic engineering is a broad term and can be divided into four categories: human, animal, plant, and microorganism. Many different specialties of genetic engineering exist, such as cancer treatment, genetic fingerprinting, the human genome project, and the genetic engineering of cheese, beer, crops, and animals. In forensic science, genetic engineers use the DNA within sample body cells to identify people. In agriculture, genetic engineers create crops with higher nutritional values. For example, some scientists have created sweet potatoes with high protein content using an artificial gene (Schoofs). With human genes, scientists developed pigs with livers that can be transplanted into humans for medical purpose (Schoofs).

## GEOLOGICAL ENGINEERING

Geological engineering, as the name implies, involves land and water. Many new tasks for geological engineers are constantly being thought of, so members of the field usually cannot count on a specific task. Rather, they perform various tasks for public and private entities. Geological engineers can be called upon to investigate sites for major land-related projects such as bridges, dams, and tunnels. They also work on efforts to mitigate toxic waste or salinization-related soil and land contamination. They use physics to map and predict the flow of water in rivers, above and underground. They can take measures to help prevent landslides, build structures in rock, and create and use detailed computer models of geological features. They build and maintain earth-related power sources, such as hydroelectric dams and geothermal plants, or work on oil drilling sites.

A geological engineer would find it difficult to secure a good position with only four years of college. Most employers prefer that their geological engineers go to graduate school, exploring earth science, geophysics and geology in greater depth, while also learning the principles of construction and design.

Geotechnical engineers' salaries according to the National Society of Professional Engineers:

| Average in 2003 | 25%tile | 75%tile | Average in 2000 | 25%tile | 75%tile |
|---|---|---|---|---|---|
| $66,162 | $45,150 | $91,740 | $66,779 | $42,500 | $75,869 |

# INDUSTRIAL ENGINEERING

Industrial engineering is described by the Institute of Industrial Engineers as "the design, improvement and installation of integrated systems of people, material and energy." Industrial engineering is an interdisciplinary field that involves the integration of technology, mathematical models, and management practices. Traditional industrial engineering is done on a factory floor. However, the skills of an industrial engineer are transferable to a host of other applications. As a result, industrial engineers find themselves working within many industries. Four of the main areas of emphasis for industrial engineers are production, manufacturing, human factors, and operations research.

The production area includes functions such as plant layout, material handling, scheduling, quality control, and reliability control. An industrial engineer would examine the entire process involved in making a product and optimize it by reducing cost and production time and by increasing quality and reliability. In addition to factory layout, industrial engineers use their expertise in other ways. For example, with an amusement park an industrial engineer would analyze the flow of people through the park looking for ways to reduce bottlenecks and provide a pleasant experience for the patrons.

Manufacturing differs from production in that it addresses the components of the production process. While production concerns are on a global scale, manufacturing concerns address the individual production station. The industrial engineer optimizes the actual material processing, such as the machining process.

The human factors area involves the placement of people into the production system. An industrial engineer in this area studies the interfaces between people, machines, and objects in the system. These may include production machinery, computers, and even office chairs and desks. The industrial engineer considers ergonomics in finding ways to improve the interfaces. He or she looks for ways to improve productivity while providing a safe, comfortable environment for workers.

Operations research is concerned with the optimization of systems. This involves mathematically modeling systems to identify ways to improve them. Project management techniques such as critical path identification fall under operations research. Often, computer simulations are required to model the system or to study the effects of system changes. These systems may be manufacturing systems or other organizations of assets. The optimization of sales territories for a pharmaceutical sales force provides a non-manufacturing example.

The Institute for Industrial Engineering (IIE) is one of the most prominent professional organizations for industrial engineers. It is organized into three societies, ten technical divisions and eight interest groups. The following technical divisions show the breadth of industrial engineering:

- Aerospace and Defense
- Energy, Environment and Plant Engineering
- Engineering Economy
- Facilities Planning and Design
- Financial Services
- Logistics Transportation and Distribution
- Manufacturing
- Operations Research
- Quality Control and Engineering Reliability
- Utilities

Industrial engineers' salaries according to the National Society of Professional Engineers:

| Average in 2003 | 25%tile | 75%tile | Average in 2000 | 25%tile | 75%tile |
|---|---|---|---|---|---|
| $82,290 | $56,027 | $107,900 | $72,972 | $45,100 | $78,318 |

## MANUFACTURING ENGINEERING

The Society of Manufacturing Engineer's Directory of Manufacturing Education (published in 1992) indicated that manufacturing engineering is the following: ". . . that specialty of professional engineering which requires such education and experience as is necessary to understand, apply and control engineering procedures in manufacturing processes and methods of production of industrial commodities and products and requires the ability to plan the practices of manufacturing, to research and develop the tools, processes, machines and equipment, and to integrate the facilities and systems for producing quality products with optimal expenditures."

The National Center for Education Statistics gave a view of an academic program in the field where it indicates that manufacturing engineering is a program that prepares individuals to apply scientific and mathematical principles to the design, development, and implementation of manufacturing systems. This includes instruction in materials science and engineering, manufacturing processes, process engineering, assembly and product engineering, manufacturing systems design, and manufacturing competitiveness.

Two definitions (Manufacturing Technologist and Manufacturing Engineer) are used by SME's Certification program.

A manufacturing technologist is one who carries out manufacturing projects in a responsible manner, using proven techniques known by those who are technically trained in the field of manufacturing. The second definition is that a manufacturing technologist is capable of performing job assignments which may involve the following: working on design, development and implementation of engineering plans; drafting; the erection of manufacturing engineering equipment; estimating; inspection and testing of materials and components. In performing these functions, the candidate must apply sound knowledge and understanding of materials, manufacturing processes and people.

In carrying out these duties, the competent supervision of skilled craftsmen may be necessary. The techniques employed demand acquired experience and knowledge of manufacturing technology, combined with the ability to organize the details of a job using established practices.

A manufacturing engineer professional acquires such education and experience as is necessary to understand, apply, and control engineering procedures in manufacturing processes and methods of production of industrial commodities and products. This engineer acquires the ability to plan the practices of manufacturing to research and develop the tools, processes, machines, and equipment and to integrate the facilities and systems producing quality products with optimal expenditure.

Manufacturing engineers make things. Everything that manufacturing engineers do is tied to the production of goods. Almost everything that we use at home, at work, and at play is manufactured. By its official professional definition, manufacturing occurs when the shape, form or properties of a material are altered in a way that adds value. Manufactured goods are everywhere—aircraft structures, machinery, electronics, medical devices, automobile parts, household products, toys, textiles and clothing, cans and bottles—almost everything we use.

Manufacturing engineers earn their living by making decisions about technology, machinery, people, and money. They make those decisions from a background of knowledge of mathematics and scientific principles, of people and organizations, and of the processes and machinery required to produce high-quality goods when people want them and at an affordable price.

Every day of their working careers, manufacturing engineers are called upon to design a process or a system or to solve a problem. They must know how to use many different kinds of resources, including machines, robots, people, computer-based controls and analysis tools, information networks and money. Individuals develop expertise in materials and manufacturing processes, process and assembly engineering, product engineering and realizations, manufacturing competitiveness, quality engineering, manufacturing systems design, and other specializations. Career paths would be as operations integrators or manufacturing strategists, defining relationships between technology and organizations.

Manufacturing engineers often work in concurrent engineering teams for launching production of new products. In this environment, they work in partnership with mechanical and electrical design engineers, applications engineers and marketing specialists, purchasing agents and supply chain managers, human resources specialists, and accountants and financial managers. All of these occupations within a business enterprise support the production of the company's products. The manufacturing engineer is the final link. Manufacturing is where the action is.

Modern corporations involved in production succeed or fail on the strength of their manufacturing. If a company can produce high-quality goods at competitive prices and in quantities and at times when people want to buy them, it will prosper and provide employment and reward to its workforce. Since manufacturing engineers work at the core of industrial companies, they learn about the vital interests of the company and can readily grow into management positions. The path to the executive suite in many industrial companies runs through manufacturing.

North Dakota State University (Web site 2003)

Manufacturing engineers' salaries according to the National Society of Professional Engineers:

| Average in 2003 | 25%tile | 75%tile | Average in 2000 | 25%tile | 75%tile |
|---|---|---|---|---|---|
| $86,307 | $58,125 | $97,000 | $73,455 | $48,497 | $85,250 |

## MARINE AND OCEAN ENGINEERING

Nearly 80 percent of the Earth's surface is covered with water. Engineers involved in marine and ocean engineering are concerned with the exploration of the oceans, the transportation of products over water, and the utilization of resources in the world's oceans, seas and lakes.

Marine engineers focus on the design, development and operation of ships and boats. They work together with naval architects in this capacity. Naval architects are concerned with the overall design of the ship. They focus on the shape of the hull in order to provide the appropriate hydrodynamic characteristics. They (who? The Naval Architect or the Marine Engineer?) are also concerned with the usefulness of the vessel for its intended purpose, and with the design of the ship's subsystems, which would include ventilation, water, and sanitary systems, thus permitting the crew to work efficiently.

The marine engineer is primarily concerned with the subsystems of the ship that allow the ship to serve its purpose. These include the propulsion, steering, and navigation systems. The marine engineer might analyze the ship for vibrations or stability in the water. The ship's electrical power distribution and air conditioning fall under the responsibility of marine engineers. They also might be involved in the analysis and design of the cargo handling systems of the ship.

The responsibilities of an ocean engineer involve the design, development and operation of vehicles and devices other than boats or ships. These include submersible vehicles used in the exploration of the oceans and in obtaining resources from their depths. He or she might be involved in the design of underwater pipelines or cables, offshore drilling platforms, and offshore harbor facilities.

Ocean engineers are involved with the interaction of the oceans and things with which they contact. They study wave action on beaches, docks, buoys, moorings, and harbors. Ocean engineers design ways to reduce erosion while protecting the marine environment. They study ways to protect and maintain marine areas that are critical to our food supply. Ocean engineers become involved with the control and treatment of pollution in the sea and seeking alternative sources of energy from the ocean.

One of the professional societies to which these engineers may be involved is the Society of Naval Architects and Marine Engineers. The society is subdivided into the following nine technical and research committees:

- Hull Structure
- Hydrodynamics
- Ship's Machinery

- Ship Technical Operations
- Offshore
- Ship Production
- Ship Design
- Ship Repair and Conversion
- Small Craft

## MATERIALS SCIENCE ENGINEERING

The origins of materials engineering can be traced to around 3000 B.C. when people began to produce bronze for use in creating superior hunting tools. Since that time, society has developed materials to meet its needs. Materials engineers develop these new materials and the processes to create them. The materials may be metals or non-metals, e.g., ceramics, plastics, and composites. Materials engineers are concerned with four facets of materials: structure, properties, processes, and performance.

Materials engineers study the structure and composition of materials on a scale ranging from the microscopic to the macroscopic. They are interested in the molecular bonding and chemical composition of materials. The materials engineer is concerned with the effect of grain size and structure on the material properties.

The properties in question might include strength, crack growth rates, hardness, and durability. Numerous technological advances are impeded by a lack of materials possessing the properties required by the design engineers. Materials engineers seek to develop materials to satisfy these demands.

A given material may have different properties depending on how the material is processed. For example, cooling can affect steel's properties dramatically. Air-cooled steel will have different properties than liquid-cooled steel. The composition of a material can affect its properties. Materials such as metallic alloys contain trace elements that must be evenly distributed throughout the alloy to achieve the desired properties. If the trace elements are not distributed well or form clumps in the metal, the material will have different properties than the desired alloy. This could cause a part made with the alloy to fail prematurely. Materials engineers design processes and testing procedures to ensure that the material has the desired properties.

The materials engineer works to ensure that a material meets the performance demands of its application by designing test procedures to ensure these requirements are met. Destructive and nondestructive testing techniques are used to accomplish this.

Materials engineers develop new materials, improve traditional materials, and produce materials reliably and economically through synthesis and processing. Subspecialties of materials engineering, such as metallurgy and ceramics engineering, focus on classes of materials with similar properties.

Metallurgy involves the extraction of metals from naturally occurring ore for the development of alloys for engineering purposes. The metallurgical engineer is concerned with alloy composition, properties, and performance. Detailed investigation of a component failure often identifies design flaws in the system. The materials engineer can provide useful information regarding the condition of materials to the design engineer.

Ceramics is another area of materials engineering. In ceramic engineering, the naturally occurring materials of interest are clay and silicates, rather than an ore. These non-metallic minerals are employed in the production of materials that are used in a wide range of applications, including the aerospace, computer and electronic industries.

There are other subspecialties within materials engineering and they focus on polymers, plastics, and composites. Composites are different kinds of synthesized materials used to create a material to meet some specific demands. Materials engineers are involved in biomedical applications, e.g., the development of artificial tissue for skin grafts or bone replacement materials for artificial joints.

One of the professional societies to which materials engineers may belong is the Minerals, Metals and Materials Society. It is organized into the following five technical divisions:

- Magnetic and Photonic Materials
- Extraction and Processing
- Light Metals
- Materials Processing and Manufacturing
- Structural Materials

Materials science engineers' salaries according to the National Society of Professional Engineers:

| Average in 2003 | 25%tile | 75%tile | Average in 2000 | 25%tile | 75%tile |
|---|---|---|---|---|---|
| $86,790 | $67,700 | $110,672 | ————No Data———— | | |

# MECHANICAL ENGINEERING

Mechanical engineering is one of the largest and broadest of the engineering disciplines. It is second only to electrical engineering in the number of engineers employed in the field. Fundamentally, mechanical engineering is concerned with machines and mechanical devices. Mechanical engineers are involved in the design, development, production, control, operation, and service of these devices. Mechanical engineering is composed of two main divisions: design and controls, and thermal sciences.

The design function is the most common function of mechanical engineering. It involves the detailed layout and assembly of the components of machines and devices. Mechanical engineers are concerned with the strength of parts and the stresses to which the parts will be subjected. They work closely with materials engineers to ensure that correct materials are chosen. Mechanical engineers must ensure that the parts fit together by specifying detailed dimensions.

Another aspect of the design function is the design process itself. Mechanical engineers develop computational tools to aid the design engineer in optimizing a design. These tools speed the design process by automating time-intensive analyses.

Mechanical engineers are interested in controlling the devices they design. Control of mechanical devices can involve mechanical or hydraulic controls. However, most modern control systems incorporate digital control schemes. The mechanical engineer will model controls for the system and programs, or design the control algorithm.

The noise generated from mechanical devices is a concern, so mechanical engineers are involved in acoustics and the study of noise. The mechanical engineer works to minimize unwanted noise by identifying the source and then designing ways to minimize it without sacrificing a machine's performance.

In the thermal sciences, mechanical engineers study the flow of fluids and the flow of energy between systems. Mechanical engineers in theses sciences deal with liquids, gases, and two-phase flows, which are combinations of liquids and non-liquids. Mechanical engineers here might be concerned about how much power is required to supply water through piping systems in buildings. They might also be concerned with aerodynamic drag on automobiles.

The flow of energy due to a temperature difference is called heat transfer, which is a thermal science area in which mechanical engineers are involved. They study and predict the temperature of components in their operational environment. Modern personal computers have microprocessors that require cooling. Mechanical engineers design the cooling devices that allow the electronics to function properly within the optimal temperature range.

Mechanical engineers design and develop engines. An engine is a device that produces mechanical work. Two examples are internal combustion engines used in automobiles and gas turbine engines used in airplanes. Mechanical engineers are involved in the design of the mechanical components of the engines as well as the overall cycles and efficiencies of these devices.

Performance and efficiency are concerns for mechanical engineers involved in the production of power in large power generation systems. Steam turbines, boilers, water pumps, and condensers are used to generate electricity. Mechanical engineers design these mechanical components needed to produce the power that operates the generators. Mechanical engineers are involved in developing alternative energy sources, which include solar power, hydroelectric power, alternative fuel engines, and fuel cells.

Another area in the thermal sciences is heating, ventilating, and air conditioning (HVAC). Mechanical engineers are involved in the climate control of buildings, which includes cooling and heating the air in buildings, as well as controlling humidity. In doing so, they work closely with civil engineers in designing buildings to optimize the efficiency of these systems.

Mechanical engineers are involved in the manufacturing processes of many industries. They design and develop the machines used in these processes and develop more efficient processes. Often, this involves automating time-consuming or expensive procedures within a manufacturing process. Mechanical engineers are involved in the development and use of robotics and automated processes.

In the area of biomedical engineering, mechanical engineers help develop artificial limbs and joints that provide mobility to physically impaired individuals. They develop mechanical devices used to aid in the diagnosis and treatment of patients.

The American Society of Mechanical Engineering (ASME) is one of the most prominent professional societies for mechanical engineers. It is divided into 35 technical divisions, indicating the diversity of this field. These divisions are the following:

- Advanced Energy Systems
- Aerospace Engineering
- Applied Mechanics
- Basic Engineering Technical Group
- Bioengineering
- Design Engineering
- Dynamic Systems and Control
- Electrical and Electronic Packaging
- FACT
- Fluids Engineering
- Fluids Power Systems and Technology Systems
- Heat Transfer
- Information Storage/Processing
- Internal Combustion Engine
- Gas Turbine
- Manufacturing Engineering
- Materials
- Materials Handling Engineering
- Noise Control and Acoustics
- Non-destructive Evaluation Engineering
- Nuclear Engineering
- Ocean Engineering
- Offshore Mechanics / Arctic Engineering
- Petroleum
- Plant Engineering and Maintenance
- Power
- Pressure Vessels and Piping
- Process Industries
- Rail Transportation
- Safety Engineering and Risk Analysis
- Solar Energy
- Solid Waste Processing
- Technology and Society
- Textile Engineering
- Tribology

Mechanical Engineers salaries according to the National Society of Professional Engineers:

| Average in 2003 | 25%tile | 75%tile | Average in 2000 | 25%tile | 75%tile |
|---|---|---|---|---|---|
| $86,819 | $64,000 | $102,280 | $79,434 | $56,000 | $90,945 |

## MINERAL AND MINING ENGINEERING

Modern civilization requires a vast quantity of products made from raw materials such as minerals. The continued production of these products using raw materials helps to keep our civilization functioning. Mining engineers are responsible for maintaining the flow of these raw materials by discovering, extracting and processing minerals for the products civilization requires.

Discovering the ore involves exploration in conjunction with geologists and geophysicists. The engineers combine the utilization of seismic, satellite, and other technological data, utilizing knowledge of rocks and soils. The exploration may focus on land areas, the ocean floor, or even below the ocean floor. In the future, mining engineers may explore asteroids, which are rich in mineral deposits.

Once mineral deposits are identified, they may be removed. One way minerals are removed is by mining tunnels. The engineers design and maintain the tunnels and the required support systems, which include ventilation and drainage. In other instances, minerals are removed from open pit mines. Again, the engineers analyze the removal site and design the procedure for removing the material. The engineer develops a plan for returning the site to a natural state. Mining engineers use boring, tunneling, and blasting techniques to create a mine. Regardless of the removal technique, the environmental impact of the mining operation is taken into account and kept to a minimum.

The mining engineer is involved in the processing of the raw minerals into usable forms. Purifying and separating minerals involves chemical and mechanical processes. Though mining engineers may not be involved in making a finished product that consumers recognize, they must understand the forms of material their customers can use and design processes to transform the raw materials into these usable forms.

Mining engineers are involved in the design of the specialized equipment required by the mining industry. The design of automated equipment capable of performing the most dangerous mining jobs helps to increase safety and productivity. Since mines are established in remote areas, mining engineers are involved in the transportation of minerals to the processing facility.

The expertise of mining engineers is not used exclusively by the mining industry. The same boring technology used in developing mines is also used to create subway systems and railroad tunnels, such as the one under the English Channel.

Mining engineers' salaries according to the National Society of Professional Engineers:

| Average in 2003 | 25%tile | 75%tile | Average in 2000 | 25%tile | 75%tile |
|---|---|---|---|---|---|
| $102,155 | $64,444 | $114,625 | $128,224 | $84,370 | $152,000 |

## NUCLEAR ENGINEERING

Nuclear engineers study nuclear energy and radiation and their beneficial uses. Usually they work at nuclear plants specifically to design, run, observe, and improve the reactors to increase energy efficiency and secure their safe operation. They handle the production and use of nuclear fuel, as well as the safe disposal of the radioactive waste byproducts. Some engineers work to develop the nuclear power sources for spacecraft and satellites. Others research and develop radioactive materials for industrial and medical uses, such as X-ray and other types of radiation used to examine and cure patients' medical problems.

Nuclear engineers are concerned primarily with the use and control of energy from nuclear sources. This involves electricity production, propulsion systems, waste disposal, and radiation applications.

The production of electricity from nuclear energy is one of the most common and apparent applications of nuclear engineering. Nuclear engineers focus on the design, development, and operation of nuclear power facilities. This involves using current fission technology as well as the development of fusion, which would allow seawater to be used as fuel. Nuclear energy offers an environmentally friendly alternative to fossil fuels. A current barrier to nuclear facilities production is the high cost of construction. This barrier provides a challenge for design engineers to overcome. Research is currently being performed on the viability of smaller, more efficient nuclear reactors.

Nuclear power is used in propulsion systems and provides a power source for ships and submarines, allowing them to go years without refueling. It is used as a power source for satellites. In the realm of interplanetary travel nuclear-powered engines are being examined as an alternative to conventional fossil-fueled engines, thus making interplanetary travel possible.

One of the main drawbacks of nuclear power use is the production of radioactive waste. This problem, however, presents opportunities for nuclear engineers to develop safe and reliable means to dispose of spent fuel. Nuclear engineers develop ways to reprocess the waste into less hazardous forms.

Another area in which nuclear engineers are involved is the use of radiation for medical or agricultural purposes. Radiation therapy has proven effective in treating cancers, and radioactive isotopes are used in diagnosing various diseases. Irradiating foods can eliminate harmful bacteria and help ensure a safer food supply.

Due to the complex nature of nuclear reactions, nuclear engineers are at the forefront of advanced computing methods. High-performance computing techniques, such as parallel processing, constitute research areas vital to nuclear engineering.

Nuclear engineers, who deal with reactor fuels and radio isotopic materials, are by necessity quite cautious because a mistake can be catastrophic, e.g., nuclear waste pollution or the explosion of a reactor. Often they need to wear special safety glasses and clothing to protect the workers' health, as well as thermoluminescent dosimeters, film badges, or other devices for recording cumulative radiation exposure time. Much stress for them comes from the need to prevent the escape of radioactive materials into the natural environment and the quick repair of any defective equipment. They are required to be able to *shift their work* at anytime to meet production schedules. Nuclear engineering is a career for people who can identify hazards and are aware of the tension between risks and benefits.

The American Nuclear Society (ANS) is one of the professional societies to which nuclear engineers belong. It is divided into these 16 technical divisions:

- Biology and Medicine
- Decommissioning, Decontamination and Reutilization
- Education and Training
- Environmental Sciences
- Fuel Cycle and Waste Management
- Fusion Energy
- Human Factors
- Isotopes and Radiation
- Materials Science and Technology
- Mathematics and Computations
- Nuclear Criticality Safety
- Nuclear Operations
- Nuclear Installations Safety
- Power
- Radiation Protection and Shielding
- Reactor Physics

Nuclear engineers salaries' according to the National Society of Professional Engineers:

| Average in 2003 | 25%tile | 75%tile | Average in 2000 | 25%tile | 75%tile |
|---|---|---|---|---|---|
| $92,499 | $78,000 | $104,000 | $90,950 | ————No Data———— | |

## PETROLEUM ENGINEERING

Petroleum and petroleum products are essential components to our civilization. Petroleum engineers maintain the flow of petroleum in a safe and reliable manner. They are involved in the exploration for crude oil deposits, the removal of oil, and the transporting and refining of oil.

Petroleum engineers work with geologists and geophysicists to identify potential oil and gas reserves. They combine satellite information, seismic techniques, and geological information to locate gas and oil deposits. Once a deposit has been identified, it can be extracted. The petroleum engineer designs, develops, and operates the needed drilling equipment and facilities. Such facilities may be located on land or on offshore platforms. The engineer is interested in removing the oil or gas in safely and reliably for the people involved as well as for the environment. Extraction of oil is done in stages, with the first stage using conventional means. Oil deposits are often located in sand. A significant amount of oil remains coating the sand after the initial oil removal. Recovery of this additional reserve requires the use of secondary and tertiary extraction techniques utilizing water, steam, or chemical means.

Transporting oil or gas to a processing facility is another challenge for the petroleum engineer. This requires the design of a heated pipeline, such as the one in Alaska to carry oil hundreds of miles over frozen tundra. In other instances, this requires transporting oil in double-hulled tankers from an offshore platform near a wildlife refuge. Such problematic situations require precautions be taken to ensure that wildlife is not endangered.

Once the oil or gas arrives at the processing facility, it must be refined into usable products. The petroleum engineer designs, develops and operates the equipment to chemically process the gas or oil into such end products. Petroleum is made into various grades of gasoline, diesel fuel, aircraft fuel, home heating oil, motor oils, and a host of consumer products from lubricants to plastics.

Petroleum engineers salaries' according to the National Society of Professional Engineers:

| Average in 2003 | 25%tile | 75%tile | Average in 2000 | 25%tile | 75%tile |
|---|---|---|---|---|---|
| $113,203 | $80,125 | $135,025 | $147,561 | $90,805 | $148,000 |

## ROBOTICS AND AUTOMATED SYSTEMS ENGINEERING

One of the newest and most exciting fields is that of robotic engineering. A robotic engineer requires many skills. He has to be competent in many programming languages, including C++, and be familiar with UNIX operating systems. A robotic engineer must be able to work in a team setting and be able to communicate his ideas effectively to his fellow team members.

The robotic engineer will mainly be programming his project to do different tasks. He will meet with the team to discuss ideas and hold brainstorming sessions for other robots. In an industrial setting, the robots he designs will spot weld in car manufacturing, load and unload pallets of goods, or do other monotonous tasks formerly done by people.

At the most basic level, a Bachelor's degree in chemical engineering, computer science, biochemistry, or chemistry is required for entry-level jobs. Lab experience at a college or being an understudy in a lab for a company helps finding a position. If robotic engineers have a Master's degree in any of the previously listed subjects, they can expect their salary to be at least $10,000 a year higher than those with a Bachelor's degree.

Many robotic engineers go on to become team leaders and senior engineers within their firms. This requires experience, a Master's degree, and probably a Doctorate to attain such positions.

As our civilization increasingly explores outer space, robots will be needed more. Every probe we send to explore space is a robot. The Pathfinder project that landed on Mars on July 4, 1997, was an outstanding show of engineering ingenuity. We were able to design and build a robot that could avoid large objects and navigate in an extra-terrestrial environment. Not only could it navigate with little human intervention, it was able to do so in spite of the 20-minute timeframe for radio communications between Earth and Mars. Robotic engineers will play an increasingly important role in continuing our interplanetary exploration successes, as they did with this one.

Manufacturers need robotic engineers to design more efficient and skilled robots to assemble more complex products. Space scientists will rely on the creativity and skill of the robotic engineer to design spacecraft to operate autonomously. This will become more important as the distance we send spacecraft from earth increases. Robots will be in the home as well helping us through life. The prospect of Artificial Intelligence in robots is not a new one, but it is slowly becoming a reality through the work of robotic engineers.

## SOFTWARE ENGINEERING

When most people turn on a computer, they see the splash screen or prompt come up and they think of it as an annoyance. They want to get on quickly, not wait two to three minutes to do so. However, few people realize what goes on in this short period: data are processed, files are updated, and tables are initialized. They fail to understand the importance of what goes on behind the scenes on their computers. The millions of lines of code that form what some would call a basic system. From the Internet browser to the solitaire game, every program on the computer has been painstakingly coded and revised to present users with a simple, friendly environment so they do not have to think about the program. Software engineers are the people who do this coding. Not to be confused with computer programmers, who focus more on actual coding, software engineers are in charge of solving problems that often appear in program development.

Two types of software engineers exist: computer applications software engineers and computer systems software engineers. Computer applications software engineers focus on actual programming of utilities and programs often for internal use in an office, whereas the computer systems software engineer coordinates technical systems and growth within the company.

Software engineers can function differently depending on their work setting as well. While working for companies that specialize in computers and technology, they can act as technical support, program developers, or help customers install software produced by their company. However, if on a marketing team, a software engineer would work to make sure that all systems were operating correctly within the team and solve other problems involving programs or programming. Software engineers usually work in a large office setting whereas other engineers may work primarily in a garage, workshop, or studio.

Potential software engineers must meet rigorous requirements to be hired. Not only do they need a Bachelor's degree in a computer or technology field to get hired, but they also need a broad knowledge of computers and technology. This knowledge can be acquired in a number of ways. One way would be through a company-sponsored internship. Another way would be to attend classes or to teach oneself through reading and experience. Many positions without a degree requirement do insist on training courses on different operating, security, and information systems (such programs are offered by most large companies such as Novell, Microsoft, Oracle, etc.). Many companies also require certification from these companies, which must be acquired through an exam.

Due to the rapid worldwide expansion of technology into the everyday working world, software engineering is one of the fastest growing professions in the United States. Most companies need at least a handful of software engineers at each office to manage their systems. Therefore, this profession is becoming more prevalent and by 2010 is expected to be a common occupation among the working class. College graduates with computer science or computer-engineering degrees are expected to have little trouble finding a good job with decent pay. In 2000, there were 697,000 software engineers employed in America. This statistic alone demonstrates how much demand there is for this type of engineering, especially when one considers that the profession did not even exist in any practical sense in 1985.

Software engineers' salaries according to the National Society of Professional Engineers:

| Average in 2003 | 25%tile | 75%tile | Average in 2000 | 25%tile | 75%tile |
|---|---|---|---|---|---|
| $83,449 | $53,000 | $112,000 | ————No Data———— | | |

## STRUCTURAL ENGINEERING

Structural engineers help prolong our lives and increase our efficiency in global economy by creating safer structures and fitting more people and objects per square inch into these structures.

Structural engineers analyze and design almost any structure imaginable, including skyscrapers, houses, bridges, wharves, retaining walls, towers, tunnels, canals, space platforms, etc. They do this in association with other engineers and specialists. The design of these structures includes determining the most suitable structural system, sizes of columns, beams, walls, staircases, foundation, etc., and the amount and type of reinforcement each element requires. A structural engineer must design these structures so that they will support their own weight and withstand natural forces such as gravity, wind, earthquakes, hurricanes, and floods. They prepare a detailed structural drawing in accordance with standard specifications, and may also be asked to supervise the construction of the structure.

During a workday, structural engineers might be in the office laying out plans on a computer, meeting with clients and drawing by hand, and on site supervising construction. Other typical tasks and duties could include the following:

- preparing sketches, specifications, and working drawings from the site and client needs

- analyzing tests from the site, e.g. soil samples
- performing calculations to test designs
- meeting with mechanical engineers, architects, and others in construction to discuss plans
- connecting with local authorities to obtain permits
- visiting a site regularly while under construction
- appearing as an expert witness at a court hearing.

Structural engineers' salaries according to the National Society of Professional Engineers:

| Average in 2003 | 25%tile | 75%tile | Average in 2000 | 25%tile | 75%tile |
|---|---|---|---|---|---|
| $89,499 | $54,750 | $98,125 | $80,113 | $51,435 | $90,620 |

The following salary information is for engineering disciplines that did not appear in detail in this section. The absence of detailed information on these careers does not mean that they are less important than others listed. Students are encouraged to investigate and report on these careers.

Architectural Engineers' salaries according to the National Society of Professional Engineers:

| Average in 2003 | 25%tile | 75%tile | Average in 2000 | 25%tile | 75%tile |
|---|---|---|---|---|---|
| $94,077 | $64,000 | $101,922 | ————No Data———— | | |

Control Systems Engineers' salaries according to the National Society of Professional Engineers:

| Average in 2003 | 25%tile | 75%tile | Average in 2000 | 25%tile | 75%tile |
|---|---|---|---|---|---|
| $87,329 | $63,463 | $92,500 | $73,199 | $54,750 | $86,989 |

Cost Management Engineers' salaries according to the National Society of Professional Engineers:

| Average in 2003 | 25%tile | 75%tile | Average in 2000 | 25%tile | 75%tile |
|---|---|---|---|---|---|
| $116,235 | $69,800 | $121,000 | ————No Data———— | | |

Facilities Engineers' salaries according to the National Society of Professional Engineers:

| Average in 2003 | 25%tile | 75%tile | Average in 2000 | 25%tile | 75%tile |
|---|---|---|---|---|---|
| $87,135 | $71,000 | $103,158 | ————No Data———— | | |

Forensic Engineers' salaries according to the National Society of Professional Engineers:

| Average in 2003 | 25%tile | 75%tile | Average in 2000 | 25%tile | 75%tile |
|---|---|---|---|---|---|
| $124,065 | $66,800 | $145,000 | | —No Data— | |

HVAC and Refrigeration Engineers' salaries according to the National Society of Professional Engineers:

| Average in 2003 | 25%tile | 75%tile | Average in 2000 | 25%tile | 75%tile |
|---|---|---|---|---|---|
| $96,622 | $62,500 | $109,250 | | —No Data— | |

Petroleum Engineers' salaries according to the National Society of Professional Engineers:

| Average in 2003 | 25%tile | 75%tile | Average in 2000 | 25%tile | 75%tile |
|---|---|---|---|---|---|
| $113,203 | $80,125 | $135,025 | $147,561 | $90,805 | $148,000 |

Quality Assurance Engineers' salaries according to the National Society of Professional Engineers:

| Average in 2003 | 25%tile | 75%tile | Average in 2000 | 25%tile | 75%tile |
|---|---|---|---|---|---|
| $84,855 | $50,125 | $85,750 | | —No Data— | |

Safety Engineers' salaries according to the National Society of Professional Engineers:

| Average in 2003 | 25%tile | 75%tile | Average in 2000 | 25%tile | 75%tile |
|---|---|---|---|---|---|
| $88,253 | $52,998 | $121,688 | | —No Data— | |

Sanitary Engineers' salaries according to the National Society of Professional Engineers:

| Average in 2003 | 25%tile | 75%tile | Average in 2000 | 25%tile | 75%tile |
|---|---|---|---|---|---|
| $78,537 | $61,250 | $89,000 | $84,715 | $53,539 | $98,630 |

Systems Engineers' salaries according to the National Society of Professional Engineers:

| Average in 2003 | 25%tile | 75%tile | Average in 2000 | 25%tile | 75%tile |
|---|---|---|---|---|---|
| $101,562 | $81,040 | $112,648 | | —No Data— | |

Transportation Engineers' salaries according to the National Society of Professional Engineers:

| Average in 2003 | 25%tile | 75%tile | Average in 2000 | 25%tile | 75%tile |
|---|---|---|---|---|---|
| $82,004 | $55,600 | $74,536 | ——————No Data—————— | | |

# SALARY INFORMATION FOR ENGINEERS PROVIDED BY THE NATIONAL SOCIETY OF PROFESSIONAL ENGINEERS

Exhibits 3 and 9 appear on page 98.

## EXHIBIT 14. Median Income by Gender vs. Highest Degree Earned and Length of Experience Among Full-Time Salaried Engineering Employees

| | Male Respondents | | Female Respondents | |
|---|---|---|---|---|
| | Number Reported | Median | Number Reported | Median |
| *BS Degree (engineering)* | | | | |
| Under 1 year | 34 | $46,085 | 17 | $47,913 |
| 1 year | 33 | $44,000 | 15 | $45,177 |
| 2 years | 65 | $47,500 | 19 | $47,000 |
| 3 years | 74 | $50,950 | 31 | $48,576 |
| 4 years | 91 | $53,000 | 25 | $47,000 |
| 5–9 years | 480 | $62,000 | 101 | $56,172 |
| 10–14 years | 383 | $75,000 | 42 | $66,536 |
| 15–19 years | 406 | $85,000 | 42 | $77,838 |
| 20–24 years | 440 | $90,000 | 33 | $90,000 |
| 25– 29 years | 417 | $94,087 | 8 | $83,559 |
| 30 years or more | 633 | $97,000 | | |
| *MS Degree (engineering)* | | | | |
| 1 year | 13 | $50,500 | | |
| 2 years | 16 | $47,000 | | |
| 3 years | 18 | $53,745 | 5 | $50,200 |
| 4 years | 25 | $52,000 | 7 | $46,800 |
| 5–9 years | 162 | $65,000 | 29 | $62,000 |
| 10–14 years | 147 | $75,100 | 19 | $73,000 |
| 15–19 years | 130 | $84,150 | 22 | $83,500 |
| 20–24 years | 162 | $91,027 | 14 | $86,250 |
| 25–29 years | 188 | $98,031 | 12 | $85,250 |
| 30 years or more | 296 | $100,000 | | |
| *Doctorate (engineering)* | | | | |
| 5–9 years | 11 | $81,052 | | |
| 10–14 years | 15 | $92,000 | | |
| 15–19 years | 16 | $89,887 | | |
| 20–24 years | 25 | $104,000 | | |
| 25–29 years | 22 | $135,854 | | |
| 30 years or more | 100 | $116,500 | | |

# EXHIBIT 3. Income by Level of Education—2003

| Variable and Category | All Members Responding | | | | | | | Full-Time Salaried Only | |
|---|---|---|---|---|---|---|---|---|---|
| | Number Reported | Average | 10th Percentile | 25th Percentile | Median | 75th Percentile | 90th Percentile | Number Reported | Median |
| *Highest Degree Earned* | | | | | | | | | |
| Less than a BA/BS Degree | 58 | $79,644 | $43,815 | $60,000 | $73,750 | $95,250 | $115,403 | 48 | $78,432 |
| BA Degree | 28 | $83,075 | $44,750 | $57,900 | $76,150 | $108,000 | $131,200 | 23 | $75,900 |
| BS Degree (non-engineering) | 60 | $90,875 | $50,550 | $59,700 | $80,324 | $107,750 | $148,900 | 49 | $72,000 |
| BS Degree (engineering) | 3,439 | $88,733 | $48,000 | $60,000 | $78,000 | $100,000 | $134,000 | 3,046 | $76,818 |
| MA/MS Degree (not MBA or engineering) | 143 | $91,006 | $56,408 | $67,500 | $81,000 | $104,000 | $130,000 | 123 | $82,000 |
| MBA Degree | 358 | $103,761 | $62,450 | $76,000 | $92,000 | $112,550 | $150,000 | 326 | $92,170 |
| MS Degree (engineering) | 1,294 | $94,522 | $52,327 | $65,491 | $82,500 | $107,000 | $144,500 | 1,135 | $83,000 |
| MBA and an MA/MS Degree | 191 | $115,888 | $67,200 | $81,000 | $100,900 | $132,100 | $178,220 | 161 | $96,000 |
| Doctorate (non-engineering) | 20 | $124,047 | $62,382 | $84,000 | $120,250 | $133,950 | $173,800 | 15 | $125,000 |
| Doctorate (engineering) | 212 | $112,297 | $54,773 | $78,000 | $101,600 | $139,758 | $174,470 | 116 | $101,850 |
| Other | 59 | $88,197 | $44,560 | $66,000 | $82,023 | $95,000 | $148,000 | 54 | $83,500 |
| | 5,862 | | | | | | | 5,096 | |

# EXHIBIT 9. Income by Registration Status

| Variable and Category | All Members Responding | | | | | | | Full-Time Salaried Only | |
|---|---|---|---|---|---|---|---|---|---|
| | Number Reported | Average | 10th Percentile | 25th Percentile | Median | 75th Percentile | 90th Percentile | Number Reported | Median |
| *Registration Status* | | | | | | | | | |
| No professional licensing or certification | 198 | $75,061 | $40,407 | $50,908 | $64,640 | $84,000 | $119,128 | 171 | $66,609 |
| Engineer-in-Training/ Engineer Intern | 739 | $60,382 | $40,400 | $45,500 | $52,900 | $69,330 | $90,000 | 692 | $52,658 |
| Professional Engineer (PE) | 4,188 | $96,670 | $56,000 | $68,378 | $85,000 | $108,000 | $145,000 | 3,645 | $84,200 |
| PE and certification in Environmental Engineering | 126 | $102,269 | $58,910 | $71,000 | $95,000 | $129,194 | $155,120 | 108 | $95,000 |
| PE and certification in Forensic Engineering | 33 | $150,920 | $60,020 | $81,000 | $120,000 | $185,700 | $301,336 | 12 | $115,625 |
| PE and certification in other engineering specialty | 184 | $97,074 | $56,800 | $72,025 | $89,175 | $109,625 | $144,440 | 154 | $90,300 |
| PE and Professional Surveyor (PS) or Land Surveyor (LS) | 225 | $107,387 | $60,000 | $78,000 | $90,000 | $119,718 | $161,200 | 168 | $89,952 |
| PE and other professional licensing | 127 | $118,514 | $61,600 | $74,300 | $92,549 | $114,400 | $203,400 | 105 | $90,000 |
| | 5,622 | | | | | | | 4,884 | |

## EXHIBIT 15   Income by Origin

Several times before, in alternate years, attempts were made to analyze income versus origin of respondents, even though not enough data permitted statistical validity. To the degree that the data in this year's survey permit, some comparisons will be made.

The sample was limited to respondents, who are employed full-time as salaried employees, and analyzed by length of experience and level of education simultaneously versus origin of respondent. Resulting data are displayed in Exhibit 15, horizontally by decreasing size of sample.

Asian/Pacific Islander respondents have higher median incomes than White respondents in six of their fourteen matching cells.

Hispanic respondents have the same or higher median incomes than White respondents in five of their ten matching cells, and higher median incomes than Asian/Pacific Islander respondents in five out of nine matching cells.

Black respondents have higher median incomes than Hispanic respondents in two of three matching cells.

The sample sizes for American Indian/Alaskan Native respondents were too small to produce an education versus length of experience cell with more than four respondents.

With the primary exception of White respondents, the sample sizes for cells of origin versus level of education length of experience are too small to permit any statistically valid conclusions.

See Exhibt 15.1 on page 100.

## ENGINEERING TECHNICAL SOCITIES

The following is a list of many of the engineering technical societies available to engineers. They can be a tremendous source of information for you, and many have student branches that allow you to meet engineering students and practicing engineers in the same field.

American Association for the
    Advancement of Science
1200 New York Avenue, NW
Washington, DC 20005
(202) 326-6400
www.aaas.org

American Association of Engineering
    Societies
1111 19th Street, NW
Suite 403
Washington, DC 20036
(202) 296-2237
www.aaes.org

American Ceramic Society
735 Ceramic Place
Westerville, OH 43081-8720
(614) 890-4700
www.acers.org

American Chemical Society
1155 16th Street, NW
Room 1209
Washington, DC 20036-1807
(202) 872-4600
www.acs.org

**EXHIBIT 15.1. Median Income by Origin vs. Highest Degree Earned and Length of Experience among Full-Time Salaried Engineering Employees**

| | White (not Hispanic) Respondents | | Asian/Pacific Islander Respondents | | Hispanic Respondents | | Black Respondents | |
|---|---|---|---|---|---|---|---|---|
| | Number Reported | Median | Number Reported | Median | Number Reported | Median | Number Reported | Median |
| *BS Degree (engineering)* | | | | | | | | |
| Under 1 year | 41 | $45,670 | | | | | | |
| 1 year | 43 | $44,000 | | | 4 | $47,125 | | |
| 2 years | 77 | $47,500 | 4 | $47,806 | 4 | $48,770 | | |
| 3 years | 90 | $49,751 | 5 | $51,000 | | | | |
| 4 years | 110 | $51,975 | | | | | | |
| 5–9 years | 537 | $61,000 | 9 | $65,300 | 16 | $57,500 | 6 | $60,000 |
| 10–14 years | 396 | $74,246 | 6 | $81,630 | 11 | $87,600 | | |
| 15–19 years | 423 | $85,227 | 6 | $62,726 | 8 | $95,000 | | |
| 20–24 years | 439 | $90,000 | 10 | $78,422 | 5 | $75,000 | 4 | $77,605 |
| 25–29 years | 398 | $94,800 | 6 | $62,290 | 11 | $82,000 | | |
| 30 years or more | 587 | $97,000 | 10 | $83,000 | 13 | $111,636 | | |
| *MS Degree (engineering)* | | | | | | | | |
| Under 1 year | 5 | $45,500 | | | | | | |
| 1 year | 15 | $49,691 | | | | | | |
| 2 years | 17 | $47,500 | | | | | | |
| 3 years | 19 | $52,490 | | | | | | |
| 4 years | 28 | $52,250 | | | | | | |
| 5–9 years | 167 | $65,000 | 7 | $68,000 | 8 | $56,430 | | |
| 10–14 years | 150 | $75,000 | 6 | $60,000 | | | | |
| 15–19 years | 132 | $84,750 | 7 | $90,550 | | | 4 | $80,000 |
| 20–24 years | 164 | $90,469 | 6 | $90,250 | | | | |
| 25–29 years | 187 | $97,000 | 6 | $91,000 | 4 | $119,000 | | |
| 30 years or more | 278 | $100,000 | 7 | $80,000 | | | | |
| *Doctorate (engineering)* | | | | | | | | |
| 5–9 years | 11 | $75,712 | | | | | | |
| 10–14 years | 15 | $92,000 | | | | | | |
| 15–19 years | 16 | $86,887 | | | | | | |
| 20–24 years | 27 | $94,100 | | | | | | |
| 25–29 years | 23 | $136,707 | | | | | | |
| 30 years or more | 89 | $120,000 | | | | | | |

American Concrete Institute
38800 Country Club Drive
Farmington Hills, MI 48331
(248) 848-3700
www.aci-int.org

American Congress on Surveying and
    Mapping
5410 Grosvenor Lane
Suite 100
Bethesda, MD 20814-2122
(301) 493-0200

American Consulting Engineers
    Council
1015 15th St., NW, Suite 802
Washington, DC 20005
(202) 347-7474
www.acec.org

American Gas Association
1515 Wilson Blvd.
Arlington, VA 22209
(703) 841-8400
www.aga.com

American Indian Science and Engi-
    neering Society
5661 Airport Blvd.
Boulder, CO 80301-2339
(303) 939-0023
www.colorado.edu/AISES

American Institute of Aeronautics and
    Astronautics
1801 Alexander Bell Drive, Suite 500
Reston, VA 20191-4344
(800) NEW-AIAA or (703) 264-7500
www.aiaa.org

American Institute of Chemical
    Engineers
345 East 47th Street
New York, NY 10017-2395
(212) 705-7000 or (800) 242 4363
www.aiche.org

American Institute of Mining, Metallur-
    gical and Petroleum Engineers
345 East 47th Street
New York, NY 10017
(212) 705-7695

American Nuclear Society
555 North Kensington Avenue
La Grange Park, IL 60526
(708) 352-6611
www.ans.org

American Oil Chemists Society
1608 Broadmoor Drive
Champaign, IL 61821-5930
(217) 359-2344
www.aocs.org

American Railway Engineering
    Association
8201 Corporate Drive
Landover, MD
(301) 459-3200

American Society of Agricultural
    Engineers
2950 Niles Road
St. Joseph, MI 49085-9659
(616) 429-0300
www.asae.org

American Society of Civil Engineers
1801 Alexander Bell Drive
Reston, VA 20191-4400
(800) 548-2723 or (703) 295-6000
www.asce.org

American Society of Naval Engineers
1452 Duke Street
Alexandria, VA 22314
(703) 836-6727
www.jhuapl.edu/ASNE

American Society for Engineering
    Education
1818 N St., NW, Suite 600

Washington, DC 20036-2479
(202) 331-3500
www.asee.org

American Society for Heating, Refrigeration and Air Conditioning Engineers
1791 Tulie Circle NE
Atlanta, GA 30329
(404) 636-8400
www.ashrae.org

American Society of Mechanical Engineers
345 East 47th Street
New York, NY 10017-2392
(212) 705-7722
www.asme.org

American Society of Plumbing Engineers
3617 Thousand Oaks Blvd.
Suite 210
Westlake Village, CA 91362-3649
(805) 495-7120
www.aspe.org

American Society of Nondestructive Testing
1711 Arlingate Lane
P.O. Box 28518
Columbus, OH 43228-0518
(614) 274-6003
www.asnt.org

American Society for Quality
611 East Wisconsin Avenue
Milwaukee, WI 53202
(414) 272-8575
www.asq.org

American Water Works Association
6666 West Quincy Avenue
Denver, CO 80235
(303) 443-9353
www.aws.org

The Architectural Engineering Institute
1801 Alexander Bell Drive, 1st Floor
Reston, VA 20191-4400
(703) 295-6370  Fax (703) 295-6132
http://www.aeinstitute.org

Association for the Advancement of Cost Engineering International
209 Prairie Avenue
Suite 100
Morgantown, WV 26505
(304) 296-8444
www.aacei.org

Board of Certified Safety Professionals
208 Burwash Avenue
Savoy, IL 61874-9571
(217) 359-9263
www.bcsp.com

Construction Specifications Institute
601 Madison Street
Alexandria, VA 22314-1791
(703) 684-0300
www.csinet.org

Information Technology Association of America
1616 N. Fort Myer Drive, Suite 1300
Arlington, VA 22209
(703) 522-5055
www.itaa.org

Institute of Electrical and Electronics Engineers
1828 L Street NW, Suite 1202
Washington, DC 20036
(202) 785-0017
www.ieee.org

Institute of Industrial Engineers
25 Technology Park
Norcross, GA 30092
(770) 449-0461
www.iienet.org

Iron and Steel Society
410 Commonwealth Drive
Warrendale, PA 15086-7512
(412) 776-1535
www.issource.org

Laser Institute of America
12424 Research Parkway, Suite 125
Orlando, FL 32826
(407) 380-1553
www.laserinstitute.org

Mathematical Association of America
1529 18th Street, NW
Washington, DC 20036
(202) 387-5200
www.maa.org

The Minerals, Metals and Materials
    Society
420 Commonwealth Drive
Warrendale, PA 15086
(412) 776-9000
www.tms.org

NACE International
1440 South Creek Drive
Houston, TX 77084-4906
(281) 492-0535
www.nace.org

National Academy of Engineering
2101 Constitution Avenue, NW
Washington, DC 20418
(202) 334-3200

National Action Council for Minorities
    in Engineering, Inc.
The Empire State Building
350 Fifth Avenue, Suite 2212
New York, NY 10118-2299
(212) 279-2626
www.naofcme.org

The National Association of Minority
    Engineering Program Adminis-
    trators, Inc.
1133 West Morse Blvd., Suite 201
Winter Park, FL 32789
(407) 647-8839
(407) 629-2502 Fax

National Association of Power
    Engineers
1 Springfield Street
Chicopee, MA 01013
(413) 592-6273
www.powerengineers.com

National Institute of Standards and
    Technology
Publications and Programs Inquiries
Public and Business Affairs
Gaithersburg, MD 20899
(301) 975-3058
www.nist.gov

National Science Foundation
4201 Wilson Blvd.
Arlington, VA 22230
(703) 306-1234
www.nsf.gov

National Society of Black Engineers
1454 Duke Street
Alexandria, VA 22314
(703) 549-2207
www.nsbe.org

National Society of Professional
    Engineers
1420 King Street
Alexandria, VA 22314
(888) 285-6773
www.nspe.org

Society of Allied Weight Engineers
5530 Aztec Drive
La Mesa, CA 91942
(619) 465-1367

Society of American Military
  Engineers
607 Prince Street
Alexandria, VA 22314
(703) 549-3800 or (800) 336-3097
www.same.org

Society of Automotive Engineers
400 Commonwealth Drive
Warrendale, PA 15096
(412) 776-4841
www.sae.org

Society of Fire Protection
  Engineers
7315 Wisconsin Avenue
Suite 1225W
Bethesda, MD 20814
(301) 718-2910

Society of Hispanic Professional
  Engineers
5400 East Olympic Blvd.
Suite 210
Los Angeles, CA 90022
(213) 725-3970
http://www.eng.umd.edu/SHPE/index.
  htm

Society of Manufacturing Engineers
One SME Drive
P.O. Box 930
Dearborn, MI 48121-0930
(313) 271-1500
www.sme.org

Society for Mining, Metallurgy
  Exploration, Inc.
8307 Shaffer Parkway
Littleton, CO 80127
(303) 973-9550
www.smenet.org

Society of Naval Architects and Marine
  Engineers
601 Pavonia Avenue

Jersey City, NJ 07306
(800) 798-2188
www.sname.org

Society of Petroleum Engineers
P.O. Box 833836
Richardson, TX 75083-3836

Society of Plastics Engineers
14 Fairfield Drive
Brookfield, CT 06804-0403
(203) 775-0471
www.4spe.org

Society of Women Engineers
120 Wall Street
11th Floor
New York, NY 10005
(212) 509-9577
www.swe.org

SPIE International Society for Optical
  Engineering
P.O. Box 10
Bellingham, WA 98227-0010
(360) 676-3290
www.spie.org

Tau Beta Pi
508 Dougherty Engineering Hall
P.O. Box 2697
Knoxville, TN 37901-2697
(423) 546-4578
www.tbp.org

Women in Engineering Initiative
University of Washington
101 Wilson Annex
P.O. Box 352135
Seattle, WA 98195-2135
(206) 543-4810
www.engr.washington.edu/~wieweb.
  com.

## LEARNING ACTIVITIES

### Activity 5.1    Design and Materials Evolution

Time:    4–6 hours

## LEARNING ACTIVITY

Designers learn to see new products in relation to their predecessors and nearest competitors. This is not just to avoid reinventing the wheel but also to learn from the work of other designers. Such product analysis leads to a deeper understanding of the design requirements the product must meet. Part of this analysis involves examining the materials used in the manufacture of the product.

In this activity, students will trace changes in materials used in the manufacture of everyday products over the past 50 years. 1

- Discuss reasons why different materials are used now and give examples.
- Examine in detail how the products' materials have changed.
- Evaluate the changes in materials and designs for these products:
  - Electric kettle
  - Folding Chair
  - Product of your own choice

## RESOURCES

http://www.jkcc.com/galaxy/ Peter's Radio Galaxy website. Many pictures of old
    products ranging from radios to miscellaneous toys.
http://www.nara.gov/ This is the site for the National Archives.
1. Advanced Design and Technology, Second Edition. (1995). Norman, E., Cubitt,
    J., Urry, S., and Whittaker, M. Longman Group, Pg 32, 47
Singer, L. and Ritz, J. (1995) Industrial Design: Enhancing Our Technological
    World. In *The Technology Teacher,* 55 (3), 19-26.

### Activity 5.2    Waste Management Systems

Time:    5–10 hours

## CASE STUDY

Municipal Solid Waste (MSW)—more commonly known as trash or garbage—consists of everyday items such as product packaging, grass clippings, old furniture, clothing, bottles, food scraps, newspapers, appliances, paint, and batteries. In 1999, U.S. residents, businesses, and institutions produced more than 230 million tons of MSW, which is approximately 4.6 pounds of waste per person per day, up from 2.7 pounds per person per day in 1960.

Several management practices, such as source reduction, recycling, and composting have been instituted to combat the growth of MSW.. *Source reduction* involves altering the design, manufacture, or use of products and materials to reduce the amount and toxicity of waste. *Recycling* diverts items, such as paper, glass, plastic, and metals from the waste stream. These materials are sorted, collected, processed, and re-manufactured into new products. *Composting* decomposes organic waste, such as food scraps and yard trimmings, with microorganisms (mainly bacteria and fungi), producing a humus-like substance for enriching the soils where needed.

Other practices address those materials that require disposal. *Landfills* are engineered areas where waste is placed into the land. Landfills usually have liner systems and other safeguards to prevent groundwater contamination. *Combustion* is another MSW practice that has helped reduce the amount of landfill space needed. Combustion facilities burn MSW at a high temperature, reducing waste volume and generating electricity.[1]

In this case study, your team's objective is to design and build a model of a waste management system for a human settlement on the Moon. Here are the procedures for the case study:

1. Your school, in many ways, is like a miniature town. It has systems for governance, health care, traffic control, recreation, and waste disposal. To get a better idea of how much waste your school generates each week, find out how many students, teachers, administrators, and other staff (and animals, if any) are typically in the buildings.

2. Next, interview the cafeteria staff and the custodial staff for the answers to these questions:
   a. What gets thrown away?
   b. How many pounds get thrown away every week? Calculate how many pounds of trash this is per person in the school.
   c. Are there any items that can be recycled before disposal? If yes, what are the recycled items?
   d. Which items are biodegradable?
   e. Into what types of containers is the garbage/trash packed for removal?
   f. Where is it taken?
   g. According to building codes, how many toilets must there be to accommodate all the people?

3. Waste is a hot topic in our culture. Why? Discuss what you know about the following phrases: Excessive packaging, landfills, toxic waste, disposable plastic goods, non-degradable material, water pollution, and air pollution.

4. In movies like those starring Indiana Jones, well-preserved, ancient artifacts are often found in the desert. Scientists also find preserved artifacts in polar ice-for example, mastodons or ancient people. Why have they not decayed?

5. Review the Moon ABCs Fact Sheet. (http://www.spacegrant.hawaii.edu/class_acts/MoonFacts.html)

   The Moon Base must be an enclosed, self-sustaining settlement. Just like your school, it must perform the basic functions of a town. Project teams are also responsible for designing and constructing several other types of systems (air supply, communications, electricity, food production and delivery, recreation, temperature control, transportation, and water supply). Your team's job is to dispose of the waste that could be generated by these other systems. Design a waste disposal system for the Moon Base. Be sure to decide what importance, if any, will be given to biodegradable materials, recycling, and the Moon outside of the constructed settlement.

6. Construct a model of this system based on your design. It must include the application of at least four facts from the "Moon ABCs Fact Sheet." For example, how will the Moon's gravity affect the design of your system? Maybe your system will be heavy but still portable by a few Moon base workers because the Moon's gravity is 1/6th of Earth's gravity.

7. Make a detailed and labeled sketch of the model.

## RESOURCES

*Recycling Household Waste: The Way Ahead,* Association of Municipal Engineers of the ICE

1991, *American Society of Civil Engineers* (Thomas Telford, Ltd.) This is a good book and its price is greatly reduced through the ASCE website ($7 versus $52 from other sources)

http://www.spacegrant.hawaii.edu/ Hawaii's Space Grant College, Hawaii's Institute of Geophysics and Planetology, University of Hawaii, 1996

http://www.challenger.org The Challenger Center is committed to providing teachers with cutting-edge techniques to get students enthused about science and technology. They offer easily downloadable clipart and publications that can add visual excitement and hands-on activities to your lesson plan. When the subject is space, this is the place for news and resources to make every classroom minute count.

*Exploring the Moon Teacher Guide,* A Teacher's Guide with Activities for Earth and Space Science, GRADES 4–12

1. http://www.epa.gov/epaoswer/non-hw/muncpl/facts.htm United States Environmental Protection Agency website.

## Activity 5.3    Engineering Career Investigation

Time:     5–7 hours

## LEARNING ACTIVITY

You will be spending several days in the library, on the Internet, and reading selected materials to research the engineering career that most interests you. Use your time wisely and ask for help if you cannot find what you are looking for. The report should be at least three pages of text and information, include a cover page with a graphic, and a bibliography citing all resources used, including websites. The report should include the following:

a. The nature of the work (what does this person do?)
b. Job responsibilities (types of things done on a daily basis)
c. Training necessary to enter this field (two-year degree, four-year degree, internship, etc.)
d. Advancement possibilities (will they get promoted, are there supervisory positions?)
e. Future outlook for this career (will it be around in two, five, ten years?)
f. Average salary (starting)
g. Education requirements (do they have to go to college or attain licensing?)

## RESOURCES

Burghardt, M.D. Introduction to the Engineering Profession, 2nd Edition. New York: Harper Collins College Publishers, 1995.

Garcia, J. Majoring, (Engineering). New York: The Noonday Press, 1995.

Grace, R., and J. Daniels. Guide to 150 Popular College Majors. New York: College Entrance Examination Board. , 1992, pp. 175–178.

Irwin, J.D. On Becoming An Engineer. New York: IEEE Press, 1997.

Kemper, J.D. Engineers and Their Profession, 4th Edition. New York: Oxford University Press, 1990.

Landis, R. Studying Engineering: A Road Map to a Rewarding Career. Burbank, CA: Discovery Press, 1995.

Smith, R.J., B.R. Butler, W.K. LeBold. Engineering as a Career, 4th Edition. New York: McGraw-Hill Book Company, 1983.

Wright, P.H. Introduction to Engineering, 2nd Edition. New York: John Wiley and Sons, Inc., 1994.

## REFERENCES

## Acoustical Engineering

http://www.arce.ukans.edu/wwwvl/acoustic.html
Microsoft Encarta—"Loudspeakers"

B&W Loudspeaker Corporation—http://www.B&W.com
Stereophile Magazine—July 1997 issue

## Automotive Engineering

i The World Book Encyclopedia, 1993
ii World Book Online Edition
iii World Book Online
iv World Book Online Edition
v www.executivesearchgroup.net
Occupations Handbook 1998–1999
www.sae.org/automag/worldchallenges/us.htm, Society of Automotive Engineers
www.magnaint.com, Auto Components Manufacturer
Moving Objects 1999, (Catalogue from Auto design exhibit)

## Aerospace Engineering (Aero or Astronautical)

http://stats.bls.gov/oco/ocos028.htm "Occupational Outlook Handbook 2000–01
    Edition"
http://.stats.bls.gov/oco/ocos027.htm "Occupational Outlook Handbook 2000–01
    Edition"
http://.stats.bls.gov/oco/ocos20016.htm "Occupational Outlook Handbook 2000–
    01 Edition"
"Discover Engineering Online" Aerospace
http://www.discoverengineering.org/eweek/Engineers/aerospace_engineering.
    htm

## Chemical Engineering

http://www.engineeringnet.com
http://www.asee.org
University of Wisconsin-Madison School of Education, Occupations Handbook
    1998–99, University of Wisconsin Board of Regents, 1998
U.S. Department of Labor, Occupational Outlook Handbook, Just Works, Inc.,
    1998
vi Obtained from Career Vision 2000 program.
vii Obtained from www.tandf.co.uk/journals/carfax/14636778.html

## Genetic Engineering

Schoofs, Mark. "The End of Nature: From Banana Vaccines to Fast Growing Fish:
DNA Lets Scientists Play God." *The Village Voice*. Dec. 1997.

*Chapter 6*

# Profiles of Engineers and Engineering Students

## Introduction

This chapter contains a collection of profiles of engineering graduates to let you read first-hand accounts of what it is really like to be an engineer. Each engineer wrote his or her own profile.

Our intent is to provide you with a glimpse of the diversity of the engineering workforce. Engineers are people just like you, and had varying reasons for pursuing engineering as a career. Engineering graduates also take very diverse career paths.

In order to capture a flavor of this diversity, each engineer was asked to address three areas:

1. Why or how they became an engineer
2. Their current professional activities
3. Their life outside of work

This is not a comprehensive survey of the engineering workforce. To truly represent the breadth of engineering careers and the people in those careers would take several volumes, not just one chapter. This is meant to be only a beginning. It is also arranged simply to show you the wide range of engineering careers that are possible. To keep the wide-ranging feel, we've avoided ranking the profiles by subject, and instead present them alphabetically. This makes it easier to see the common bonds across all the disciplines.

We recommend that you follow up and seek out other practicing engineers or engineering students, to get more detailed information about the specific career path you might follow. As with the information we have collected in other chapters of this text, our goal is to assist you in finding that career path which is right for you. You are the one who must ultimately decide this for yourself.

The following Table 6.1 summarizes those who provided profiles.

## TABLE 6.1 Summary of Profiled Engineers

| Name | BS Degree | Graduate Degree | Current Job Title |
|---|---|---|---|
| Sue Abreu | BS/E/BioMed | MD | Lt. Col and Medical Director |
| Moyosola Ajaja | BS/EE | | Software Engineer |
| Patrick Rivera Antony | BS/Aero | | Project Manager |
| Artagnan Ayala | BS/Aero | | Combustion Engineer |
| Sandra Begay-Campbell | BS/CE | MS/CE | Executive Director of AISES |
| Raymond C. Barrera | BS/EE | MS/Software | Computer Engineer |
| Linda Blevins | BS/ME | MS/ME & Ph.D. | Mechanical Engineer with NIST |
| Timothy Bruns | BS/EE | | Software Manager |
| Jerry Burris | BS/EE | MBA | General Manager |
| Bethany Fabin | BS/ABE | | Design Engineer |
| Bob Feldmann | BS/EE | MS/Comp, MBA | Director, Tactical Aircraft Systems |
| Steven Fredrickson | BS/EE | Ph.D. | Project Manager, NASA |
| Myron Gramelspacher | BS/ME | | Manufacturing Manager |
| Edgar Hammer | BS/ME/Mat | | President, SSI Consulting Group |
| Karen Jamison | BS/IE | MBA | Operations Manager |
| Beverly Johnson | BS/ME | MS/EM & MBA | Supervisor |
| James Lammie | BS/CE | MS/CE | Member of Board of Directors |
| Ryan Maibach | BS/CEM | | Project Engineer |
| Mary Maley | BS/AgE | | Product Manager |
| Jeanne Mordarski | BS/IE | MBA | Sales Manager |
| Mark Pashan | BS/EE | MS/EE & MBA | Director, Hardware Operations |
| Douglas Pyles | BS/E/Mgt | MS/CEM | Engineering Consultant |
| James Richard | BS/CEM | | Director of Mechanical Systems |
| Patrick Shook | BS/ME | MS/ME | Senior Engineer |
| David Swain | BS/AAE | | Executive Vice-President |
| Nana Tzeng | BS/ME | MS/ME | Design Engineer |
| Patrice Vanderbeck | BS/ME | | Electronics systems Engineer |
| Jack Welch | BS/ChE | MS/ChE & Ph.D. | Chief Executive Officer (retired) |
| Shawn Williams | BS/EE | | Product General Manager |
| Adel Zakaria | BS/ME | MS/IE & Ph.D. | Sr. VP Engr & Manufacturing |

## PROFILE OF A BIOMEDICAL ENGINEER: SUE H. ABREU, FT. BRAGG, NORTH CAROLINA

### Occupation

Lieutenant Colonel, Medical Corps, United States Army
Medical Director, Quality Assurance, Womack Army Medical Center

### Education

IDE (BSE, Biomedical Engineering), 1978
MD, Uniformed Services University of the Health Sciences, 1982

### Studying Engineering

As I started college, I was planning to be a teacher. Because of taking an elective class in athletic training, I developed an interest in sports medicine. I ended up taking most of my classes in aeronautical engineering so I could study the lightweight structures and materials that could be used to design artificial limbs or protective equipment for sports. Late in college, I decided to go to medical school and ended up graduating from college with an interdisciplinary engineering degree.

### Career Life

After medical school, I specialized in nuclear medicine. In nuclear medicine we use small amounts of radioactive compounds to see how things work inside of people. By using special cameras that detect radiation and computers that help gather the data, we can watch how various organs function. We can do three-dimensional studies and quantify results. In nuclear medicine, I am a consultant to other physicians: I help them decide what tests might be helpful and discuss the meaning of the results of the nuclear medicine procedures we do for their patients.

I ended up in a field I never had heard of when I started college, but I found it as I kept exploring areas that intrigued me. I tried new classes and looked for opportunities that interested me, even if they didn't fit the paths most students followed. As a result, I found a specialty I enjoy, and I'm now doing a great deal of teaching within my specialty of nuclear medicine and in my current work in quality assurance.

So, be sure to follow your dreams—if you can take something you love doing and find a way to earn a living doing it, you will end up much happier than if you set money or prestige as your goals.

### Life Outside of Work

Outside of work I enjoy skydiving. I volunteer as the team doctor for the U.S. Parachute Team and have traveled all over the world with them. I currently live in a large house on six acres in the country, not far from an airport with a parachuting center. I share the house with an airline pilot and an artist—both expert skydivers—who help make it a great place to live. Although I was married, I chose not to have children; but the dog and cats help keep us company here.

## PROFILE OF A COMPUTER ENGINEER: MOYOSOLA O. AJAJA, CHANDLER, ARIZONA

### Occupation

Software Engineer at Intel Corporation

### Education

BS, Computer and Electrical Engineering, 1997

### Studying Engineering

I came into engineering the easy way—by excelling in math and physics in high school. Deciding to enroll at Purdue and pursue a dual degree in computer and electrical engineering was a little more complicated. I wanted to learn more about computers, and I wanted to seek my fortune in a distant land. I picked the U.S. and justified the 7000-mile journey from Lagos, Nigeria, where my family lived, to West Lafayette, Indiana, where Purdue is located, with the phrase "dual degree." (I understand that degree option is no longer available. Fellow adventurers will have to justify their journeys with a different explanation.)

During my first year at Purdue I set two goals for myself: first, find a scholarship to fund my education, and second, gain useful work experience. I applied for dozens of scholarships. I was partial to those offered by engineering firms that provided internships, since internships for first-year engineering students were very scarce. In addition, I attended every resume or interview preparation workshop offered during that year. My efforts paid off. I was invited to join the cooperative education program with a summer placement with Intel Corporation, and later I was awarded an Intel Foundation scholarship which paid my tuition.

### Career Life

Today, I work as a software engineer with Intel in Arizona. I develop hardware emulation units and validation test suites for new processors. What that means in plain English: I take descriptions of hardware features and functions and write software programs that behave like the hardware should. This is cheaper than fabricating silicon devices each time a change is made during design. As an Intel engineer with access to the latest and greatest technologies, I am constantly challenged to learn new things to remain at the leading edge of computer technology.

### Life Outside of Work

I have tried to maintain a balance between my work and my non-work activities. My weekend mornings are spent running with my dog or hiking up Camelback Mountain in Phoenix. The evenings are spent in classes like dog training, theology, or photography. My real passion is traveling, and through engineering school, internships, and my current assignment, I have met people who helped fuel my interest in increasingly diverse destinations.

I've discovered that engineering is a discipline, not just a major. The distinction here is that a discipline involves the development of the faculties through instruction and exercise, while a major is simply a field of study, an area of mental focus, or a concentration. For me this means the qualities of an engineer should be apparent in all I do. The guide I use is the Code of Ethics approved by the IEEE, which is presented in the Ethics chapter.

## PROFILE OF AN AEROSPACE ENGINEER:
## PATRICK RIVERA ANTONY

### Occupation

Project Manager, Boeing Space Beach

### Education

BS, Aerospace Engineering

### Studying Engineering

When Apollo 11 landed on the moon I was eight years old. I still remember to this day how excited I was and exactly where I was and how I felt when I saw Neil Armstrong step off the ladder of the lunar module and become the first man to walk on the moon. Math and science was always my favorite subject in school, so it gave me a good background to get started in my college studies.

Along with space travel, rockets and airplanes have always fascinated me. The whole concept of being able to fly and be free is amazing. I also knew that if I wanted to become an astronaut, I would have to learn to fly. So I got my pilot's license and found that the fundamental principles of flying are exactly what you learn in the courses taken for Aerospace Engineering.

Now, I would probably put more emphasis on my non-engineering classes. To be a good engineer now, you have to be able to communicate effectively, work in a team environment, and know different aspects of the business. Being good technically is not enough anymore. I would take more English, Business, and Social Science classes to make me a more well rounded individual and more valuable employee.

### Career Life

Currently I am the Project Manager of the Performance Management and Continuous Improvement team for Boeing Space Beach, HB/SB Host Site Engineering. My responsibilities include developing and maintaining the management system used for organizational oversight, planning and operational assessment. I coordinate company policy deployment of business goals and strategies from the Vision Support Plan to the Engineering functional organization. Additionally, I interface with Program Management and Technical Managers to define performance requirements and develop innovative process and organizational solutions.

In my current management role, I spend a lot of time in planning meetings, so every day is different. I also do not often have an opportunity to use my engineering education, but as a leader of an organization, my engineering background allows me to communicate competently on technical matters related to the business.

### Life Outside of Work

Most of my free time I try to spend with my wife and two children. But when I am not with them, I am studying. I am currently pursuing a Doctorate degree in Organizational Psychology with an emphasis in Change Management. With what little free time I have after that, I enjoy golf, hiking, fishing, and biking.

## PROFILE OF AN AEROSPACE ENGINEER: ARTAGNAN AYALA, GILBERT, ARIZONA

### Occupation

Combustion Engineer, Diversity Organizations Manager, Honeywell

### Education

BS, Aeronautical and Astronautical Engineering, 1995
MS (in progress), Mechanical & Aerospace Engineering

## Studying Engineering

I have been interested in space since I was very young, I always wanted to be an Astronaut. This was my motivation to become an engineer. I figured that, if one day I was to climb on a rocket and go to outer space, I'd better know how it works. I am still working towards that goal.

My education has definitely met my expectations. I have applied what I learned in my job, and some of it to life in general. Engineering is not only a field you go into, but also a way of thinking. You are thought to solve problems, which can be applied to everything you do.

If I could start over, I would interact more with professors, take better notes, and learn more about statistics.

### *Career Life*

Currently I am a Combustion Engineer, a Six Sigma Plus Black Belt, and the Diversity Organizations Manager for Society of Hispanic Professional Engineers at Honeywell Engines & Systems.

As a combustion engineer I design and develop combustion systems that are installed in Auxiliary Power Units and Industrial Power Generators. I have finished the development on one system, designed a technology demonstrator, and am currently designing a premixed fuel delivery system.

As a Six Sigma Plus Black Belt I apply statistical tools to improve all sorts of processes, from combustion system development and manufacturing.

As Diversity Organizations Manager, I am responsible for the company's contact and participation with the Society.

What I like the most about what I do is the diversity of my responsibilities. I get to apply my engineering skills every day, and I get to learn more skills. This has allowed me to receive my Black Belt certification, and participate in the Honeywell Quest for Excellence, a company event where teams with outstanding results present their work in a competition to win the Premier Achievement Award, the biggest team honor. I am particularly proud of my participation as presenter in 2 events, and making it to the finals in one of them.

### *Life Outside of Work*

My wife Laura, a Graphic Designer, and recently we expanded our family with the arrival of our first daughter, Deanna Isabella. I devote most of my free time to my family and some to my studies.

Before we had a baby, I participated in a volleyball league, went dancing at clubs and concerts on weekends, and traveled outside of Arizona.

**PROFILE OF A CIVIL ENGINEER:**
**SANDRA BEGAY-CAMPBELL, BOULDER, COLORADO**

**Occupation**

AISES Executive Director

**Education**

BSCE, 1987; MS, Structural Engineering, 1991

## Studying Engineering

I am a Navajo and the executive director of the American Indian Science and Engineering Society (AISES), which is a non-profit organization whose mission is to increase the number of American Indian scientists and engineers. I am the third executive director in the Society's twenty-year history and the first woman to serve in this position. I manage the Society's operations and educational programs. For more AISES information, check out www.aises.org.

In 1987, I received a BSCE degree from the University of New Mexico. I worked at Lawrence Livermore National Laboratories before I earned a MS, Structural Engineering degree from Stanford University. I also worked at Los Alamos National Laboratory and Sandia National Laboratories before accepting my current leadership position. Within AISES, I served as a college chapter officer, a national student representative, and board of directors member. I was the first woman AISES board of directors Chairperson.

In the sixth grade, I was very interested in architecture, but I knew I was not an artist. I also enjoyed math and solving problems so I looked into the engineering profession. I attended a "minority introduction to engineering" program as a high school junior and I discovered that civil engineers worked on a variety of interesting public projects, which included work with architects. This program solidified my decision to become an engineer.

### Career Life

One of the earliest challenges I faced was in continuing my structural engineering studies following the 1989 San Francisco Bay-Area earthquake. I was a first quarter graduate student at Stanford when the earthquake hit. Through prayer and reflection, I understood my unique role as an American Indian engineer. I must use my best knowledge to design structures for earthquake resistance, but my cultural heritage taught me the wisdom that engineers ultimately cannot control Nature and that we have to accept the consequences from natural phenomena.

### Life Outside of Work

Life outside of work is difficult to describe at this point in time. With the re-building of the AISES organization and relocation of the offices, I don't have much time for outside activities. I have also been commuting between Boulder, Colorado, and Albuquerque, New Mexico. In brief, my hobbies are watching college basketball, watching movies, and working on my home's backyard. My husband and I have two dogs and a cat.

## PROFILE OF A COMPUTER ENGINEER:
## RAYMOND C. BARRERA, GAITHERSBURG, MA

### Occupation

Computer Engineer, Advanced Concepts and Engineering Division, Space and Warfare
Systems Command Systems Center, San Diego

### Education

BS, Electrical and Computer Engineering 1989
MS, Software Engineering 1999

### Studying Engineering

I was very fortunate during high school to work for an archaeologist and her husband who
were great mentors. To me archaeology is like detective work—finding bits of information
here and there and putting them together to form the big picture. Dr. Bernice McAllister
taught me the scientific methodology an archaeologist needs to base sound conclusions
on evidence. I think I would be happy had I become an archaeologist, but I really enjoy
building things. My dad's training as an electronic technician had gotten me interested in
electronics when I was very young. That, with some encouragement from Dr. McAllister's
husband, Capt. James McAllister, USN (ret) helped convince me to select Electrical Engineering as my specialty.

### Career Life

I work at a research, development, test and evaluation laboratory for the US Navy. I am involved in testing and system engineering of command and control systems. Command and
control systems are used by tactical commanders for decision making and direction. I
began working here in 1989 so I was here during Desert Storm. Perhaps even more important than the technical work is the ability to communicate. Not very many engineers work
alone. A former Navy Admiral, Grace Hopper (who is said to have coined the computer term
"bug") used a length of wire to describe a nanosecond to programmers. It was about a foot
long, the distance that electricity could travel in one billionth of a second. But then she
showed a microsecond—a coil of wire almost a thousand feet long. She was trying to convince programmers not to waste even a microsecond. Often the most difficult engineering
challenge is to share an idea with others in oral and written presentations, but that is the
only way these ideas can come to life.

### Life Outside of Work

My wife Martha and I spend most of our time outside of work with our new daughter Laura.
I do have some flexibility on my work schedule so I can spend more time with her. I've been
able to select job assignments that don't require too much travel. Since this is a research
laboratory there are always new things to do. In the over ten years I've been here, no two
have been the same. In this command alone there are engineers working with supercom-
puters, lasers, networking, marine mammals, 3-D displays, simulators, and sensors.

## PROFILE OF A MECHANICAL ENGINEER:
## LINDA G. BLEVINS, GAITHERSBURG, MARYLAND

### Occupation

Mechanical Engineer, National Institute of Standards and Technology

### Education

BSME, 1989; MSME, 1992; PhD, 1996

### Studying Engineering

During high school I discovered that I enjoyed mathematics. I learned about engineering when I participated in a six-week summer honors program at the University of Alabama before my senior year in high school. I took college calculus that summer, and I was hooked. I chose to study mechanical engineering because the course subjects are diverse and the industrial demand for mechanical engineers remains steady. As a co-op at Eastman Chemical Co., I worked on engineering problems in power and chemical plants. The concepts that I learned in classes came to life during the alternate semesters that I worked, and the money I earned helped pay for school. After earning a BS degree from the University of Alabama, I obtained an MS degree from Virginia Tech, and a PhD degree from Purdue University. I never would have set or achieved these goals without encouragement and advice from faculty members. Because these mentors played such valuable roles in my life, I would advise college students to get to know their professors well. These personal investments will be rewarding for years to come.

### Career Life

I am a mechanical engineer in the Building and Fire Research Laboratory at the National Institute of Standards and Technology (NIST), a national research laboratory operated by the U.S. Department of Commerce, located in Gaithersburg, Maryland. Our goals are to study the ways that fires ignite, spread, and extinguish so that our nation can minimize the loss of lives and property to fires. My primary job function is to improve the accuracy of measurements made during fire research. A few things routinely measured are toxic gas concentration, temperature, and heat intensity. I spend my time developing laser-based instrumentation, devising computer (math) models of instrument behavior, designing laboratory equipment, tinkering with electronics, publishing papers, writing and reviewing research proposals, and presenting talks at conferences. In addition, I work on a project funded by the National Aeronautics and Space Administration (NASA) to study fires in space. Working in a research laboratory ensures that I am constantly learning and growing, and I realize every day how lucky I am to be here. My job is exciting, fun, and rewarding.

### Life Outside of Work

During my free time, I enjoy hiking, rollerblading, and reading. I participate in a weekly bowling league and I manage a softball team each summer. I also volunteer as a member of the Mechanical Engineering Advisory Board at the University of Alabama. This allows me to travel home to Alabama (and visit my family) several times a year. Finally, I volunteer regularly to educate children and community members about the excitement of engineering.

## PROFILE OF AN ELECTRICAL ENGINEER: TIMOTHY J. BRUNS, ST. LOUIS, MISSOURI

### Occupation

Software Manager at Boeing Co.

### Education

BSEE, 1983

## Studying Engineering

I became interested in electronics at a young age by building electronic kits from companies like Radio Shack and Heathkit. As a teenager, I became very active in local citizen's band (CB) radio groups. It was an easy decision for me to pursue a degree in engineering. The technology has changed so much since I graduated, and I have needed to stay current with the latest technology and to find ways to apply it to my line of work. If you are just starting out in engineering, I encourage you to apply yourself and do your very best in all your classes. When I arrived at Purdue I felt as if I was the least prepared of any of my classmates, but I worked hard and did very well. Some of the better prepared students did not apply themselves from the beginning and suffered as a result. One thing I would have done differently is to get to know my professors and teaching assistants better. In large universities and organizations it is easy to get lost in the crowd, and I wish that I had formed better friendships and relationships with my instructors.

### Career Life

I am the software manager for a team of 15 developers that is creating a Windows NT application. This application uses the latest technologies such as MFC, COM and ActiveX. A typical day is spent reviewing the technical work of the team, along with reviewing schedules and making estimates for future work. I often meet with customers of our product and suppliers of our software development tools. Since our program is just getting started, I have been spending a lot of time interviewing people who would like to join our team. It is difficult to say how I apply my engineering training directly to my current job. I know that my engineering degree has given me the ability to plan and organize the work of our team, and to solve the many problems that come up. The thing I like best about my job is the wide variety of assignments I have had in my 15 years with Boeing. Working in a large company gives me the ability to have several "mini-careers," all while working for the same company. A significant accomplishment that I have made while working at Boeing is the introduction of new tools and technology into the software development process. One tool that we have introduced automatically produces source code from a graphical representation. This tool enables us to bypass much of the labor-intensive and error-prone aspects of software design.

### Life Outside of Work

In the engineering field, particularly in electrical and computer engineering, you will find that the technology changes very rapidly. In my case, I stay abreast of the latest technologies by enrolling in evening computer classes through the local universities. I enjoy home "engineering" projects such as designing a new deck. My wife, Donna, and I keep very busy raising our two sons, Garrett and Gavin.

## PROFILE OF AN ELECTRICAL ENGINEER: JERRY W. BURRIS, LOUISVILLE, KENTUCKY

### Occupation

General Manager of Refrigeration Programs for General
Electric Appliances

### Education

BSEE, 1985; MBA, 1994

### Studying Engineering

I have always had a curiosity about how things work (especially electronic devices). My parents recognized this at a very early age. They encouraged me to think about becoming an engineer. I was the child in the family who was always asked to fix the TV or electronic games. This continued through high school, where I excelled in math and science.

Purdue University was a natural choice for me, not only for its reputation for engineering excellence, but also due to the added bonus of having Marion Blalock and her Minority Engineering Program. This program has served as a recruiting magnet for Purdue and also has served as a mechanism for helping retain and matriculate students of color at Purdue.

While at Purdue, I was active in many extracurricular activities including leadership roles with NSBE and Kappa Alpha Psi fraternity. My early involvement in academics and extracurricular activities led to a full scholarship, which I received from PPG during my freshman year. Summer internships with PPG and IBM were invaluable in terms of giving me insight into what career path I wanted to pursue (design, manufacturing, or sales/marketing).

### Career Life

I chose the technical sales and marketing route with General Electric's Technical Leadership Program. This premier program gave me advantages over direct hires in terms of exposure and training. After working six years I earned an MBA from Northwestern's Kellogg School of Management; I focused on global business, teamwork, and marketing.

My career has taken me from a role as a sales engineer, calling on industrial and OEM customers, to branch manager, with profit and loss responsibility, leading a team of nine people; to general manager of Refrigeration Programs at GE Appliances, where I now manage a $2 billion refrigeration product line.

I have been blessed with a lovely wife, who is also a Purdue and Northwestern graduate. We have two active children—Jarret, who is 7, and Ashlee, 4. We are managing dual careers at GE. This comes with significant challenges. However, GE has been very supportive of both of our careers.

### Life Outside of Work

Life can not be all about work! You have to strive for balance. I have sought to keep God first in my life. I enjoy coaching my children in soccer, baseball and basketball, and I try to stay active with my own personal sporting activities. My favorite activities are listening to jazz music, traveling to exotic locations, managing our investment portfolio, and improving my golf game.

## PROFILE OF AN AGRICULTURAL ENGINEER: BETHANY A. ELKIN FABIN, WATERLOO, IOWA

### Occupation

Design Engineer, 8000 Chassis Design Team—John Deere
  Waterloo Works

### Education

BS, Agricultural and Biological engineering

### Studying Engineering

When I began to explore career options, I was told that an engineering degree was the ticket to achieving success in a variety of fields. I investigated the Agricultural and Biological Engineering program at Penn State and discovered therein the opportunity to examine many aspects of engineering and agriculture under one discipline. I found my niche. This major provided the chance to "sample" many engineering topics and thus make knowledgeable decisions on what areas I wanted to pursue in future jobs. My Business Management minor also afforded many opportunities, and I would recommend that every engineer take at least a few business classes. I would also recommend getting involved in professional societies whenever possible. They provide many networking opportunities and a good preview of the job market. If I were to start my schooling over, I would take more of the hands-on classes. Also, I cannot begin to convey the importance of an internship or some kind of related work experience. Having the opportunity to work for a variety of companies in a variety of positions has helped me greatly in my career.

### Career Life

In my current position as a chassis design engineer for John Deere, I work with others to design parts for tractor frames, coordinate homologation and standard reviews for update programs, and coordinate projects with supporting teams. In the latter role, I develop general specifications to ensure that we meet customer requirements and implement verification processes.

In a typical day of work, I spend a couple of hours working on Pro/E software designing and modeling parts. I also spend time working with suppliers and purchasing personnel to get parts quoted and ordered. In addition, I spend some time in our shop checking on prototype builds or test procedures, and some time in meetings working with different groups to keep people informed. The thing I like best about my job is the freedom I have to work on a variety of projects. It's nice to work for a company that has developed a strong name for itself and works diligently to stand behind their products.

### Life Outside of Work

Outside of work, I welcome every opportunity to travel with my husband and play host to out-of-state friends, relatives, and foreign exchange students. MBA classes are taking up much of my time off the job currently, but in my free time I find I enjoy music, sports, rowing, training my dog, remodeling my house, and gardening. My membership in the local chapter of American Society of Agricultural Engineers also keeps me busy with meetings and seminars.

## PROFILE OF AN ELECTRICAL ENGINEER:
## BOB FELDMANN, ST. LOUIS, MISSOURI

### Occupation

Director, Tactical Aircraft Mission Systems, The Boeing Company

### Education

BSEE, 1976; MS, Computer Science, 1980; MBA, 1999

### Studying Engineering

My interest in engineering evolved naturally from my lifelong interest in science. Mathematics, while not my life's ambition, was interesting and satisfying. High school offered me the opportunity to enjoy learning about science, and I knew that I wanted to explore it even more in college.

In the mid-1970s, computers were not commonplace except in colleges, and I was hooked with my first FORTRAN programming class as a freshman. From that point on, I wanted to learn more about the hardware and software that made computers work. I oriented my electrical engineering curriculum toward digital electronics and used every elective I could to take software or software theory courses. While in college, I began a four-quarter stint as a cooperative engineering student at McDonnell Douglas in St. Louis. As a co-op, I was able to design software in the flight simulators (McDonnell was a world leader in simulation) and to work on a research design team for advanced flight control systems. Each semester when I went back to school, I would be at the library when Aviation Week magazine arrived, and I would read it cover to cover. When I graduated, I started my career as a software designer. My first day on the job, I was told that I would be responsible for the design of the software that controls the automatic carrier landing system on the F/A-18 aircraft. Ever since that first day, I have never been disappointed with the technical issues that have challenged me.

### *Career Life*

Today I am leading a team of over 800 engineers in the design and production of Mission Systems (also known as avionics) for the F-15, the F/A-18, the AV-8B, and the T-45 aircraft. My role as team leader for the organization is to ensure that the various product teams are providing outstanding value to our customers with the quality of our designs. I no longer write software, but I interact with the technical teams, coaching and guiding them through the difficult challenges of today's technically-exploding world. My proudest recent accomplishment was leading a team of engineers in the design and flight test of a reconnaissance system for the F/A-18. That system will provide the United States with its first manned tactical reconnaissance capability since the mid-1980s.

### *Life Outside of Work*

My life outside the office centers around outdoor activities and my family. My wife and I have three sons, all of whom play soccer and baseball (I have coached each one at various times). I really enjoy golf, bike riding, and other outdoor activities. My wife and I receive great pleasure from watching our sons grow up.

## PROFILE OF A COMPUTER ENGINEER:
## STEVEN E. FREDRICKSON, HOUSTON, TEXAS

### Occupation

Project manager of the Autonomous Extravehicular Robotic
   Camera for NASA; Electronics Engineer, NASA Johnson Space
   Center, 1995–present

### Education

BS, Computer and EE 1992; PhD, Engineering Science, 1995

## Studying Engineering

To prepare for a leadership role in the emerging information society, I studied electrical and computer engineering as an undergraduate. At Purdue I supplemented engineering studies with non-engineering courses and extracurricular activities, and sought experiences to develop practical business skills. One highlight was the Cooperative Education Program. Three "Co-op" tours at NASA introduced me to software design, robotic control systems, and neural networks. This early work experience intensified my interest in advanced study of electrical engineering and robotics. To simultaneously satisfy my desires to engage in advanced academic research and to gain personal international experience, I pursued an engineering doctorate program in the Robotics Research Group at Oxford University.

I am extremely pleased with the universities I attended and the fields of study I completed to prepare for my current career. I would offer three recommendations to anyone pursuing an engineering path: 1) participate in Co-op or similar programs, 2) develop effective oral and written communications skills, 3) explore opportunities to study abroad.

### *Career Life*

When I returned to NASA as a robotics research engineer, I transitioned from specialized research in artificial neural networks to broadly focused applied engineering. As project manager of the Autonomous Extravehicular Robotic Camera (AERCam) project, I have led a multidisciplinary team of engineers in development of a free-flyer robotic camera to provide "bird's eye" views of the Space Shuttle or International Space Station. Despite this deliberate transition to a project leadership role, it has been imperative for me to maintain my core technical skills. To ensure continued technical proficiency, I participate in several training courses and technical conferences every year.

### *Life Outside of Work*

As much as I enjoy working at NASA, I believe it is essential to maintain outside interests. For me, that starts by spending time with my wife, Becky. Since Becky is pursuing a joint engineering and medical career, it can be demanding at times. The key for us has been to develop outside activities that we can enjoy together. Currently these include teaching Sunday school, participating in Bible study, attending concerts and plays, jogging, lifting weights, climbing at an indoor rock gym, and traveling. In addition, we allow each other time to pursue individual interests, which for me include reading, aviation, and golf.

## PROFILE OF A MECHANICAL ENGINEER:
## MYRON D. GRAMELSPACHER, HARTLAND, WI

### Occupation

Manufacturing Manager, General Electric

### Education

BSME, 1989

### Studying Engineering

I started at Purdue University in August 1985 in the engineering program. My interests in math and science were what really drove me to initially pursue opportunities in the field of engineering. My initial focus was in civil engineering, since I liked the concept of being able to work on roads, bridges, and outdoor structures. By learning more about the various engineering disciplines through seminars during my freshman year, I changed my mind and decided to pursue mechanical engineering. I felt that a degree in mechanical engineering would allow me more versatility and options in the workplace. Looking back on my college days, I wish I had taken courses in both business and foreign language to supplement my technical background.

### *Career Life*

I graduated from Purdue University in 1989 with a degree in mechanical engineering, and started with General Electric (GE) as part of the Manufacturing Management Program. This program provided me with an opportunity to have six-month rotational assignments in two GE businesses. My first year was with GE Transportation Systems in Grove City, Pennsylvania, followed by a year with GE Aircraft Engines in Cincinnati, Ohio. In 1991, I transferred to the GE Medical Systems division in Milwaukee, Wisconsin. Since that time, I have held various positions in the Sourcing group, including supplier quality engineer, buyer, and team leader of the mechanical sourcing department. I also had the opportunity to live in Paris, France, for a year, heading up an Eastern European initiative. During that time, my efforts focused on the identification and qualification of suppliers in Eastern Europe. This position required that I travel throughout Europe, making it possible for me to experience different cultures and surroundings. This was a truly challenging and rewarding experience, both for my wife and me.

I currently hold the position of a Black Belt in GE's Six Sigma quality program. I utilize the Six Sigma tools and methodology to drive both process and product improvements that reduce costs, and ultimately impact our customers. The analytical skills and systematic problem solving techniques that I gained through my undergraduate engineering courses have greatly contributed to the many opportunities and successes I have had in my professional career.

### *Life Outside of Work*

I now am attending Marquette University, working toward my MBA. An MBA will complement my technical background and enable me to strengthen my overall business knowledge. Outside work, I enjoy making landscaping improvements around the house and tackling various wood-working activities. My wife, Kim, and I enjoy traveling in our spare time. I also enjoy playing golf, tennis, and softball.

## PROFILE OF AN ENGINEER:
## EDGAR T. HAMMER, GREENSBURG, PENNSYLVANIA

### Occupation

President, SSI Consulting Group

### Education

BS Mechanical/Metallurgical Engineering

### Studying Engineering

I enjoy solving problems, and once classes started in my engineering program, I never considered pursuing any other major. It wasn't always easy, but solving problems is what engineers do and I love it. My hardest choice was selecting between mechanical and metallurgical engineering majors. After switching back and forth a few times, I discovered that many colleges would allow you to design your own "interdisciplinary major." I did just that and now probably have the distinction of being the only graduate from Grove City College with a BS in Mechanical/Metallurgical Engineering.

### Career Life

I started working as a practicing metallurgical engineer doing testing in a quality assurance department. I picked up industrial statistics along the way, which came easily with my engineering math background. All engineers learn a little computer programming (when I started, personal PCs had 4 Kb—not Mb—of RAM). I was able to write some useful applications, combining my engineering, statistics and programming knowledge, that caught my employer's attention, and eventually I was promoted to head of the department.

I found that I really enjoy programming and there is a nice market for technical software. In engineering you'll write equations, create flow charts, and draw charts and graphs to illustrate your data. That all translates directly to software development. My engineering training has literally equipped me for technical programming better than computer science would have.

Today, I have found my niche in the quality assurance industry as a consultant, trainer, and software developer. I am president of my own company and enjoy the perks of being my own boss. I rarely practice my pure engineering disciplines, but I rely on the foundations of them daily.

### Life Outside of Work

More important than my career is my family and deep faith in God. Through engineering, I have been blessed with a career that lets me keep my family a priority, allowing me to spend considerable time with them and to provide well for them. At times, engineering has also been a stumbling block to my faith. Mankind sometimes becomes prideful in his ability to explain his world. But focus on what we don't know and you will see that we understand only a tiny portion of the universe. I am convinced that the complexity and order we see reveals the hand of the master engineer Himself. I guess that puts us in good company!

## PROFILE OF AN INDUSTRIAL ENGINEER: KAREN JAMISON, DAYTON, OHIO

### Occupation

Operations Manager, Jamison Metal Supply, Inc.

### Education

BSIE, 1988; MBA, 2000

### Studying Engineering

I didn't grow up knowing I wanted to be an engineer, but luckily my high school guidance counselor recognized my science and math abilities and encouraged me to try engineering. I firmly believe that engineering is a wonderful career in and of itself, and that it can be an excellent stepping stone for any other career you may wish to pursue in the future.

I chose industrial engineering because I am highly interested in improving the processes people use to do their work. Industrial engineering provides both technical challenges and the opportunity to work with all kinds of people.

If you are just starting to think about engineering or are trying to choose a specific discipline, talk to as many practicing engineers and professors as you can. Become involved in organizations on campus that will let you interact with other engineering students and practicing engineers.

I also highly recommend the co-op program. I had over two years of work experience when I graduated, and I knew what types of work I would enjoy. It is definitely to your advantage during interviews to know what type of job will best suit you, and to be able to speak intelligently on that subject.

Finally, remember that grades aren't everything but that your education is invaluable. If I were to do one thing differently, I would study to truly learn and understand the content instead of with the goal of getting a good grade in the class.

### Career Life

Until last year, I was a consultant focusing on process improvement and business process re-engineering. Now I am learning to run Jamison Metal Supply, which is a business my parents founded 25 years ago. My job includes anything and everything that needs to be done. My primary responsibilities are overseeing operations to ensure quality products and timely deliveries, ordering steel for inventory and special orders, and pricing the material we sell.

I use my engineering training in all kinds of ways. I am working on updating our physical inventory system to better utilize warehouse floor space; I schedule customer orders to meet promised delivery times; and I am updating our computer system. Most importantly, engineering has taught me how to approach solving a problem and how to manage my time.

### Life Outside of Work

My time outside of work is concentrated on completing my MBA degree, but I do find time for having fun as well. One of my favorite hobbies is crewing for a hot air balloon. I also teach a sign language class at the University of Dayton, and am vice president of the Purdue Club of Greater Dayton, Ohio. I think engineering is a very flexible field that allows individuals to prioritize their lives any way they wish.

## PROFILE OF A MECHANICAL ENGINEER: BEVERLY D. JOHNSON, WATERLOO, IOWA

### Occupation

Supervisor in Wheel Operations at John Deere Waterloo Works

### Education

BSME; MS, Engineering Management

### Studying Engineering

My education includes a BS in Mechanical Engineering from the United States Military Academy, an MS in Engineering Management from the University of Missouri, Rolla, and my current study in the Executive Master's Degree Program at Northwestern University, Evanston, IL.

I think engineering is a very rewarding career because you can see the results of your effort every day. Engineering offers opportunities to create, build, design, and sometimes even destroy. Also, the analytical tools you develop in your engineering coursework make studying other subjects easier, and they are applicable to everyday life.

I truly enjoy my career in engineering. It is a dynamic career field that has taken me to many different jobs and many different places. I have done everything from constructing buildings and roads in Germany and the Hawaiian Islands to my current work as a supervisor in the wheel operations for the John Deere Waterloo Works.

### *Career Life*

I have been with the John Deere Waterloo Works for two years, working in various engineering assignments such as quality engineering, project management, and process redesign. My current assignment as a supervisor in Wheel Operations is focused in production. I am responsible for the assembly processes pertaining to the tires and wheels for the 7000 and 8000 series tractors. I am also responsible for the daily supervision of the wage department personnel. Although my job is sometimes hectic, it is also very rewarding as I watch what our department is able to accomplish every day.

Prior to joining John Deere I spent nine years as a military officer in the U.S. Army Corps of Engineers. My primary responsibilities included the construction of buildings and roads, and the development and training of other engineers. My work with the military allowed me to live in, and travel throughout, Europe and the Pacific Islands.

### *Life Outside of Work*

Although I chose engineering over journalism, my favorite pastimes are reading and writing. I also exercise regularly and compete in sports. I volunteer my time to the Boys and Girls Club of Waterloo, the American Red Cross, and a local university. However, my most important responsibility, and the most enjoyable, is the time I commit to the care and development of my two children, Colbert, 6, and Randy, 3.

## PROFILE OF A CIVIL ENGINEER: JAMES L. LAMMIE, NEW YORK

### Occupation

Board of Directors, Parsons Brinckerhoff Inc.

### Education

BS, Civil Engineering, 1953; MS, Civil Engineering, 1957

### Career Life

When I grew up, my father worked in a steel mill in Pittsburgh, the City of Bridges. I was fascinated with the many different bridges and what could be done with steel. I knew that I wanted to build things. I was fortunate to win an appointment to West Point, which was founded as the first engineering school in the U.S.

After graduation, I spent 21 years in the Army Corps of Engineers working on a wide variety of military and civil engineering projects all over the world. After retiring from the Army I knew I wanted to be a Project Manager on big projects, so I joined Parsons Brinckerhoff, Inc. and spent seven years as a consultant Project Manager for design and construction on the Metropolitan Atlanta Rapid Transit project (MARTA), the most rewarding period of my professional career. Today, my grandchildren ride what I helped build—a most rewarding feeling.

After MARTA, I had the pleasure of serving as the CEO of Parsons Brinckerhoff, Inc., the largest transportation design firm in the U.S., for the next fourteen years. Today, as a member of the Board of Directors of our employee-owned firm, I am still involved in some of our mega projects: the Central Artery Highway project in Boston, the new Taiwan High-Speed Rail system, the Bay Area Rapid Transit extension to the San Francisco Airport, and many others. The high point of my job is getting involved in critical project decisions and being able to "kick the tires" of work under construction.

## *Life Outside of Work*

Thanks to my varied career in engineering, construction and management, I am also able to participate in a variety of outside activities: the Transportation Research Board (TRB), the Institute for Civil Infrastructure Systems (ICIS), the Engineering Advisory Board at Purdue University, and the National Academy of Engineering (NAE). I also teach and lecture on Project Management, Leadership, and Engineering Ethics. During my career, the high points have been presenting proposals and winning major jobs, election to the National Academy of Engineering, and receiving an Honorary Doctorate at Purdue University.

I always enjoyed participating in a variety of sports, until my knees gave out. The most personally rewarding aspect of my life over the years has been the companionship of my wife, three children (all in the medical profession, thanks to my wife's nursing career), and my eight grandchildren (with three going to colleges close by, permitting frequent visits).

## PROFILE OF AN ELECTRICAL ENGINEER: RYAN MAIBACH, FARMINGTON, MICHIGAN

### Occupation

Project Engineer at Barton Malow Company

### Education

BS-CEM (Construction Engr. and Management), 1996

### Studying Engineering

I think I have construction in my blood. My family has been involved in construction for four generations, and I was always surrounded by it while growing up. I can remember walking along beams with dad at a very young age, much to the irritation of my mother. So when it came time to choose a career, construction seemed to be the obvious answer.

Knowing what I wanted to study before I entered college was very nice, but it did not make my freshman year any easier. The first year of engineering school requires that you gut-out the tough prerequisites. During my college days, two extra-curricular activities were particularly rewarding to me. The first was my involvement in student organizations, which helped me to develop my leadership skills. The second was taking advantage of summer job internships, which provided practical experience by bringing to life the theoretical class-room education. Looking back, I should have taken more technology-related classes. I have found that technological skills are highly sought after in the job market.

### Career Life

After graduation, I went to work for Barton Malow Company, a national design and construction services firm, and was made a field engineer on a basketball arena construction project. Presently, I am a project engineer working on a hospital expansion project. In my position I have a great deal of flexibility in terms of how I use my time throughout the day. I have worked on design development, purchasing contracts, scheduling, and subcontractor supervision.

Construction has been just as exciting as I had hoped. Every day when I leave my project, I can see what new accomplishments have been made—always, more steel has been hung or walls erected since I arrived. An interesting aspect of construction work is learning about our clients' businesses. It helps to have an understanding of our their industries in order for projects to be successful. For example, in order to make the present hospital expansion successful, our project team must understand how the hospital functions, overall. Having the opportunity to work on a variety of projects allows people in the construction industry to learn a great deal about various industries throughout their career.

### Life Outside of Work

The spirit of camaraderie common to a construction site fosters friendships with co-workers that often continues outside of the work place. I have participated in a variety of community activities with my co-workers, ranging from speaking to school groups to refurbishing old homes for the elderly. Individuals looking for a challenging career which offers a lifetime of learning experience will find the construction industry very rewarding.

## PROFILE OF AN AGRICULTURAL ENGINEER: MARY E. MALEY, BATTLE CREEK, MICHIGAN

### Occupation

Product Manager, Kellogg Company

### Education

BS, Agricultural Engineering (food engineering)

### Studying Engineering

Math and science were always my favorite subjects, with the best part being the story problems where the concepts were applied. The idea of using scientific principles to solve a problem is what led me to choose engineering as a major. I would get to learn some more about math and chemistry as well as do something useful with that knowledge. That happened at college and continues to happen in my job.

### *Career Life*

Today at Kellogg Company, the work I do is varied from day to day. My role is to make sure our manufacturing facilities have all the information they need, at the right time, to bring new products to market. That means coordinating the work from many different departments and gaining a consensus on the critical tasks to meet the timeline. You might ask, "What does that have to do with engineering?" Primarily, I bring together a myriad of details into one final outcome, just as all engineers do in combining the known facts to reach a solution. I just get to add some more unknowns and assumptions, such as dealing with people and changing requirements. The biggest challenge is getting the project accomplished to meet the needs of the consumer (that's you) before any of our competitors do.

Since Kellogg Company is a global company, my work affects the entire world. These days I work on projects for North America, Mexico, and Southeast Asia. I have had the opportunity to learn about other cultures and adapt our food products to fit their lifestyles. With manufacturing being located outside the U.S. as well, I encounter the varying work procedures and government regulations of each country. It makes my job challenging and enjoyable.

### *Life Outside of Work*

Certainly, working at Kellogg's is not all that I do. My job is just one part of life. I find I need outlets for creative activities and for making contributions for the betterment of our world. Through sailboat racing I find a time of total concentration and a chance to apply aerodynamic principles. This also provides a fun way to have some competition. On the creative side, I participate in the handbell choir at my church. For me, music is a way to use my whole brain in the interpretation of notes into an emotional song of praise.

Perhaps more important is how I can give back to the community and the world in which I live. To promote cultural activities in my town, I serve on the board of directors of the Art Center. Through the Presbyterian Church, I lead the Benevolence Committee in disbursing funds and educating the congregation about the outreach in our community, the nation, and the world.

## PROFILE OF AN INDUSTRIAL ENGINEER: JEANNE MORDARSKI, ALBUQUERQUE, NM

### Occupation

Sales Manager, LightPath Technologies, Inc.

### Education

BSIE; MBA

## Studying Engineering

The engineering workload at Purdue was quite a shock to me. My first two years were a struggle and I was afraid to get involved in extracurricular activities. By my junior year, I became more concerned that I was missing out on the "college experience" than I was about my grades. I became an active member in several campus organizations—the best decision I ever made. I was forced to balance my studies and personal life. I broadened my network of friends, developed leadership skills, and learned to manage my time more effectively. As a bonus, my grades improved tremendously.

My emphasis within the IE curriculum was on Production and Manufacturing Systems. I accepted a production supervisor position with Corning Inc. after graduation. This role put me in the middle of the action, and taught me to think on my feet and make sound decisions. For eight years I worked at Corning in various engineering and manufacturing capacities. During this time, I was able to land a one-year tenure in Japan implementing Process Management Systems at our facility in Shizuoka.

For three years I took evening classes working toward an MBA from Syracuse University. In my course work, I realized how much I enjoyed the business side of things. Upon completion of my degree, I accepted a sales manager position at Corning in the telecommunications market.

### *Career Life*

I recently left Corning to work as a sales manager for a start-up company, LightPath Technologies, Inc. Working for a large company directly from college gave me invaluable experience. The structure enabled me to work more effectively. However, as I progressed through the ranks at Corning, I realized that this same structure was limiting my ability to contribute because of the many management layers. In my current role at LightPath, we are introducing new products to the telecommunications market.

### *Life Outside of Work*

I have an eclectic mix of interests outside of work. I truly enjoy exercise and the outdoors. On weekends, you'll find me skiing, camping, hiking, rock climbing, or biking. I love international travel and scuba diving, and take every opportunity I can to participate in both. I have recently taken up Latin social dance and kickboxing. I am also involved with the Purdue Alumni Association in Albuquerque. I like to keep busy and have worked very hard to strike a balance between my career and personal life. I seldom work more than 40 hours a week. I made it a goal to be more productive during work hours to minimize overtime and unnecessary stress. It usually works.

## PROFILE OF AN ELECTRICAL ENGINEER: MARK ALLEN PASHAN, RED BANK, NEW JERSEY

### Occupation

Director of Hardware Development,
    Lucent Technologies

### Education

BS; MS; MBA

### Studying Engineering

When I was in high school trying to decide which career
to pursue, I had a number of criteria: I wanted a job that I'd look forward to each day, that offered continuous learning, and that offered a reasonable level of financial stability. Engineering satisfied those criteria for me. I enjoyed math and science (the foundations of engineering) in high school, but engineering is more than number crunching. The field of engineering rewards creativity, the ability to find a better way to solve a problem. If I had to do it over again, I'd still choose engineering, but I'd also have bought more shares of Wal-Mart, Lucent, and Yahoo when they were first offered.

### *Career Life*

In my career, I have advanced through a number of levels of technical management, and currently have about 130 engineers reporting to me. My job is no longer at the level of designing integrated circuits. I guide my team's progress on a number of new product development activities. I work to make sure we have the right people working on the right things at the right time. I set priorities among the competing needs of the business, and evaluate new business opportunities. To do my job, I use a combination of business and technical judgment: what are the future customer needs, what are the available and soon-to-be-available technologies, what are my competitors doing and what may they do next, who can do the work and work well together, and can we get the work done in time and at a reasonable cost. The end results are new products introduced into the marketplace that turn a profit for the business. That goal can only be achieved through others. A good part of my job is getting my teams to achieve more than they thought possible.

This is the best time in history to be an engineer. There are more available alternatives than ever—from startup companies to large established firms, from full-time to part-time work hours. There are more opportunities for continuing education and there is the potential for significant financial reward for those willing to take a risk.

My organization is spread across three states and I have customers and suppliers all over the world. My job requires travel and long hours, and I couldn't do my job and have a family without the support of my wife, Reem (see following profile). But we do it together and the kids are a joy (even when they don't always obey). I enjoy a number of activities outside of work such as basketball, traveling, and dining out.

## PROFILE OF AN ENGINEER: DOUGLAS E. PYLES

### Occupation

Engineering Consultant, Ingersoll-Rand

### Education

BS, Interdisciplinary Engineering

## Studying Engineering

I enjoyed science in high school, especially physics. I would describe myself during that time as a loner, a trait that became a weakness for me when I went to college. If I were to give advice to a student just beginning to study engineering I would say the following: 'Don't live in a shell, become active in one or two student organizations, make friends in your classes, get help right away and when you need it, don't be afraid!' If I had it to do over again I would try to follow this advice.

### Career Life

I work for Ingersoll-Rand, at the Tool & Hoist Division Business Development Center. My title is Engineering Consultant, which is the first level on Ingersoll-Rand's technical career ladder. The career ladder is for people who, after progressing through the ranks, want to continue pursuing areas of technical expertise rather than managing people.

My primary responsibility at this time is supporting New Product Development Teams in the areas of mechanisms, design, testing, and manufacturing. I am personally responsible for insuring that portions of a system meet all performance, cost, manufacturability, and life targets. During the course of a project we travel extensively in a team-wide effort to learn the needs of the customer. I also participate with and help guide the team as we develop customer specifications for the final product.

### Life Outside of Work

As I write this, I am also running an analysis on a Unix workstation that sits next to my PC. In the past, we have had to rely on experience and a process of building and testing to perfect our products. Now we have the advantage of computer analysis. I hold several patents for designs that were developed in this way. I enjoy being able to answer complex questions and make decisions about designs before making prototypes. I also really enjoy working on projects that have the potential for increasing market share and profits for our division. This type of analysis is only one of the many tools we use to develop high-quality products for an ever more demanding and changing marketplace.

When I am not slaying dragons at work, I am likely to be with my wife and three kids. I still maintain that anyone with children has no spare time, and has to schedule all their fun. On Wednesday nights, I play music with a different group of people. We have several guitars, a bass and a keyboard, as well as dedicated vocalists. We are responsible for leading music at church most every Sunday morning. I play the twelve-string guitar. When I was in school, I never thought I would ever find myself up in front of 300 people singing and playing music! It shows me that whatever stage of life I am in I must always look for opportunities to reach and to grow.

## PROFILE OF A CONSTRUCTION ENGINEER: JAMES T. RICHARD, PE, BUFFALO GROVE, ILLINOIS

### Occupation

Director of Mechanical Systems Service, Siemens Building Technologies, Inc.

### Education

BS, 1984 and MS, 1995 in Construction Engr and Management

### Studying Engineering

When I entered college, I had no idea in which engineering field I wanted to major. After learning more about the different types of engineering majors and opportunities during my freshman year, I selected construction engineering. This program gave me the opportunity to combine civil and mechanical engineering into a single concentrated major.

Equally as important as choosing your field is taking advantage of campus activities. Social skills are critical for your career and make the rest of your life more fulfilled. I was a member of a number of student organizations, including the Purdue Varsity Glee Club, which performed throughout the United States and Europe and at President Reagan's Inaugural Ball.

### Career Life

After receiving my master's degree, I became director of Mechanical Systems Services for Siemens Building Technologies, Landis Division. I am responsible for developing the mechanical service business in North America and assisting approximately 100 branch offices. I am very proud of my department and the way we have grown.

When working with 100 branch offices, there is no typical day. What's exciting about my work is starting from ground zero on a project and giving it life. Our effort to build a world-class service organization pumps me up daily. It's exciting to answer questions such as "How can we do that better?" and "How can we make it better?" I am constantly applying the problem-solving techniques engineering teaches to take us from point A to point B.

### Life Outside of Work

My biggest accomplishment is coordinating work, home, marriage and fatherhood (two children)—and I love them all. It's quite a juggling act, but provides tremendous satisfaction. My wife is an executive with Motorola and has an equally demanding schedule. Consequently, we both put a premium on family activities.

I also try to stay involved with my profession outside of work. I'm a registered professional engineer and a member of a number of professional organizations including the American Council for Construction Education, Association of Mechanical Engineers, Society of Professional Engineers and the American Society of Heating, Refrigeration and Air Conditioning Engineers.

## PROFILE OF A MECHANICAL ENGINEER: PATRICK J. SHOOK, COLUMBUS, INDIANA

### Occupation

Senior Engineer, New Product Development, Cummins
  Engine Company

### Education

BSME, 1992; MSME, 1994

## Studying Engineering

Mr. Myers, my high school chemistry teacher, had a discussion with me one day about Purdue's co-operative education program. He could see my interest in math and science and pointed me toward a field which I knew very little about—engineering. I investigated, and with high expectations, made the decision to attend Purdue to study and to prepare for what seemed to be a very interesting career.

By the end of the first semester during my freshman year, I had decided to pursue a mechanical engineering degree. This was after many discussions with junior and senior engineering students as well as with my father and a few professors. I had grown up in a family which owned a general contracting business (house construction, remodeling, etc.) and the broad variety of topics of study within mechanical engineering seemed to fit my desires. I also signed up to become a co-op student in order to obtain valuable work experience as well as to help pay for my education.

After graduating with my BSME, I entered into a research assistantship at Purdue for an intense (but extremely rewarding) two years on the way to obtaining my MSME. The most exciting task given to me by Prof. Fleeter was to build and operate a helicopter engine compressor test stand.

After graduating with my MSME, I hired on at Cummins Engine Company and worked for four years as a mechanical development engineer. This time was filled with designing abuse tests for semi-truck engines and determining how to improve the components that wore out during those tests. Without a doubt, learning how to work with people is easily 50-percent of my job. Since I have been at Cummins, I have learned that being clear with people concerning the goals of a plan is extremely important. As in a football huddle, everyone on the "field" needs to know what the "play" is and how to execute it.

Since July 1998, I have been working in a new position which focuses on engine cycle simulation. This has been primarily computer work and has re-sharpened my skills in fluid mechanics, thermodynamics, and heat transfer. The variety in my job has been enjoyable: from defining customer requirements for specific components to maximizing work processes within the structure of a large company.

## *Life Outside of Work*

Outside of work, my wife and I do our best to serve the Lord and our church. In the past, we have both taken and taught classes on what the Bible says about marriage. We wanted to build on a good foundation and have enjoyed our marriage more and more with each year.

## PROFILE OF AN AERONAUTICAL ENGINEER: DAVID O. SWAIN, ST. LOUIS, MISSOURI

### Occupation

Executive vice president, The Boeing Company

### Education

BS, Aeronautical Engineering, 1964

### Studying Engineering

I had demonstrated strong mathematics and physics skills in high school and was drawn to engineering because I understood it was a well-paid profession. I was very interested in the space program following the Soviet Union's launch of Sputnik in 1957 and was drawn to aerospace. The field has met all my expectations.

### Career Life

If I could do everything over again, I would do everything the same. My advice for beginning students is to focus on learning and not worrying so much about their specific career path.

I joined McDonnell Douglas in 1964 as an engineer on the manned Gemini project. In 1972 I moved to Tactical Missile programs and worked on the Tomahawk and Harpoon/Standoff Land Attack Missile. In 1987 I became vice president and general manager for strategic business development for McDonnell Douglas Astronautics Company. I was named senior vice president and C-17 program manager in 1991. This was one of my most significant accomplishments in that I led the C-17 program through difficult and troubled times. It is now on the right track.

I currently am executive vice president for Phantom Works, the research and development organization of The Boeing Company. I have overall responsibility for Phantom Works and its operations in Southern California, Seattle, Mesa, St. Louis, and Philadelphia. Some of the exciting projects we are involved with include: Joint Strike Fighter, Future X Space Vehicle, Canard Rotor Wing, Space-Based Laser, Space-Based Radar, Common Support Aircraft, Uninhabited Combat Aircraft, Blended-Wing Body, Solar Orbit Transfer Vehicle, and Deepwater Ship. In a typical day, I meet with customers, suppliers, peers, direct reports, and employees. I lead, coach and teach. My normal work hours are 11–12 hours a day. I travel frequently to California, Washington state, and Washington, D.C.

### Life Outside of Work

I use my engineering training in my current job by applying logic and asking penetrating questions. I also enjoy learning from other people. What I like best about my job is the fact that I am working on projects that have significant value.

I enjoy doing things with my family. I have four children and seven grandchildren. I also like racquetball and golf. I am a trustee on the St. Louis Science Center Board and serve as an advisor for Purdue University. I also find a great deal of reward from providing good advice to interested individuals.

## PROFILE OF A MECHANICAL ENGINEER: NANA TZENG, SEATTLE, WASHINGTON

### Occupation

Design Engineer, The Boeing Company

### Education

BSME, 1997; MSME, 1998

## Studying Engineering

What I enjoyed most about being an engineering student was making stuff—in other words designing and fabricating mechanical parts. While I was at Purdue, I participated in the Solar Racing Club. As one of the few mechanical engineers on a team dominated by electrical engineers, I helped improve the braking system, performed computer aided stress analysis on the chassis, learned to weld, machined rotors and other parts, and got to be driver of the solar car. The experience was not only rewarding, it also helped me relate what I read about in textbooks with applying that knowledge. College offers many extracurricular opportunities and I would encourage any engineering student to become involved in hands-on activities and research projects.

### Career Life

My career really is rocket science! I currently work in the Instrumentation Development and Design group at Rocketdyne, the division of The Boeing Company that designs and develops rocket engines. The team I work with is responsible for all the sensors and electrical components on the Space Shuttle Main Engine. My latest project is the redesign of the spark ignition system. This involves the design of components and tooling, creating and updating of drawings, and working with the manufacturing team to improve the fabrication process. I also help the members of my team analyze sensor data from hot fire tests and space shuttle flights. Because the nature of my work is highly technical, I regularly use the knowledge and skill I gained as an engineering student. Now that I am familiar with the complexity of rocket engines and the detailed work that goes into building one, it's even more amazing when I see everything come together during launch.

### Life Outside of Work

Ever since I finished school, I have been able to develop other interests and hobbies, some of which are golf, photography and snowboarding. I also often enjoy hiking, camping, rollerblading, shopping, concerts, clubs, etc. The best advice I have to offer to students of any discipline is to keep an open mind and take advantage of your opportunities.

## PROFILE OF A MECHANICAL ENGINEER: PATRICE VANDERBECK, CEDAR FALLS, IOWA

### Occupation

Electronics systems engineer for the 6000 and 7000 series John Deere
 tractors

### Education

BSME, 1982

### Studying Engineering

I like to understand how things work. Math was my favorite subject in high
school, followed closely by the sciences. Engineering seemed like the appropriate career for me, based on my interests. My college advisor recommended I study
mechanical engineering after assessing my capabilities and interests. It sure was the right
direction. The engineering curriculum can be difficult, but an engineering degree gives you
many options. You can go into design, management, marketing, sales, law, manufacturing,
or research, to name just a few areas. If you decide you want a change, it makes it easier
for you to move on in a new direction. The engineering degree will open doors for you.

Given the opportunity, I would have changed two things about the path I took. First, I
would have developed better study skills in high school, or earlier in college. I had to do
some backtracking because of this. Also, I would have worked at a company that manufactures a product (like tractors) directly out of college instead of starting at a consulting
firm. I learned it was important for me to work where a product is manufactured.

### Career Life

I have worked at three companies since graduating in 1982. I am currently the electronics
systems engineer for the 6000 and 7000 series John Deere tractors. I make sure that the
different engineering teams within the electronics and vehicle groups are communicating
with each other. My job combines design, program management, negotiating, and communicating. I deal with current tractor-related issues and the designs for new tractor programs at the systems level. I love my job because it is never boring. I am always learning.
I work with many talented, dedicated people. When I was in a previous position at John
Deere, I had design responsibility for a device for left-hand control of the forward and reverse movement of a tractor. It had 38 subassemblies. Between the supplier and me, we
designed these subassemblies into a very small package. It took a lot of development effort to get the assembly to work perfectly within the entire system of the tractor. In the end,
the device was well received by our customers for its function and reliability.

### Life Outside of Work

I enjoy biking, skiing, weaving, entertaining, and reading. I belong to an investment club and
a reading club. I am married to another engineer. We have a vacation home on the Missis-
sippi River where we do a lot of boating and entertaining. We love to travel. My husband
and I do volunteer work with Habitat for Humanity and with our church. Our engineering
jobs allow us to live a comfortable life and to enjoy fun things like travel and boating.

## PROFILE OF A CHEMICAL ENGINEER: JACK WELCH, FAIRFIELD, CONNECTICUT

### Occupation

CEO of General Electric (Retired, 2002)

### Education

BSChE; MSChE; PhD

### *Career Life*

The man called "CEO of the century" by the editor-in-chief of Time magazine is an engineer. Jack Welch, who led General Electric's transformation over the past two decades into a global technology and services giant, started with the company as an engineer in Pittsfield, Mass. He had earned his BS ChE from the U of Mass in 1957, and followed that with an MS and PhD from the U of Illinois in 1958 and 1960.

In high school he had captained the hockey and golf teams and earned the distinction of being voted "Most Talkative and Noisiest Boy" by his classmates. "No one in my family had ever gone to college, but I had that ambition," Welch recalls. "Of course, believe me, my mother had that ambition for both of us."

"Life is a series of experiences, a series of steps if you will," he continues. "Every time you're reaffirmed, every time someone tells you you're Okay, you can go on to the next step, the next challenge. Well, my teachers in the Engineering Department told me I was Okay. In fact, they told me I was really good. A couple of them practically adopted me, and told me I had what it took to go on to graduate school. I had never even thought of graduate school. But they really believed in me."

After earning his PhD, Welch returned to Massachusetts and GE's Chemicals Division for his first job as a development specialist. It was on that first job that he demonstrated many of the leadership traits that characterize him to this day.

"I was an entrepreneur in a small business outside the mainstream of GE—the plastics business. My technician and I were partners working on the same thing. We had two people, then four people, then eight people, then 12. Today, GE Plastics is a $6 billion business. But it started that way. Everyone's involved. Everyone knows. Everyone's got a piece of the action. The organization's flat. All these things are from when I was 26 years old."

Welch's rapid rise in GE continued, and in 1981 he became the eighth chairman and CEO of the company that was founded in 1892. Although he recently retired, the organization he led was named "Most Admired" by Fortune magazine and "Most Respected" by the Financial Times.

Yet he described his job running a company with 1998 revenues of approximately $100 billion as "not rocket science." Instead he saw his key role as allocating both human and financial resources in a way that will continue GE's growth. "My job is allocating capital, human and financial, and transferring the best practices. That's all. It's transferring ideas, putting the right people in the right jobs and giving them the resources to win," he says.

Welch, now the father of four and recently a grandfather for the fourth time, continues on the golf course his winning ways that began in high school. He's twice won his club championship and has even bested well-known pros in friendly play.

## PROFILE OF AN ELECTRICAL ENGINEER: SHAWN D. WILLIAMS, TWINSBURG, OHIO

### Occupation

Product General Manager at GE Appliances

### Education

BSEE, 1985

### Studying Engineering

The opportunity to further my math and science abilities attracted me to engineering. I really enjoyed calculus, chemistry, and physics in high school, so I attended Marion Blalock's Target Cities Luncheon in Chicago. At the luncheon, Purdue engineering students spoke about engineering, and their comments opened my eyes to engineering as a possible career.

As a freshman in engineering at Purdue, I queried several juniors and seniors in various engineering disciplines on their course work. I concluded that electrical engineering would allow me to leverage my foundation in mathematics and physics. In addition, computers were becoming more popular, and I concluded that EE would provide insights into computers. My advice to students is that you remain persistent and disciplined in pursuing an engineering degree.

### *Career Life*

Currently, I am a regional sales manager for General Electric Industrial Systems. My primary job responsibilities include delivering top-line sales growth on an annual sales volume of $130 million. I lead a team of 70 employees (mostly engineers) throughout six states in the Midwest as they execute business strategies in their local trading areas.

In the course of a day, I may do such things as interview for an open sales engineer position, review a trading area strategy with a general manager, expedite a delivery with a factory for a distributor principal, and provide coaching to a new sales engineer on a project. The best aspect of my job is the variety of strategic and tactical tasks in which I engage. My engineering degree provided the solid foundation for me to understand the technical nature of the products, but the ability to handle multiple tasks is key in executive management.

### *Life Outside of Work*

Obtaining an engineering degree has assisted me in providing a lifestyle for my family that I never thought possible while growing up. It is imperative for me to balance work and family life. My wife and my son and I spend weekends visiting the zoo, seeing friends, playing board games, and attending church.

On a personal note, I am the primary provider for my family, so keeping in shape physically, mentally, spiritually, and intellectually are keys to a successful life. I accomplish this by running 3 to 4 miles three times a week; consistently working on lowering my golf handicap of 13 by playing golf with customers and friends weekly; and incessantly learning about life through Bible reading, management books, and by mentoring and coaching employees, students and family members. I can honestly say that without my engineering experiences, the life I have now would still be just a dream!

## PROFILE OF A MECHANICAL ENGINEER: DR. ADEL A. ZAKARIA, WATERLOO, IOWA

### Occupation

Senior VP, Engineering and Manufacturing, Worldwide Agricultural
    Equipment Division

### Education

BSME, Egypt, 1967; MSIE, 1971; PhD, Industrial Engineering, 1973

## Studying Engineering

Growing up in a developing country, I viewed engineering as an instrument for making genuine progress. It has offered me an opportunity to be part of moving things forward in our generation. In particular, I was interested in how things are made. Manufacturing offered a unique way to work with three domains: things, people, and systems. In today's highly technical society, engineering will offer a student an ideal foundation for several careers. A coop or summer job experience can be very helpful in getting early exposure and making one's study a lot more interesting and meaningful.

### Career Life

Within a $7 billion Worldwide Agricultural Equipment Division, I guide the work of 20,000 employees who engineer, manufacture, and provide product support to our customers.

A typical day for me in the last year might include:

- visiting with key customers and their servicing dealer in Nebraska
- chairing a Worldwide Combine Product Council in Germany
- visiting a new joint venture site in India to review progress of the factory construction
- reviewing the results of a business improvement team

I enjoy the breadth of my job, working with a variety of people, the disciplines they represent, and the constant challenge of leading change. Not only have I been able to apply engineering training, but I have had to constantly augment it by learning and expanding my knowledge in new, evolving areas.

Significant accomplishments in my career include the early development of cellular manufacturing, the development of computer aided design and manufacturing tools (CAD/CAM), guiding the development and introduction of our company's first worldwide tractor product platform, and the breakthrough in forging a new win/win labor strategy with our unions.

### Life Outside of Work

My hobbies outside of work include racquetball and photography. Our family (including our two daughters) has traveled together in over 30 countries. We enjoy learning about the world's people and cultures. We also have camped in most of the U.S. and Canadian national parks. At various times I have been active in a number of national engineering societies. I also have participated in a number of community volunteer groups, including a hospital board, United Way, and the Boys & Girls Club.

## EMERY SANFORD

### *Where did you attend college or are you currently going to college?*

University of California-Berkeley and presently the University of Sydney in Australia for study abroad. I am a mechanical engineering major.

### *What kind of high school classes helped you get to where you are today?*

High school math, especially calculus, prepared me for the university-level math and engineering courses. I never would have pursued engineering, however, had it not been for the principles of engineering courses and the hands-on case studies.

### *Describe your experiences in college.*

Classes freshman year were hard. It was like a slap in the face compared to my high school courses. Once I got through the first year, however, I figured out my learning style and grew more comfortable with the university-style classes. College has been a great experience in and out of the classroom. I think it is imperative that engineering students find something extracurricular to do with their spare time. Focusing only on school is sure to drive you crazy.

Another bit of advice for engineering students: find people to study with. Facing tough thermodynamics or mechanics problems alone can be intimidating. Having just one other person to talk to about schoolwork can make a world of a difference. Those same people will become your best friends around exam times, when it can once again become difficult to stay self-motivated.

### *What do you do for fun outside of school?*

I race competitively for the UC-Berkeley sailing team. I also ski, snowboard, and surf. I enjoy riding and working on motorcycles.

## COLIN QUINN

### *Where did you attend college or are you currently going to college?*

I am currently a junior studying mechanical engineering at University of Wisconsin-Madison.

### *What kind of high school classes helped you get to where you are today?*

The engineering classes in high school put me at least one to two years ahead of most of my classmates my first year in college. The hands-on aspect of the curriculum is phenomenal. Knowing the theory behind something is important to learn, but applying that theory in a real-world situation brings the learning to the next level. Personally, the experience I got with composite materials was invaluable down the road. I have used the knowledge I acquired in high school and applied it to building carbon fiber components for Formula 1 style race cars.

### *Describe your experiences in college.*

College has shown me the importance of hands-on experience. I have many incredibly smart people around me who ace their classes but have trouble getting jobs after graduating. Getting involved in projects is just as important, if not more important, than your grades. I am a member of the UW Formula SAE team. We build a Formula 1 style racecar around a 600cc motorcycle engine. We design and fabricate all the parts, do all the engine programming, all the CAD work, and all the suspension simulations. We compete against over 130 international teams in Detroit every spring. I have learned more in this project than in all my classes combined, plus I get to drive the fastest car I will ever drive.

### *If applicable, describe your work experiences in engineering.*

I am currently on co-op at Wolf Appliance, a subsidiary of Sub-Zero, in Madison, Wisconsin. I work with the design engineers on design and research and development projects. I have programmed and developed a steamer and fryer unit, modified drawings and models on Solidworks, done prototyping work, and done tear-downs on competitor's units. It has been a great learning experience and introduction into the professional world. I strongly recommend co-oping to anyone who wants a good job directly out of college.

### *What do you do for fun outside of work?*

I play soccer, go snowboarding and wakeboarding, play cards with friends, and race the formula car on the weekend.

# MIKE KAHAN

*Where did you attend college or are you currently going to college?*
B.S. in Computer Science—MIT, June 2003, (expecting) Masters of Engineering (M.Eng) in Computer Science—MIT, June 2004.

*What kind of high school classes helped you get to where you are today?*
Calculus (I and II) was helpful for freshman classes such as physics, chemistry, calculus, etc. Also, provided a good background for the calculus needed for introductory Electrical Engineering.

Chemistry and Physics were somewhat helpful for their college equivalents (chemistry more so than physics) and were useful for understanding things like silicon gate structure (n and p doping).

Engineering 1 and 2 (Independent Study) was my first real introduction via classes to the world of engineering. They opened my eyes to what several different types of engineering could be like and appealed to my taste for hands-on tinkering.

Science Olympiad (not a class, but an after school activity) was my first introduction to engineering via the events where you build stuff (e.g., bridges, 2L bottle rockets, small cars). The Olympiad helped me to learn the science material taught in class.

High-Mileage Vehicle: My first big TEAM project was useful for learning how to work with others, especially under deadlines. It was my first exposure to things like mechanical engineering, welding, serious planning, etc.

*Describe your experiences in college.*
I'll focus only on the school aspects here. (I could write a book about the non-school stuff, such as my Sigma Chi fraternity experiences.)

As a freshman, I was nearly blown out of the water by the caliber of people surrounding me. At the time, all freshman classes were Pass/No Record (even better than Pass/Fail!), and it was a good thing too. I felt like the dumbest kid at MIT. So many people already knew the material we were learning (calculus, physics w/calculus, chemistry, etc.) It took me nearly a year to ramp up to the college workload. MIT is notorious for working its undergrads to death (sadly, MIT has averaged something like one suicide per year over the past 12 years for a school of 4,000 undergrads and 4,000 grads). Freshman year went by quickly, probably because I didn't really like the classes I was taking (calculus, physics, chemistry, biology, etc.) and because I had my head in a book the entire time catching up to my classmates.

Sophomore year was a trial-by-fire immersion into my major (Computer Science). It was my first term on grades at MIT, a scary thing when you have been struggling to post a B average the past year. I slept four hours a day for almost the entire year, and just about quit my major because I was so miserable. I took most of the heavily math-oriented core classes during sophomore year, including those required for Electrical Engineering. (At MIT, Computer Science and Electrical Engineering are in the same department and share the same set of four core classes). Nothing struck me as all that interesting, but that was probably because I could only take one header class in theoretical Computer Science. Header classes serve as an introduction to specializations within Computer Science.

Junior year was the polar opposite of sophomore year. I slept more. I started taking classes in areas I enjoyed, such as Artificial Intelligence (A.I.) and Systems Engineering. I took fewer classes with 300-plus students (commonplace for almost all freshman and sophomore classes). In addition, I became involved in a research project related to A.I. and began looking into MIT's five-year Masters program for MIT undergrads (the M.Eng program). Life was looking up academically.

Senior year, I hit my prime. I earned my first 4.0 for a term in college during the fall. I applied and was accepted to the M.Eng program. I began taking some of my graduate classes in A.I. and continued working with my research project. The graduate classes I took were challenging but enjoyable. I was able to apply the software engineering and A.I. techniques I had learned over the past year and a half to my research project.

All in all, I learned much about engineering and life at MIT. The first two years of college were a trial-by-fire, the second two were my chance to specialize and use what I had learned. Sleep is important. Do not get fewer than six hours a night for extended periods of time. If possible, find out what you want to do early on and take classes/work in that area because it will make your life more enjoyable. Just like high school, you need to find extracurricular activities that interest you and run with them, but do not spread your time too thin.

### If applicable, describe your work experiences in engineering.
The summer after freshman year, I worked for Cisco Systems in San Jose, California. I spent the summer learning how to program in C and in Perl while writing an application that would set up and test Voice-Over-IP hardware. My group was organized into small teams of three to five people. I spent a significant amount of time communicating with members of other teams to gather the information I needed to do my job. Hardware testing introduced me to the real-world importance of fault tolerance and software modularity.

During the month of January of sophomore year, I worked for a startup company in Waltham, Massachusetts. This company was making a product that mounted inside your car and would display a browser-like application with information relevant to your current location. I worked on the hardware and software for displaying your location, heading, and speed on a map. The team I was a part of faced many technical challenges that often required solutions be created using the engineering techniques I had seen while working for Cisco.

The summer after sophomore year, I worked for Draper Laboratories in Cambridge, Massachusetts, where I developed programs for route planning in autonomous helicopters. My administrative group was less team-oriented than Cisco and more individual-oriented. The project I was working on was still in the experimental stage, so there was less focus on quality assurance (QA) and more focus on getting the software to work in the first place.

The summers after junior and senior years, I worked with my research group in A.I. We explored the use of A.I. in aiding collaboration between teams of people. In addition to the A.I. algorithms we employed, I used many of the software engineering techniques I had learned from my college classes, including principles of good software design, modularity, interfaces, design patterns and code reviews.

### What do you do for fun outside of work?
Outside of work, I was (and still am) an active member of my fraternity, Sigma Chi. For fun, I like to play video games, go camping, watch sports on TV, work on websites (mostly the fraternity's: http://sigmachi.mit.edu), ride my bike around Boston, hang out with my friends and patronize Crossroads (the local bar).

## GIDEON SORKIN

*Where did you attend college or are you currently going to college?*
Brown University, Providence, RI Class of 2006

*What kind of high school classes helped you get to where you are today?*
I took a varied selection of courses in high school, everything from Honors English to Calculus to Mr. Gomez's amazing Engineering program.

*Describe your experiences in college.*
I really enjoy my engineering classes. I have to take a lot of math, which I tend to like less than the physics and engineering but because the math is applied so quickly to my engineering work it is made more fun. My experience in engineering has been a positive one where I leave lecture feeling as if I understand the world better than before I entered lecture.

*What do you do for fun outside of work?*
I am a member of the Brown University Crew team. Between doing a varsity intercollegiate sport and my engineering course load, that is about all I do.

## EXERCISES AND ACTIVITIES

1. Select five of the engineers profiled in this chapter. Prepare a one-page summary of their current responsibilities.

2. Select five other engineers profiled in this chapter. What were some of the common factors that got them interested in engineering or helped them succeed in school?

3. Of the engineers profiled in this chapter, describe three who have unique, non-engineering jobs.

4. Make a list summarizing the current jobs and responsibilities of five of the engineers in this chapter.

5. What types of career paths are available to engineers? Make a list of seven different career paths taken by engineers in this chapter.

6. Select a historical engineering figure. Prepare a one-page profile on this person using the format used in this chapter.

7. Interview a practicing engineer and prepare a profile on him or her, similar to the format used in this chapter.

8. Select five of the engineers profiled in this chapter and list some of the challenges they had to overcome during their education and/or career.

9. Prepare a one-page paper summarizing some of the challenges engineers face today.

10. Select one of the engineers profiled in this chapter and prepare an oral presentation on that engineer's company or organization.

11. Which career path presented in this chapter sounds most appealing to you?

12. Which of the engineers' stories can you relate to the most? Why?

13. What are your goals for your life outside of work? In this chapter, did you read about anyone with similar goals and/or interests? Explain.

14. Imagine yourself five years from now. Write your profile using the same format as presented in this chapter.

15. Imagine yourself ten years from now. Write your profile from that perspective.

16. Imagine yourself 25 years from now. Write your profile from that perspective.

17. What are your thoughts about the success and fulfillment found by engineers who obtained an advanced degree outside the field of engineering?

18. Based on the profiles and your own experience, what are the advantages and disadvantages of getting a graduate degree in engineering?

19. Prepare a matrix of the bachelor's degrees the engineers in this chapter received and their current positions. How many work in jobs traditionally associated with bachelor's degrees?

20. Select your preferred major and prepare a list of potential career paths it could lead to. Provide a brief explanation for each option.

21. Today's engineering workforce is truly diverse, with both men and women from all ethnic backgrounds working together. This was not always the case. Select a historical figure who was a woman or minority engineer and prepare a one-page profile which discusses the difficulties she or he had to overcome.

22. Research a company you would be interested in working for and list the engineers in their upper-level management structure. Identify the highest level of management currently held by an engineer in that company.

23. Prepare a one-page paper on how some of the engineers profiled have addressed family responsibilities amidst an engineering career.

24. A stereotype of engineers is that they are boring loners who only care about numbers and technology. Based on the personal-interest sections of the profiles, prepare a report refuting this stereotype.

# Chapter 7

# Technical Communication

## WHY DO WE COMMUNICATE?

What is the purpose of communication? Well, there are many purposes, actually. For starters, it might surprise you to consider that we are often the most important readers of our own writing. What we write should be valuable to us as a future resource. If you keep that in mind, written communication will be easier and more rewarding.

Of course there are a number of additional reasons for communicating with excellence: obtaining good grades, creating effective reports, convincing others of your perspective—both on a professional and personal basis. Each will require that you assume a different mindset.

Effective communication:
- transfers vital information
- provides a basis for judging your knowledge
- conveys your interest and competence
- increases the knowledge of others
- identifies to you the gaps in your information
- allows you to carry out desired changes

Engineers must possess the technical skills to complete engineering analysis, evaluation, and design. Engineers must possess strong communication and teaming skills. One of the biggest complaints of employers of entry-level engineers is lack of communication and teaming skills. Possessing these strong skill sets will enable you to excel in engineering. Most engineers in industry, research, and government complete design work in collaborative teams and routinely communicate their discoveries and progress to peers, managers, other engineers, and the community.

In this chapter, you will learn about oral and written communication skills. First, two golden rules of communication for engineers:
- **Brevity is best.**
- **Using a shorter, simpler word is typically preferable to a longer, more difficult word.**

As you consider your own communication, there are a number of other things to keep in mind:

## *Is the language you are using suited to your audience?*

- Do you know how to communicate effectively to your audience?
- Do you need to expand your communication skills to communicate more effectively to your audience?

## *Is what you are saying clear to your readers?*

- Will your readers draw the same conclusions you draw?
- Do you use words, phrases, or ideas that could lead to multiple interpretations?

## *Do you know your audience?*

- Does your audience have any unique preferences?
- Do they have any hidden expectations?
- How do they feel about the things you are communicating?

## *Do you know how your audience perceives things?*

- Do all engineers see things the same way?
- If you are addressing an audience of non-engineers, how might they interpret things differently?

Typically, you will be communicating with people who are making decisions concerning your work. These decision-makers have minimal time, and you want to have a maximal impact. While other types of communication place great emphasis on complex language and nuance, technical communication is all about getting it said and done quickly and simply.

## ORAL COMMUNICATION SKILLS

In an engineering context, you will make presentations to many audiences. The purpose of this section is to share information on planning and executing an effective oral presentation.

There are two types of presentations: formal and informal. Formal presentations involve a preset time and date and usually involve a reasonable amount of preparation. They may be made to colleagues, to expert panels, at design competitions, at the conclusion of a capstone design experience, at conferences in which engineering information is shared, or in the workplace when you communicate your information to your design team, management, the public, or funding and decision makers. For formal presentations, you will likely use a computer program such as PowerPoint; you might hand out printed copies of your slides. Formal presentations usually require formal business attire.

Informal presentations tend to be shorter, and do not require as much preparation time or formal business attire. You do not usually need to prepare a Power-Point presentation either. Sometimes, you will have to make impromptu, or spontaneous presentations. You might make such presentations in a business context, technical meeting, or at a meeting involving a design team.

A good presentation renders a service to your audience. A presentation is an opportunity to take the audience from where they are to where you want them to be. The best presentations will do the following:

- give valuable information the audience probably would not have had otherwise.
- be in a form that the audience can put into immediate use.
- motivate and inspire the audience to want to put the your presented information into immediate use.

With proper preparation in advance, you can concentrate on how you deliver information and on audience impact during your presentation. More specific information regarding these sections is included below.

For formal presentations, you will likely use a an overhead projector or a computer presentation program such as PowerPoint. Each PowerPoint image or each overhead is referred to as a slide (or an overhead) in this discussion.

## If you are nervous

Being nervous about doing a presentation is normal. In fact, the two most common fears of people living in the United States are public speaking and flying in airplanes. The following list of techniques is intended to help you combat your nervousness.

- Breathing is important; when you get nervous, your respiration rate tends to speed up and your body can go into adrenaline overdrive (a fight or flight mechanism). Controlling your breathing is one way to stay calm.
  - Take three deep breaths before you start.
  - Practice square breathing, which is inhaling seven counts, holding seven counts, exhaling seven counts, and holding seven counts.
- Visualize yourself successfully presenting your talk.
- Work with your nervousness by tailoring the talk to your strengths. Most people are most nervous during the initial stages of speaking. Use one or more of these tips, which can be used throughout the presentation.
  - **Use humor:** This makes the atmosphere less formal. Start your presentation with a joke, especially one that pertains to your presentation.
  - **Use props:** It gives you something to do with your hands. A show and tell will interest the audience.

- **Use your nervous habit:** Work with that habit to your audience's bene-fit. If you like jingling change (coins) in your pockets, pull something vi-sual out of your pocket to start your presentation.
  - **Use trivia:** Focus your audience's attention or teach the audience some-thing interesting that pertains to your talk.
- **Have your presentation ready at least a day in advance.** Being finished usually helps to cut down on your nervousness. Student Alicia Abadie says, "The more you procrastinate, the more nerve-wracking it can be."

## Before the presentation

Preparation is the key to successful oral presentations. To prepare for a presen-tation, you need to do the following:

- Identify your purpose and audience
- Organize the information
- Gather the information you need to present
- Create presentation material with maximal audience impact
- Practice your presentation after it is prepared

## Identify your purpose and audience

To give a presentation with maximum impact, know your audience:

- How many people will you be speaking to?
- How well do you know your audience? If you are speaking to a small audi-ence with whom you are familiar, you can use their names during your talk. This focuses the attention of your audience.
- What is the background of your audience (age, professional background, personal background, etc.)? You must tailor your talk, using appropriate vo-cabulary and information the audience will understand and put into action.
- What is the audience expecting from you in terms of information, tone, level of formality, etc.?
- What does the audience know about your topic? You should know what they know and concentrate on the information that they do not know. The idea is to avoid being repetitive, and out of the knowledge base of your audience on the other. If you are not sure what the audience knows, ask your contact person before your presentation, so you can prepare accordingly.
- How long is your allotted time to speak? Speak for less than your limit to give the audience a chance to ask questions.

You need to keep the purpose of your talk in mind when you are preparing. In what capacity are you speaking? For example, are you giving a report to your peers? Are you presenting your final design and recommendations from your project? Focus the information in your presentation around your purpose, and your audience. This will ensure that your audience is in the best position to take your information and act accordingly.

## Organization

You should organize your presentation so the audience can digest the delivered information easily. The following tips will help you to organize your information.

- The key to a well-organized speech is realizing that even the most engaged audience will not pay attention all the time. Thus, you should limit your speech to three to five major points . Your talk will be successful if the audience understands and retains the information. How much your audience understands is more important than how much information you present.

Did you know that people speak 100–125 words per minute, but the human brain can handle up to 500 words per minute? This is how you can pay attention to what is going on when someone is speaking while your mind is on something else.

- **Have a one-page summary.** For a presentation longer than five minutes, an oral one-page summary at the beginning of the talk is useful. This lets the audience know what is coming and in what order, and will help your audience to follow your presentation.
- **Use the power of repetition.**
  - Repeat your main ideas; the audience will remember only three to five of the ideas you present
  - Repeating key points makes it easier for the audience to stay with you
  - Repeat your main message at least three times, in three different ways
  - An adage from an accomplished speaker: "Tell them what you're going to tell them, tell them, then tell them what you told them."
- **Provide a written supplement if appropriate.** A written supplement helps people to focus on what you are saying instead of writing it down. This is a security thing; if people they already have the information in front of them, they are more likely to listen and focus on your presentation. The more ways in which you present your message, the more likely the audience is going to remember it.

In terms of organizing the information itself, a sample outline of a presentation could be as follows:

- **Introduction:** Who you are and why you did this?
- **Need:** Why your engineering design needed?

- **Solution:** You have the design that addresses the need, and here it is. In- clude major design features and why you did what you did (explain how your design specifically addressed the needs).
- **Bottom line:** What are the benefits? How much it will cost?
- **Conclusion**

## Gather the information you need to present

Although this point seems self-explanatory, remember these few tips when gath- ering information for an oral presentation:

- The information you include in your presentation should pertain to three to five main points you plan to share with the audience. This information will support your main points and will help with the organization of your presen- tation.
- Make use of graphics, pictures, and photographs because they communi- cate a wealth of information in a concise way
- Make sure that you do not include too much or too little information in your presentation. You should have approximately one slide for every minute that you speak. Some speakers can include more slides, but you should not ex- ceed one slide per 30 seconds. Thus, if your presentation is ten minutes long, you should not have more than 20 slides.

## Create presentation with maximal audience impact

Your impact can be visual (what people see), auditory (what people hear), or kine- matic (what people do or hands-on experience). Remember, keep it simple. Your audience will stay with you if you use accessible language, and the audience will retain information better if you provide information in more than one of the ways mentioned above.

The following tips will assist you in maximizing the visual impact of your talk.
- Present one central idea per slide.
- Be as brief as possible; make two clear, understandable slides than one cluttered slide with too much information.
- Font size (the size of your type) is important. Use at least 18-point font size for overheads. You can use smaller font size for slides but not much smaller (12-point, the size of the printed words in this book, is too small for slides).
- Slides do not need to be complete and self-explanatory; the speaker can add details (handouts are helpful here).
- Orally identify or locate colors to help listeners who are color blind (10% of all males are color blind).
- Use high contrast colors (white and blue, yellow and navy, etc.) with slides or overheads; this helps people with color blindness.
- The more you must use a pointer, the worse the slide; keep it simple.

- If you have many numbers in a table, use multiple slides with the empha-sized number circled or highlighted. If you are discussing only five numbers, why show more than five on a slide?
- Do not use all capital letters. They are difficult to read. Use capital letters sparingly for emphasis of an important point.
- Use graphs or charts instead of tables. Most people can comprehend graphs and charts more easily than in tables.

The following tips will assist you in maximizing the auditory impact of your presentation.

- **Be an enthusiastic speaker**
  - Be enthusiastic about your message and your audience; everyone en-joys listening to speakers who care about their subject.
  - You do not have to be a cheerleader. Show with your voice, eyes, and body language what you are saying is important and meaningful.
  - Another adage: "There are no boring topics, only boring presentations."
- **Master your subject**
  - You need to be able to concentrate on reaching the audience, which means you should make eye contact with the audience, and concentrate on "reading" your audience (are they engaged, bored, etc., and what can you do to get and keep their attention?). This takes energy. You should know your material well enough so that you appear confident.
- **Never read or memorize your presentation.** Reading reduces eye con-tact and memorization makes your talk appear canned and boring
- **Use pauses to your advantage.** When you make an important point or when you have presented something complicated, give the audience time to think about and digest your point. You can build pauses into your presen-tation to slow down the pace if you tend to speak more quickly when doing public talks.

Using Kinematic Methods will engage your audience. The methods can be direct, where your audience does something, or indirect, where they watch as you do something hands-on. Make use of the following tips for kinematic methods:

- If you are demonstrating something, make sure that your audience can see the entire demonstration from their viewpoints. Make sure that you point out the major concepts or ideas.
- If you have the audience do something, especially if it involves multiple steps, make sure that the entire audience is with you as you step through the process. You can ask, "has everyone completed this step?" before pro-ceeding to the next one.

## Practicing your presentation

Practice your presentation out loud because it will give you insight into your tech-nique. First, most presentations tend to be scheduled, and you do not want your presentation to go longer than its allotted time. Long-winded speakers will often alienate their audiences or will lose their attention. Most people tend to finish their presentations ahead of time because they speak more quickly when nervous. If you practice your presentation, you will know how long it takes to complete. If you are nervous, practice will enable you to build in several places in your presentation in which you can check yourself and slow down if necessary. Second, the more you practice, the more familiar you become with the material that you are presenting. This familiarity can be important if your information is technical. Showing the audience you are familiar and comfortable with the presented information will help them (and you) gain confidence in your subject. Most likely, you will explain the concepts better as you understand them more fully.

Now that we have talked about everything you can do before a presentation to maximize its effectiveness, think about the presentation itself.

## Getting started

The first 15 to 30 seconds of a presentation are critical; establish eye contact and rapport if you are going to turn down the lights for slides or PowerPoint. You can establish a connection with the audience in several ways:

- You can start by saying, "Good morning (or afternoon or evening), my name is <name>, and I am going to discuss <subject.>"
- You can establish the level of formality. For example, you can tell the audience that it is okay to ask questions at any time.
- You can establish your tone with humor (telling a joke), or asking a trivia question with a small reward (e.g., a piece of candy) for the winner.
- Interacting with the audience is an excellent way to establish good rapport. After you introduce yourself and the subject of your talk, you may involve the audience by asking a general question. Those interested can raise their hands or interact with an answer. This information could be useful to you (you can learn something about the audience for a better presentation) and can provide you with a good lead into the rest of your talk.

During your presentation, monitor your audience. Do they look engaged, confused, or bored? If they look engaged, do not change anything that you are doing. If your audience is confused, you could ask them a question, or say, "I know that this information is difficult and some of you look confused. Can I clear up anything for you?" If your audience appears bored, you might re-engage your audience with another trivia question or by asking the audience become active.

The conclusion of your talk is important. You need to let your audience know that you have finished, e.g., "This concludes my presentation. Do you have any questions?"

## Presentation hardware

- Know how the equipment (PowerPoint, overhead projector, etc.) works beforehand. For overhead projectors, if you can read the overhead when you place it on the projector, the audience can read it. You may have to focus but your overhead will not be backward or upside down.

- Ensure the proper video projection system is available and that the computer and system are compatible. You can bring your own just in case.

## WRITTEN COMMUNICATION SKILLS

In your engineering travels, you will volunteer or be called upon to communicate your ideas, designs, findings, and concepts in writing. Many technical writing books contain detailed information about this. In fact, most of you will take an entire course in technical writing, during which time you will get experience writing technical papers, reports, cover letters, executive memos, engineering design notebooks, proposals, etc. This section of the chapter is intended to provide conceptual information with respect to the writing you will encounter in engineering, including reports, proposals, memorandums (memos for short), résumés, and cover letters. These next sections include brief descriptions on how to best craft these writing pieces.

> For more detailed information, consult a resource like S. Stevenson and S. Whitmore, *Strategies for Engineering Communication.*

Before we begin discussing the formats of written communication, recall what was said at the beginning of this chapter about the two golden rules of communication for engineers:

- **Brevity is best. Simplify whenever possible.**
- **Using a shorter, simpler word is typically preferable to a longer, more difficult word.**

And here are a couple of common grammatical issues to consider:

- Each paragraph should address a **single topic** and have a single primary focus (stated in a topic sentence) accompanied by supporting details.
- Avoid **clichés,** which tend only to confuse the reader of technical writing.
- **Its and it's:** "Its" (without an apostrophe) is used to denote possession For example: The dog chased its tail. "It's" is a contraction that means "it is." For example, It's sunny outside. If you are not sure which to use, simply insert the words "it is" in place of "it's." If the sentence makes sense, keep "it's." If not, use "its."
- **Affect and effect:** "Affect" is a verb meaning to influence, e.g., Her leadership positively affected the situation. "Effect" is a noun meaning result, e.g., One effect of securing the contract was the hiring of more student workers. The tricky part is that effect can be used as a verb meaning to bring about, e.g., He effected considerable change when he became manager.

> *Your Secret Weapon:* Get yourself a good writing guide such as *The Chicago Manual of Style* or *Webster's Concise Handbook for Writers.* These aids will answer all your grammatical questions.

## The Audience

Before you begin to write, ask yourself, "Who is going to read this, anyway?" You must consider the needs and perspectives of your audience. Their level of education or experience may dictate the language and style of your writing. They may use words or expressions that are unique to their part of the country or their part of the world.

Always remember that technical material needs to be clear and precise. You cannot simply write as you please and hope that your readers will "figure it on their own." It's as if you are the screen writer and science requires a very clear script. You must prepare your readers, give the required information to help them understand, and repeat information as necessary to bring them to full understanding.

## Technical Writing

Because your engineering career most likely will involve a great deal of technical communication, it is vital that you understand the qualities of effective technical writing. In general, technical writing:

- identifies the main premise early on
- is clear, objective, and economical
- follows a specific format (i.e., memos, lab reports, abstracts)
- takes a problem-solving approach
- involves a specialized vocabulary
- often incorporates signs, symbols, formulas, graphs, and tables
- documents completed work

Let's look at some of the unique aspects of technical writing:

Content—Technical writing is typically based on research. Therefore, it must be factual and objective.

Methodical Construction—The way technical engineering material is written is specific and consistent throughout the profession. In general, paragraphs begin with a topical sentence, which is followed by supporting statements. Technical writers like to present the most essential information first and then present supporting and clarifying information. In this way there is no mystery as to the purpose of the communication.

Clarity—Technical communications must be clear and complete.

Objectivity—Technical writing should be unemotional and unbiased. It should not convey the feelings of the writer or others.

Importance in Workplace—Technical documents provide the information necessary for the proper functioning of an engineering-related business. Such documents supply information related to design, construction, research, and operations. Improper or incomplete technical communication can lead to design flaws, hazardous developments, waste of resources, production delays, loss of competitive advantage, and other undesirable outcomes. Such failure to effectively communicate can have an adverse effect on one's career.

# FORMATS OF WRITTEN COMMUNICATION

As an engineer, your written communication will take a variety of forms. These include brief notes to a superior or to a peer, hand-written reminders about a project, informal reports, memos, formal reports that may document months of work, and visual presentations that require significant preparation, yet little text. Much of your writing will need to conform to one standard format or another. In this section, a number of these formats are presented.

## Formal Reports

The principal purpose of the formal report is to present information gathered from an experiment or simulation in a document that will be useful to your employer and colleagues.

### Title

The title should be as brief as possible, clear, and appropriate. Seven or eight well chosen words is typical. The title doesn't need to do any more than introduce the report.

### Summary

Though the Summary (or Abstract) is presented early in a formal report, it should not be written until all other parts of the report have been completed. It should state, in simple, declarative sentences, *what* was attempted, *how* it was accomplished (including a discussion of any special techniques used), and what the *implications* are. That is, it should state the results and main conclusions.

This is both the shortest and most difficult section of the report to write. It is also the most important. In technical publications, the abstract often is the only portion of the report that many people will read. Therefore, it should be written to stand alone. It must communicate all the relevant ideas and results in one to two paragraphs (250–350 words). And be sure to discuss the details, which occurred in the past, with past-tense verbs.

### Table of Contents

In the Table of Contents, list each heading found throughout your report along with its page number. However, it is customary not to list the Abstract among the contents.

## Nomenclature

List and define all symbols used in the report. They should be listed alphabetically—Arabic symbols first, then Greek. Such a listing will enable your readers to find definitions of symbols used in the Analysis portion of your report with ease. Note that abbreviations should not be listed in the Nomenclature section.

## Introduction

The essential task of the Introduction is to orient the reader to the substance of the experiment or project and the context in which it was executed. The Introduction should state the *motivation* for the experiment and any relevant *background* information. It should introduce the material contained in the report by noting what is presented in each of the sections that follow.

## Analysis

The Analysis should begin with the basic, general (and well-known) relationships and proceed to the specific formulas to be used in the interpretation of the data. It is usually appropriate to use an explanatory sketch. All the symbols used should appear in the listing of Nomenclature, described above. Analytical results which have been previously derived and are readily available (equations from a handbook, for example) can be quoted and appropriately referenced. Their derivation need not be repeated unless the derivation is critical to the engineer's understanding of the experiment.

All relevant mathematical steps should be presented in the Analysis as well. Mathematical explanations and commentary are crucial to a good Analysis. Be careful not to force the reader to consult lab handouts or textbooks to understand the specific analysis involved in the experiment. If it is important enough to mention, it is important enough to explain.

## Experimental Procedure

A schematic representation (a diagram) of the experimental equipment or simulation program, including detailed views of unusual or important components, is a valuable aid in informing the reader about the experiment. The diagram can be used to document pertinent dimensions of the apparatus, and it can be used to specify the specific experimental equipment used for the study. If the procedure used in the experiments is not a familiar one, it will be necessary to include details of the techniques used. Your information should be clear enough that someone familiar with the general area of investigation should be able to reproduce your experiments with precision, based on the information given in this section. The report can be hindered by an incomplete or incorrect discussion of the experiment's procedure and equipment.

Though you may be tempted to slip into use of the present tense when describing the procedure, be careful to describe it in the past tense.

Some major points to remember are:

- When first referencing a figure, be sure to put the figure on the same page (or the next page). Your reader shouldn't have to hunt through your report to find the figure.
- Make figures at least one-third of a page in size. Figures that are too small are harder to read.
- Figures and tables should have clear and complete titles. Beneath the title, a short explanation of the figure's contents is normally necessary.

## Results

The answers derived from the analysis of an experiment are presented in the Results section of the report. This section should contain short, declarative statements of the results which summarize specific data presented in graphs or tables. The purpose is to tell the reader (without discussion) your interpretation of the results. Keep in mind that the results could be interpreted in more than one way. Thus, the importance of your stated results.

Realize that your readers may start reading your report in midstream, meaning that they may start with the Abstract and then go directly to the Results. Never assume that the report will be read from page 1 to the end, which is seldom the case.

You also should distinguish between "Figures," which include schematic drawings, photographs, graphs, etc., and "Tables," which are tabular presentations of data or computational results. Each type of data presentation should be numbered sequentially with its title: for example, "Figure 1. Schematic of the Combustion Chamber."

Though you are to present your data without discussion, do not make this presentation without explanation. A Results section containing only tables and graphs would be difficult to understand. Remember, you want to make comprehending your report as easy as possible for your readers.

## Discussion

The Discussion section starts with a brief summary statement of the results and then proceeds to a discussion of these results. The principal task in this section is to interpret the results, to note what went "as expected," what was unexpected, and what is of technical interest. The interpretation of the results in terms of the motivation for the experiment and its correlation to the current company projects should be the focus of the discussion.

The discussion could involve a comparison with other, similar investigations or comparison with expected results. The strong points of the work should be brought out here along with any limitations, because if the writer does not point out the limitations of his work, someone else surely will later. It may also be legitimate to comment on possible future investigations. Discuss the specific results of the experiment, using references to the accuracy of the measurements. It is useful to note the estimated uncertainties and their effect on the calculated values. For example: "The pressure was 1.0 kPa and the velocity was 10 m/s. Note that the "information content" of this statement is much greater than in the statement: "The manometer liquid was fluctuating and the pressure could not be accurately measured." It is extremely important to provide specifics and avoid vague modifiers such as: greater than, about, and higher than.

Keep in mind that good paragraph construction will present a thesis statement or idea and then give supporting details for that statement. When new points need to be made, make sure you make them in new paragraphs. You should make an outline for the Discussion section (just as you should for all parts of the report) so it follows a logical progression which will support the conclusions presented in the next section, Conclusions. Carefully judge the information that you are providing to the readers to make sure it satisfies their specific expectations. For instance, they expect to find information presented in a pattern that presents what is known first and then what is new second. Sentences should begin with known information and then progress to new, related information.

## Conclusions

A typical way to begin the Conclusions section is to make this statement: "The following conclusions are supported by the results of this study," and then list the conclusions in one or more simple, declarative sentences using numbers to differentiate each separate conclusion. Remember that in your Conclusions section, engineers will be looking for concise statements that clearly communicate what your results indicate—not additional explanatory material or any plans for further investigation. They want the mass of data synthesized into the briefest conclusions you can present.

## References

References are to be listed at the end of the report. In general, be sure to list: author's last name, then first and middle initials (and then list co-authors with initials first), title of source being quoted (in quotes), edition of source, volume of source (if applicable), city of publication (from source's title page), publisher, date of publication, and pages referenced. Note the variety of possible formats in the following samples:

Walker, R.E., A.R. Stone, and M. Shandor. "Secondary Gas Injection in a Conical Rocket Nozzle," *AIAA Journal,* Vol. 1, No. 2, Feb. 1963, pp. 334–338.

Turner, M.J., H.C. Martin, and R.C. Leible, "Further Development and Applications of Stiffness Method," *Matrix Methods of Structural Analysis,* 1st ed., Vol. 1, New York, Macmillan, 1964, pp. 203–206.

Segre, E., ed., *Experimental Nuclear Physics,* 1st ed., Vol. 1, New York: Wiley, 1953, pp. 6–10.

Book, E. and H. Bratman, *Using Compilers to Build Compilers, SP-176,* Aug. 1960, Systems Development Corp., Santa Monica.

Soo, S.L. "Boundary Layer Motion of a Gas-Solid Suspension." *Proceedings of the Symposium on Interaction between Fluids and Particles,* Institute of Chemical Engineers, Vol. 1, 1962, pp. 50–63.

Always give inclusive page numbers for references from journal articles and a page or chapter number for books. Each reference in the text must be cited.

## *Appendices*

Lengthy calculations and side issues that are not really related to the main theme of the report should be placed in the Appendix. To decide whether something should go in the body of the text or in the Appendix, think about whether or not its inclusion in the main body of the report required as part of the description of the investigation. If the answer is No, the item either should be left out or put in the Appendix.

## Communications Checklists

One way to make sure that you have successfully completed an assignment is to evaluate your work with a checklist. You can create custom checklists for each of the various communications formats you use. These checklists will guarantee that you have included everything that is required to inform the reader.

## Formal Report Checklist

### *Title Page*

1. Title of paper
2. Course/project
3. Date due
4. Section meeting time
5. Name

### *Abstract*

6. Purpose of the lab/experiment
7. How the lab was performed
8. What was discovered, achieved, or concluded
9. Use of past tense
10. Reference to experiment, not paper
11. No personal pronouns (I, we)

### *Nomenclature*

12. Listed in alphabetical order
13. Upper case, then lower case (A a B b c G g1 g1a)
14. Arabic and Greek listed separately (Arabic first)
15. Only symbols are listed

## Table of Contents

16. All sections represented
17. Abstract and Table of Contents are not listed
18. Observations as headings; analysis, equipment, procedures, results as sub-headings
19. All columns aligned

## Introduction

20. Motivation for the experiment stated
21. Sufficient information to orient reader to the substance of experiment
22. Sufficient information to excite reader
23. Listing of sections which follow

## Observations

24. Mathematical model used to predict system behavior presented with ample explanation and lead-in
25. Equations numbered
26. Proper punctuation in equations
27. Equations properly spaced
28. Schematic of equipment used
29. Tables oriented correctly, labeled and referenced
30. Figures oriented correctly, labeled and referenced
31. Highlights of equipment used
32. Highlights of the procedure (not specific steps)
33. Data presented with clear indication of what it applies to
34. Clear indication of what data refers to
35. Trends in data stated (to be discussed in the Discussion section)
36. Clear indication of what reader should observe in the data

## Discussion

37. Complete discussion of the results
38. Connection of data and objectives is clearly stated
39. Comparison to similar experiments presented
40. Strong points of the study given
41. Weak points of the study given
42. Statements are specific
43. Logical progression which supports Conclusions that follow

## Conclusions

44. Begins with, "The following conclusions are supported by this study"
45. Conclusions are numbered
46. Conclusions are concise and highly specific
47. Vague statements are not used
48. Conclusions flow directly from Discussion

## References

49. Initials for first names
50. All necessary information included
51. References numbered in text like this: [1]

## A Sample Abstract

An Abstract is a statement summarizing the important points of a given text. The following abstract satisfies the checklist features listed in the previous section. By making sure that the required items are included for each part of your report, you have a chance to review what you've included and what still needs to be addressed before the paper is submitted. (Note: The numbers that appear in the following abstract refer to the preceding checklist. References aren't numbered in a real abstract.)

### *Abstract*

Experiments {10} were conducted {9} to assess the appropriateness of digital signal analysis to the design, development, and testing of Whirlwind Corporation's new light aircraft gas turbine. Such analysis could be used to monitor the transient and steady state property variation of the new power plant, predict potentially catastrophic failure, and pinpoint sources of extraneous noise generation {6}.

Several elementary sinusoidal and square wave signals were generated {9} by a commercially available Waveteck function generator. These signals were then converted {9} to the frequency domain by LabView's "Spectrum Analyzer" via the Fast Fourier Transform. Various combinations of sampling frequency and sample size were investigated {9}. When deemed appropriate, these signals were {9} also low-pass filtered {7,10}.

The Discrete Fourier Transform accurately represented {9} only components less than half the sampling frequency. Higher frequency components were reflected {9} across the Nyquist frequency or its integer multiples. This aliasing was eliminated {9} by low-pass filtering, but on occasion, important signal components were discarded {9}. Whenever the input signal contained components that were not integer multiples of the frequency resolution, the magnitude of the corresponding spectrum peaks were diminished {9}. This leakage was reduced {9} by increasing the frequency resolution by increasing the sample size {8,10}. These signal analysis techniques proved {9} their utility and applicability to the new gas turbine project.

In this abstract, the author has addressed the important elements of the assignment, which was to tell the reader why the experiment was performed, how it was done, and what the principal results were.

Notice some of the methods that the writer has used to make the text flow smoothly. Line 8 begins the paragraph with "Several elementary sinusoidal and square wave signals," which prompts the reader to wonder what is to follow. Line 25 shows the writer relating "These signal analysis techniques" back to the previous sentence. Be sure to help your readers by presenting a logical, flowing report.

## OTHER TYPES OF COMMUNICATION

### Executive Summaries

Some assignments may require that you submit an Executive Summary instead of a full formal report. This type of report condenses the work you have done into the briefest document possible. It assumes a managerial perspective instead of a technical one, and will contain enough information that an individual with limited technical expertise will be able to make clear decisions based on your report. The Executive Summary presents just the simple facts and describes the key elements of your work in non-technical language—concise and straight to the point. A good Executive Summary contains the following information:

- the background of the situation or problem
- cost factors
- conclusions
- recommendations

If you have constructed Abstracts before, you will understand that they are similar to the Executive Summary. Both are quite brief. They focus on what was investigated, the conclusions your work has produced, and the recommended course of action. In some instances, this Summary will be all that is read before a decision is made either to proceed or to halt further action.

### Proposals

Proposals are documents that address a specific need and usually describe the need or problem at issue, define a solution, and request funding or other resources to solve the problem. Proposals can be solicited (as when an organization asks for proposals) or unsolicited (as when you send in a proposal and ask for funding or for a project to be implemented). Solicited proposals involve a Request For Proposal (RFP) with the following guidelines:

- what the proposal should cover
- what sections it should have
- when it should be submitted
- to whom it should be sent
- how it will be evaluated with regard to other proposals

For Proposals, follow the ABC format:

**A:** The **Abstract** gives the summary or big picture for those who will make decisions about your proposal. The abstract usually includes some kind of hook or grabber, which will interest the audience to read further. The abstract will include the following:

- the purpose of the proposal
- the reader's main need
- the main features you offer and related benefits
- an overview of proposal sections to follow

**B:** The **Body** provides the details about your proposal. Your discussion should answer the following questions:

- What do you want to solve and why?
- What are the technical details of your approach?
- Who will do the work, and with what?
- When will it be done, and how long will it take?
- How much will it cost?

Typical sections in the body portion of the report will be given to you in the RFP, or else you will have to develop them yourself, making sure you address the following:

- a description of the problem or project and its significance
- a proposed solution or approach
- personnel
- schedule
- cost breakdown of funds requested

Give special attention to establishing need in the Body. Why should your proposal be chosen? What makes it unique? How will your design or recommendations help the community?

**C:** The **Conclusion** makes your proposal's main benefit explicit and will make the next step clear. This section gives you the opportunity to control the reader's final impression. Be sure to:

- emphasize a main benefit or feature of your proposal
- restate your interest in doing the work
- indicate what should happen next

## Memos

The *Memo* provides a convenient, relatively informal way to communicate the existence of a problem, propose some course of action, describe a procedure, or report the results of a test. Though informal, memos are not to be carelessly written. They must be carefully prepared, thoughtfully written, and thoroughly proofread for errors.

Memos contain introductory information that summarizes the memo's purpose. The memo's conclusion restates the main points and contains any recommendations. They begin by clearly presenting all relevant information right up front, as follows:

**To:**       **name, job title**
             **department**
             **organization**
**From:**     **name, job title**
             **department**
             **organization**
**Subject:**  **(or "Re:") issue addressed in the memo**
**Date:**     **date**

The following may be used as well, if applicable:

**Dist:**     **(Distribution) list of others receiving the memo**
**Encl:**     **(Enclosure) list of other items included with the memo**
**Ref:**      **(Reference) list of related documents**

The format of the text of the memo is also simple and contains the following information:

*Foreword*—statement of the problem or important issues addressed
*Summary* (or *Abstract*)—statement of results, findings, or other pertinent information
*Discussion or Details*—technical information, discussion of the problem or issues, or support for the claims in the Summary

## Activity

Prepare an answer to the following memo. Keep in mind the appropriate format and approach for a good memo.

*Situation*—Assume that the following memo has been received in your office at Engo-Tech Consultants (an engineering consulting firm specializing in technology compatible with the request in the memo). Please read it and address its concerns.

**To:**     Engo-Tech Consultants
**From:**   Michelle De LaBaron, General Manager, Euro-Flags
**Date:**   August 1, 2002
**Re:**     Increased activity for Euro-Flags
           To ensure the continued success of our operation at Euro-Flags
           we believe it is essential that we add a new technical emphasis
           to the variety of entertainment opportunities we currently offer.

Our mission includes more than simply existing as a provider of "fun and games." We also want to stimulate our patrons intellectually, and thrill them with the wonder of current technology.

With this in mind, we ask that you present us with a proposal which outlines how your firm could assist us in this endeavor. Please explain not only what your firm can do, but also how your proposed plans would appeal to our patrons.

## Email

Electronic mail, or e-mail, has become a part of life in almost every office. Messages gather in computer mailboxes at a phenomenal rate, and the time required to respond to these e-mails continues to grow. It is, therefore, vitally important that we look closely at what we receive and what we send. E-mail must be read carefully and the responses we send must be well written. Just as with more formal reporting, if your communication includes spelling errors, poor punctuation, and sloppy construction, it will likely hamper your reader's comprehension. And it projects general incompetence—even if you are quite competent technically. Besides, e-mail provides an easily preserved record of your communication, so it's worth doing right.

E-mail's instant transmission and spontaneous feel can tempt us into serious communications errors. It's easy to get sloppy. Given the format, e-mail can be easily misunderstood—which is one reason why the variety of "smiley face" punctuation tags evolved. And once we have sent a message, it is out of our hands and we can do nothing to change the content or presentation. Here are a few simple rules to follow:

- Use proper grammar—appropriate sentence structure, subject/verb agreement, spelling, flow, etc.
- Carefully proofread, edit, and spellcheck your e-mail. Are there gaps in your train of thought, or opportunities for confusion, that may cause a reader to misunderstand your message?
- Have you thought about your response before you hit "Send"? Once it is sent, it can't be retrieved. With professional e-mail, it's often best to wait as long as possible to send your reply, even if you only have an hour. Let it rest, then re-read your reply before sending.

E-mail is a fantastic tool, but it has limitations. Never use it when a face-to-face dialog actually is needed. Do not use it when a formal document is required. E-mail does not carry the status of a formal report. And remember that e-mail is not quite as private as you think. A message may be downloaded and left on the screen of your recipient, in plain view of anyone entering his office. And in truth, many e-mail logs and data archives are more public than you may realize. It is vital, therefore, that you make every effort to use the tool wisely, and avoid letting it embarrass you or cause you undue grief.

## Progress Report

At some point in every job, you will be asked to indicate how you are progressing toward some goal. Managers, co-workers, stockholders—even you yourself—will want to know what you've accomplished. The Progress Report provides the means by which you can report your status.

The reports need to be attractive and easy to understand. Commonly, the progress report will have sections much like any other report prepared for management:

- Introduction
- Project description
- Summary of what you have done

Each section is discussed briefly below:

Introduction—Here you capture the interest of the readers by informing them of what you are going to do in the report. You expose readers to the scope of the work being done, the purpose of the work, and any major changes that have been required in the project.

Project Description—Here the reader is acquainted with the time-frame of the project and the current progress. Phases that have been completed and the time it took to complete them are presented in this section. It is also important to explain the tasks that remain to be completed.

Summary—This is where you draw the whole report together. Summarize the main points of action and reiterate where you are along the road to completion.

## Problem Statements

Problem solving is a critical aspect of engineering. Before you begin to work on a project you may be asked to present a Problem Statement pertaining to the project which explicitly states the problem to be investigated and outlines the course you intend to take. Correctly defining the problem is vital to the solution. If you are not on the right track from the beginning, you may well fail to obtain the appropriate solution. So here are some basics:

- Work with data that you already have collected—evaluate all reasonable avenues of pursuit in finding a solution as you define your problem.
- Consult your contacts who have knowledge of the problem.
- Investigate the problem first-hand.

Here's a silly anecdote which may help you remember to properly identify a problem.

A man who had a chronically sore neck went to the doctor. He told the doctor of his suffering, that he was unable to turn his head without severe pain. Even visiting the doctor caused him pain. The doctor asked him what he had done to try to solve the problem. The man listed a variety of attempted remedies—none of which had worked. The doctor looked at the man, analyzed the situation, and calmly said, "When you get up in the morning, try taking the hanger out of the shirt you're going to wear before you put it on."

Thus the problem was properly identified and solved.

## Cover Letters

Cover Letters will be especially important as you attempt to land your first job in engineering. Such letters are sent along with résumés and transcripts. They "cover" your other material—thus the name. You should highlight information about yourself that separates you from the rest of the applicants and demonstrates your knowledge of and interest in the company. A great Cover Letter can open doors, but a poor one can slam them shut.

The Cover Letter functions as your introduction to a prospective employer. This one-page document introduces your to the company in a more conversational manner. In many instances, they will represent the only personal impression you will get to make on a potential employer. View the letter as an opportunity to communicate what you have to offer them. If they don't like what they read, you may never get the chance to talk to them in person.

In the first paragraph, introduce yourself and describe how you became familiar with the position for which you are applying. In the next paragraph(s), detail your qualifications and what makes you qualified for the job. You should include how your personal character and interests will enable you to excel at the job and to fit in with the company. Your concluding paragraph should include a request for an interview (or for the company to follow up), your contact information, and a thank you for the company's time and consideration.

### Specifics you should you include in a Cover Letter

- the date, your address and phone number, and the name and address of the person to whom you are writing
- 1st paragraph—the reason you are writing the letter, the source of your information about the employer, and what you would like to do for the employer (the position for which you would like to apply)
- 2nd paragraph—a brief discussion of your resume, hitting the highlights
- 3rd paragraph—"current" information which may not yet be appropriately included in your resume (courses you are currently taking, research you may be involved in, activities you are about to engage in)
- 4th paragraph—summary paragraph in which you thank the reader for his or her consideration "in advance" (You may also include your phone number, and mention that you will call to confirm receipt of the letter.)

Remember, you want your Cover Letter to present the most positive impression possible. A good deal of time should be spent perfecting it. Cover Letters and Resumes should be printed on the same kind and color of paper. Always use a high-quality laser printer to print these items.

## Sample Cover Letter

August 10, 2002

Ms. Lydia Baron
Human Resources Director
Harrigan Corporation
120 Rollaway Road
Chicago, IL 60606

Dear Ms. Baron:

I am very interested in the summer internship program offered at Harrigan Corporation. I know from my research that Harrigan's rapidly expanding automotive parts operation has an outstanding reputation. The opportunity to gain valuable experience through working in that division would be of tremendous benefit to me in my professional development.

My interest in engineering began at an early age, and I look forward to a lifetime of continual engineering education. I am completing my junior year at Michigan State University, having taken a number of courses toward my degree in mechanical engineering. As my resume shows, I have concentrated a great deal on developing my computer skills as part of my engineering background.

This semester I am taking two courses—Advanced Mechanics and Finite Element Analysis—which have been particularly valuable in preparing me for working at Harrigan Corporation. I have studied the current work schedule of the Brighton facility and understand that these courses fit well with your plans for the new turbo charger assemblies.

A copy of my resume is enclosed for your consideration. I will contact you during the week of August 20 to see if you require any additional information. If you have any questions for me in the interim, please contact me at (517) 555-1212, or at smith@pilot.msu.edu. I look forward to discussing with you the job opportunities at Harrigan Corporation.

Sincerely,


Jason D. Smith

222 Landon Hall
Michigan State University
East Lansing, MI 48824
(517) 555-1212
smith@pilot.msu.edu


Enclosure

## Résumés

A résumé is one of the most important documents you will ever create. It sells you and your qualifications. There are two main types of résumés: skills résumés and experience résumés. A skills résumé is for people who have not yet completed significant work experience. The skills résumé highlights the skills and talents to benefit the potential employer, even if the applicant has little or no technical work experience. Experience résumés highlight prior work experience related to the job for which a person is applying. In general, most engineering students in the first two years of study and those who graduate without technical work experience will use a skills résumé; engineering students with co-op or internship experience in an engineering context would use an experience résumé.

There is a wide variety of appropriate formats for résumés, so ultimately you have to decide upon a format that you will be happy with. Regardless of the specific format, every résumé must present your name, address and phone number; your educational background; your previous employment (if any); and extracurricular activities. The following simple format is merely one of many. Key elements are highlighted in bold type.

Make sure that you get as many people as you can to read and critique your résumé. Listen to and evaluate all the comments you get, and then customize the résumé to fit your needs.

## Résumé Checklist

### Information

- Objective is clearly stated.
- Clearly convey how your education, experience, activities, and honors support the objective.
- Experience, education, and/or skills segments are effective.
- All activities, honors, and other data are appropriate for the employment and the reader.

### Organization

- Name and key headings stand out.
- Information within each heading is ordered from most to least important.
- Experience segment is arranged to highlight your strengths and career objective.

### Style

- Language is simple, direct, and precise.
- Noun phrases are consistently used for headings.
- Strong verb phrases or clauses are consistently used to describe experience, skills, and activities.
- Parallel structure is used effectively.
- Have no errors in grammar, punctuation, or spelling (no typos whatsoever).

## Important things to remember about résumés

- A résumé should be crafted for a specific position or job. This shows the employer that you care about the position. Your résumé's objective section should reflect some specificity as a result.
- You may need to change your résumé for "scannable format," if you apply for jobs with larger engineering companies. These companies use a computer to scan each résumé submitted to them for keywords. If so, you should consult your career services office or look at the company's recruitment information to determine these keywords and use them on your résumé. The scanner will target you and your résumé as a potential employee more quickly.
- If possible, keep your résumé to one page. Many employers frown upon résumés longer than one page.
- The format of a résumé is important. Your information should be arranged to be visually pleasing and readable. Use the following strategies in terms of format:
  - List dates in reverse chronological order (starting with your most recent information first)
  - Use at least 11-point font size (preferably 12-point) and typeface such as Times New Roman
  - Use one-inch margins around the page
  - Headings such as Education, Work History, Skills, Honors and Activities, etc., should be in bold, can be in capital letters, and should be the only information on that line of the page
  - Your résumé should be printed using a high-quality laser printer on excellent bond paper (at least 40-pound ivory-colored or off-white paper)

## Here is a list of recommended action verbs you may want to use, categorized by topic:

### Human Relations

| | | |
|---|---|---|
| worked with | taught | helped |
| volunteered | served | assisted |
| interacted | counseled | trained |
| sponsored | directed | guided |

### Research and Design

| | | |
|---|---|---|
| researched | experimented | observed |
| analyzed | tested | assessed |
| solved | verified | designed |
| discovered | devised | created |
| investigated | evaluated | |

### Communications

| | | |
|---|---|---|
| drafted | edited | addressed |
| designed | revised | interpreted |
| composed | prepared | lectured |

|            |             |           |
|------------|-------------|-----------|
| published  | taught      | conducted |
| presented  | instructed  | published |

**Management**

|             |             |           |
|-------------|-------------|-----------|
| managed     | contracted  | saved     |
| rated       | maintained  | scheduled |
| evaluated   | established | sold      |
| devised     | negotiated  | verified  |
| planned     | controlled  | produced  |
| organized   | purchased   | improved  |
| coordinated |             |           |

(See actual Sample Résumé on next page.)

## Thank-You Letters

After interviewing for a new job or co-operative position, you should send Thank-You Letters to the individuals who interviewed you. You shouldn't wait more than 48 hours to send these letters. They should be businesslike and concise. And remember that a strong follow-up letter may be just the thing that puts you ahead of the competition.

Here's what you should include in your Thank-You Letter:

*1st paragraph*
— Thank the interviewers for their time, and reiterate your interest in working for the company.

*2nd paragraph*
— Briefly restate your qualifications. This is the time to address any positive qualities you may have failed to mention during the interview.

*3rd paragraph*
— Close the letter with a final thank-you, and express your interest in hearing back from the interviewer. You may even want to give a specific time frame within which you will follow up with a phone call. Provide the interviewer with phone numbers where you can be reached and your e-mail address, if you have one.

*Additional advice:*
— Write or modify each Thank-You Letter separately. Don't use generic form letters. Try to personalize your summary of the interview in each letter.
(Even if you are rejected for the position, sending a Thank-You Letter may open up other doors or future opportunities with that company.)

## Relevant Readings

Here are just a few of the many technical communication books that are on the market. They all contain pretty much the same information, presented in varying formats. Having such a resource on your shelf will be of great value.

## Sample Résumé Format

**Your Name** (centered)

**College Address**                                      **Permanent Address**
**College Phone Number**                                 **Permanent Phone Number**

**Objective:**              An entry level position in (state field or position sought).

**Education:**              • your degrees
                           • your school
                           • graduation date (or anticipated date)
                           • grade point average and basis (e.g., 3.4/4.0)

**Work Experience:**        • your job titles (list most recent first)
                           • the companies for which you worked
                           • dates for which you worked for those companies
                           • responsibilities in those positions

**Activities:**            • meaningful activities in which you have participated

## Sample Actual Résumé

**Jason D. Smith**

222 Landon Hall                                          37 Dilgren Avenue
Michigan State University                                Port Anglican, VT 12225
East Lansing, MI 48824                                   (234) 555-1212
(517) 555-1212
smith@pilot.msu.edu

**Objective:**             An internship in the manufacturing area of a mid-sized
                          company

**Education:**             Bachelor of Science in Mechanical Engineering
                          Michigan State University, East Lansing, Michigan
                          Expected Date of Graduation: May 2003; GPA: 3.6/4.0

**Computer Skills:**       MS Office 2000, Lotus, C++, MATLAB

**Employment:**            May 2002—August 2002; Production Engineer
                          Labardee and Sons, Aberdeen, Texas
                          • Developed new production plans for the Labardee
                            grinding machine
                          • Consulted with primary Labardee clientele

                          June 2001—August 2001; Machinist
                          Rawlins Machining, Deer Park, New York
                          • Produced updated schematics for all standing jobs
                          • Instituted changes in assembly line procedures

**Honors/Activities:**     Ames Scholarship, Dean's List (4 semesters), MSU Swim
                          Team, racquetball

**References:**            Available upon request

Fear, D., *Technical Writing. Glenview,* Illinois: Scott, Foresman and Company. 1981.

(For the writer who wants instruction in clear, concise steps.)

Eisenberg, A., *Effective Technical Communication,* 2nd Edition. New York: Mc-Graw-Hill. 1992.

(General overall coverage of the major topics of technical writing; includes proposals, letters, and reports.)

Hirschorn, H., *Writing for Science, Industry, and Technology.* New York: D. Van Nostrand. 1980.

(Particularly helpful appendix. An approach to writing that takes the writer from the beginning of the process to the end product.)

Houp, K. and T. Pearsall, *Reporting Technical Information,* 6th Edition. New York: Macmillan. 1988.

(The best of the bunch, both for future use and present needs. Gives lots of examples, and helps greatly in the writing process.)

Huckin, T., and L. Olsen, *English for Science and Technology.* New York: McGraw-Hill. 1991.

(Good book for the non-native English speaker about the process of report construction.)

Mathes, J. C., and D. Stevenson, *Designing Technical Reports.* New York: Macmillan. 1991.

(A process approach to the writing of a technical presentation—written from the information perspective, not from the "form of the report" perspective.)

Michaelson, H., *How to Write and Publish Engineering Papers and Reports.* Phoenix: Oryx Press. 1990.

(Primarily aims at the writing of papers for publication. Does a good job of looking at quality of writing and the concern for the reader.)

Miles, J., *Technical Writing—Principles and Practices.* Chicago: SRA Research Associates. 1982.

(A general text to help with basic problems. Includes a great deal on the process of getting started.)

Turner, M., *Technical Writing.* Reston, Virginia: Reston Publishing. 1984.

(A teaching text with a good visual format. Nice section on memo writing.)

## Also

Beer, D. A Guide to Writing as an Engineer. New York: John Wiley & Sons. 1997.

Eisenberg, A. A Beginner's Guide to Technical Writing. New York: McGraw Hill. 1998.

Finklestein, L. Pocket Book of Technical Writing. New York: McGraw Hill. 2000.

Pfeiffer, W. Proposal Writing. New Jersey: Prentice-Hall. 2000.

Pfeiffer, W. Technical Writing. New Jersey: Prentice-Hall. 2000.

## EXERCISES AND ACTIVITIES

1. Prepare a five-minute presentation to be given to a group of your fellow students concerning an engineering topic of interest to you. Prepare overhead or PowerPoint slides for your presentation. You may not need more than one slide per minute of presentation.

2. Prepare a presentation, similar to the one above, to be given to a group of elementary students. What information did you change?

3. List at least five clichés that you have used or heard that could make comprehension difficult for a foreign student.

4. Write a memo informing your fellow students of a seminar you are presenting.

5. Write a memo to your supervisor requesting that your work activity be reviewed for a possible raise.

6. Write a formal report about a lab experiment that you have completed, using the suggestions presented in this chapter.

7. Write an Executive Summary of a lab experiment that you have completed.

8. Write an e-mail to your local newspaper responding to an issue (deer hunting, guns, smoking, abortion, air bags, etc.) for the "Letters to the Editor" section.

9. Engineering is not recognized as a profession by federal agencies since it does not have any entrance requirements. In fact, a college degree is not even required. Write an e-mail to your state representative supporting or opposing this position of the federal agencies.

10. Write your resume.

11. Write a cover letter to the OLG Manufacturing Company seeking employment in the Engineering Department.

12. Write a thank-you letter to a Dr. Jan Harris, as though she interviewed you for an engineering position.

# Chapter 8

# Succeeding in the Classroom

## Introduction

As an engineering student, an important goal is to succeed in the classes needed to attain an engineering degree. How this is done depends great deal on the individual student's style, temperament, and strengths. This chapter presents strategies to help students succeed. As with any tips on success, you need to look at these techniques and decide if you can use them. If you are unsure about some of the suggestions, experiment with them to see if they fit you.

The three components of succeeding in academics are ability, attitude and effort. No book can address the first item. Though you need ability to succeed in engineering, many students had the ability but failed because of poor attitudes or poor or inefficient effort. Many students had limited natural ability, but because of their positive attitude and extraordinary effort, are excellent engineers. This chapter focuses on techniques to improve the components for success that you control.

## Attitude

Your attitude is the most important thing you control when taking on any challenge, including studying engineering. You can expect success or failure. ALWAYS look for the positives; NEVER dwell on the negatives.

Approaching your classes and teachers with a positive attitude is the first key to succeeding in your engineering studies. Each class should be viewed as an opportunity to succeed. Learn to confront difficulties along the way and move ahead. Many students will decide that a certain class is too difficult or that the teacher is horrible and expect to do poorly. Many fulfill their own prophecy by failing. Other students look at the same situation, overcome the obstacles, and excel.

## Goals

In setting goals, you have two essential elements: height and time. You might ask, "How high do I set my goals?" This will vary from individual to individual, but goals should be set attainably high. A common management practice is for a team to decide on a reasonable goal to accomplish. Then the bar is raised and a stretch goal is defined that encourages the team to produce more. This stretch goal is designed to push the group farther than they would have gone with only the original goal. Sometimes this stretch goal looks reasonable and other times it looks unreachable. What often happens, though, is that the stretch goal is met. As an individual, setting stretch goals helps you grow. When setting goals for yourself, look at what you think you can do, which are your base goals. Then establish stretch goals. You may reach those higher goals.

The second element in goal setting is time. When will the goal be attained? Have long-term goals and short-term goals. Long-term goals help you get where you want to go and answer what you want to do with your engineering degree. An appropriate long-term goal as a freshman would be a particular entry-level job. This will help answer the question "Why am I doing this?" when your classes get tough.

Short-term or intermediate goals should be stepping stones to the achievement of your long-term goals. The academic setting's advantage is it has natural breaks, semesters or quarters, for the establishment of goals within specific time frames. Goals can be set for the academic year, and for each semester or quarter. These provide intermediate goals.

Short-term goals need to be set and rewarded. Rewarding short-term goals keeps motivation high in the quest for long-term goals. Many students lose sight of their long-term goals because they do not set short-term rewardable goals. Short-term goals may be an "A" on an upcoming quiz or test, or it may be completing a homework assignment before the weekend.

## Example

| | |
|---|---|
| Long-term goal | Job as a design engineer for a major manufacturer of microprocessors |
| Intermediate goals | 3.7 GPA after sophomore year<br>3.5 GPA at graduation<br>Hold an office in the local chapter of IEEE<br>Internships after sophomore and junior years |
| Short-term goals | "B" on first math test<br>"A" in calculus<br>3.7 GPA after first semester<br>Join engineering club and attend meetings<br>Complete chemistry lab by Friday |

In the list of short-term goals above, there are opportunities for rewards. Perhaps there's a new CD or video game you want. Set a goal of a certain grade on that calculus test. If you meet the goal, you can go buy the CD or game.

Clearly define your goals by writing them down. It is much more powerful to write down your goals than to think about them. Some people post them where they can see them daily. Others put them in a place where they can retrieve and review them. Both methods are effective. For short-term goals, establish weekly goals. Select a day, probably Sunday or Monday, to establish the goals for the week. This is a great time to establish the rewards for your weekly goals. Plan something fun to do on the weekend if you meet those goals.

At the end of each semester, examine your progress toward your long-term goals. Every semester gives you an opportunity to assess your progress and start afresh in new classes. Take advantage of this. Examine your long-term goals and make alterations if needed.

Guidance counselors or advisors work with student placement and are excellent resources to get information on what employers are looking for. What credentials do you need to be able to get the kind of job you desire? If you know this as a freshman, you can work toward obtaining them. You can also go directly to the employers and ask these same questions of them.

## Keys to Effectiveness

Once you have set goals regarding grades, the next step is achieving those grades. Effort and effectiveness may be the most important components of your success as a student. Numerous well-equipped students end up failing. Some students come to the university poorly prepared or begin badly but eventually succeed. The difference is in their effort and effectiveness. The following suggestions contain strategies to improve personal effectiveness in your studies.

### Take Time to Study

One of the strongest correlations with a student's performance is the time he or she spends studying. Developing the study habits and discipline to spend the needed time studying is a prime factor in separating successful and unsuccessful students.

### Go to Class, and Be on Time

Grades are highly correlated with class attendance and performance. This may seem obvious, but you will run into many students who believe class attendance does not matter. In the classroom, important information will be presented or discussed. If you skip class, you may miss this information. By skipping class, you miss an opportunity to be exposed to the necessary material, and you will have to make up that time on your own. Missing class only delays the expenditure of time you will need to master the material. Frequently, you will need more than an hour of independent study for every hour of missed class time.

### Make Class Effective

The first component of making class effective is sitting where you can get involved. If the classroom is large, can you see and hear in the back? If not, move to the front. Do you need to feel involved in the class? Sit in the front row. Do you fall asleep in class? Identify why. Ask yourself, "Am I sleeping enough at night?" If the answer is no, get more sleep.

The second component for making class time effective is to prepare for class. Learning is a process of reinforcing ideas and concepts. Use the class time to reinforce the course material by reviewing material the day before. Take time before the class to skim over the relevant material. Skimming means reading it through but not taking time to understand it deeply. This will make the class time more interesting and understandable and will make note taking easier.

After class, reread the assigned sections in your textbook. Concentrate on the sections highlighted by the lecture. By doing so, you will not waste time understanding the nonessential parts of the book.

## Keep Up with the Class

Class is most effective if you keep up. An excellent short-term goal is to master the material of each course before the next class. Most courses are structured in such a way that you can master one concept and then move to the next concept during the following session. Often in science and engineering classes, the concepts build upon one another. If you fall behind, you will not be able to master concepts that depend on previous lectures, and you will not understand the current lectures.

Some of your classmates may insist that you do not need to keep up so diligently since some course majors cover material that is more conceptually based. That is, once you understand the concept, you have the course mastered. Engineering, math and science courses are not this way. To master them, you must study extensively and prepare over an extended period. In some courses, you will master only a few basic concepts even though the entire course is dedicated to the application of these few concepts.

## Take Effective Notes

A main component in keeping up with your classes is taking effective notes for each class. Effective notes capture the key points of the session in a way that allows you to understand them when you are reviewing for the final exam three months later.

Suggested note-taking strategies:

1. Skim the assigned reading prior to class to help identify key points.
2. Take enough notes to capture key points but do not write so much that you fail to hear the presentation.
3. Review your notes after class to interpret them, filling in gaps so they will still make sense later in the semester.
4. Review your notes with other students to ensure you have captured the key points.
5. Review your notes early in the semester with your teacher to be certain you are capturing the key points.

In class, your job is to record enough information to allow you to flesh out your notes later. You do not have to write everything down. Getting together with classmates after class helps interpret your notes and gives you different perspectives on the main points. If you really want to make sure you have captured the key points, ask your teacher. Doing so early in the semester will not only set you up for successful note taking throughout the semester but will also allow you to get to know your teacher.

## Group Studying

Studies have shown that more learning takes place in groups than when students study by themselves. Retention is higher if a subject is discussed, rather than just listened to or read. If you find yourself doing most of the explaining of the ideas to your study partners, take heart! The most effective way to learn a subject is to teach it. Anyone who has taught can confirm this; they really learned the subject the first time they taught it.

> *"The most important single ingredient to the formula of success is knowing how to get along with people."* —Theodore Roosevelt

If you are not totally convinced of the academic benefits of group studying, consider it a part of your practical engineering education. Engineers today work in groups more than ever before. Being able to work with others is essential to being an engineer. If you start studying in groups sooner, working this way will be second nature by the time you enter the workforce.

Studying with others will also make it more tolerable, even fun, to study. There are many stories about students preparing for exams and spending the whole day on a subject. It is difficult to stick with one subject for an entire day by yourself. With study partners, you find yourself sticking to it and maybe even enjoying it.

Studying with others makes it easier to maintain your study commitments. Something we all struggle with is self-discipline. It is much easier to keep a study commitment if others are depending on you. They will notice if you are not there and studying.

Group studying is more efficient if properly used. Not everyone in the group will be stuck on the same problems. So, you will have someone who can explain your particular problem. The problem areas common to everyone before the group convenes can be tackled with the collective knowledge and group perspectives. Solutions are found more quickly if they can be discussed. More efficient study time will make more time for other areas of your life.

Choosing a group or partner may be hard at first. Try different study partners and different group sizes. Ray Landis, Dean of engineering at California State University, Los Angeles, and a leading expert on student success, suggests studying in pairs. "That way each gets to be the teacher about half the time." The trick is to have a group that can work efficiently. Too large a group will degenerate into a social gathering and not be productive in studying. Whichever group size you choose, here are some basic tips for group studying:

1. Prepare individually before getting together.
2. Set expectations for how much preparation should be done before getting together.

3. Set expectations on what will be done during each group meeting.
4. Find a good place to convene the group that will allow for good discussion without too many distractions.
5. Hold each other accountable. If a group member is not carrying his or her own weight, discuss it with that person and get him or her to comply with the group's rules.
6. If a member continues to fail to carry his or her own weight or comply with the group's rules, remove that person from the group.

## Define a Study Spot

Depending on your own learning style and personality, you may need a certain kind of environment to study efficiently. Some resources attempt to describe the perfect study environment. In reality, no perfect study environment or studying method exists. What is crucial is that you know what you need and find a place that meets those needs. For some, it involves total quiet while sitting at a desk. Others may prefer some noise, such as background music, and prefer to spread out on the floor. Whatever environment you decide is the best for you, pick a place where you can be effective.

Cynthia Tobias describes the conflict she has with her husband's view of a proper study environment in her book *The Way They Learn:*

> *I have always favored working on the floor, both as a student and as an adult. Even if I'm dressed in a business suit, I close my office door and spread out on the floor before commencing my work. At home, my husband will often find me hunched over books and papers on the floor, lost in thought. He is concerned.*
>
> *"The light is terrible in here!" he exclaims. "And you're going to ruin your back sitting on the floor like that. Here, here! We have a perfectly clean and wonderful roll top desk." He sweeps up my papers and neatly places them on the desk, helps me into the chair, turns on the high-intensity lamp, and pats my shoulder.*
>
> *"Now, isn't that better?" he asks.*
>
> *I nod and wait until he is down the hall and out of sight. Then, I gather all my papers and go back down on the floor. . . . It does not occur to him that anyone in his or her right mind could actually work better on the floor than at a desk, or concentrate better in 10-minute spurts with music or noise in the background than in a silent 60-minute block of time.*

Different people have different ways in which they learn most efficiently. You need to discover yours so that you can learn in the most efficient way possible for yourself. If you are unsure of what works best for you, try some different options. Try various study environments and track how much you are able to get done. Stick with the one that works best for you.

If you and your brother, sister, or roommate have different styles and needs, then you will need to negotiate to decide how the room will be used. Both of you will have to be considerate and compromise because it is shared space.

One item that is a distraction for almost everyone is television. Because television is visual, it makes concentrating on homework impossible although many students swear they are still productive. A few might be, but most are not. We strongly recommend against having a television set turned on in your bedroom or study area. It is not even good for study breaks, which should be shorter than the typical 30-minute or 60-minute program time slots on TV. Forgoing television may be one temporary sacrifice you have to make to be successful.

## Test-taking

Most courses use written, timed exams as the main method of evaluating your performance. To excel in a course, you need to know the material well and be able to apply it in a test situation. The early part of this chapter provided tips on how to improve your understanding of the course material. The second part of the chapter is about test-taking skills. Ask any student who has been away from school for a few years. Taking tests is a skill that needs to be practiced. Just like a basketball player who will spend hours practicing lay-ups, you must practice taking tests. A great way to do this is to obtain past exams. A couple of days before the test, block out an hour, sit down, and practice taking one of the old tests. Use only the materials allowed during the actual test. If it is a closed-book test, do not use your book. In this way, you are accomplishing two things. The first is assessing your preparation. A good performance on the practice test indicates that you are on the right track. The second thing you are doing is practicing the mechanics of taking the test. Again, with the basketball analogy, players will scrimmage in practice to simulate game conditions. You should simulate test conditions. What do you do when you reach a problem you cannot solve? Most people will start to panic. You must stay calm and reason your way through whatever is preventing you from solving the problem. This may mean skipping the problem and coming back to it or looking at it in a different way. In any case, you will run into these pitfalls on tests, and by simulating test conditions, you will take tests better.

Before the first test in the course, visit your teacher and ask him or her how to assemble a simulated test. The old exams may be the solution. Instructors do not like to hear, "What is going to be on the test?" They are more receptive to requests for guidance in your preparation. The instructor's suggestions will focus your studying and make it more efficient.

### Taking the Test

The first guideline for test taking is to come prepared. Do you have extra pencils? Can you use your calculator efficiently? Breaking your only pencil or having your calculator fail during the test is stressful and will prevent you from doing your best even if you are given a replacement. Proper planning for the test is the first step toward success.

The second step is to skim over the entire test before beginning to work the problems. This is to ensure that you have the complete test. It also gives you an overview of the test so you can plan to attack it efficiently. Take note of the point weighting for each problem and the relative difficulty.

The next step is to look for a problem that you know you can solve easily. Do this problem first. It will get you into the flow of doing problems and give you confidence for the rest of the test. In addition, doing the problems you know you can tackle will ensure that you get full credit for these before moving on to the ones that are more problematic.

## Watch the Clock

Always keep track of the time during a test. Tests are timed, so you need to use time efficiently. Many students will complain they ran out of time. While this may prevent you from finishing the test, it should not prevent you from showing what you know. Look at the number of points on the test and the time you are allotted to take the test. This will help you estimate how long to spend on each problem. If there are 100 points possible on the test and 50 minutes to complete it, each point is allotted 30 seconds. Therefore, a 10-point problem should only take five minutes. A common mistake is wasting too much time struggling through a difficult problem and not even getting to others on the test. Pace yourself so you can get to every problem even if you do not complete all of them. Most instructors will give partial credit. If they do, write down how you would have finished the problem and go on to the next when the allotted time for that one is up. If you have time at the end, you can go back to the unfinished problem. Nevertheless, if not, the instructor can see that you understood the material and can award partial credit.

Remember that the goal of a test is to get as many points as possible. A teacher cannot give you any credit if you do not write anything down. Make sure you write down what you know and do not leave questions blank. Another common mistake occurs when students realize they have made a mistake and erase much of their work. Often, the mistake they made was minor and partial credit could have been given, but they received no points because they erased everything.

Concentrate on the problems that will produce the most points. These are either problems you can do relatively quickly and get full credit for or ones worth a large portion of the total points. For instance, if there are four problems, three worth 20 points and one worth 40 points, concentrate on the 40-point problem.

Leave time at the end, if possible, to review your answers. If you find a mistake and do not have enough time to fix it, write down that you found it and would have fixed it if you had time. If you are taking a multiple-choice test, be careful about changing your answers. Many studies find that students will change right answers as often as they change wrong answers. If you make a correction on a multiple-choice test, make certain you know your answer was wrong. Otherwise, your initial answer may have been correct.

Think. This may sound basic but is important. What is the question asking? What does the instructor want to see you do? When you get an answer, ask yourself if it makes sense. If you calculate an unrealistic answer, comment on it. Often a simple math error produces a ridiculous answer. Show that you know it is ridiculous but that you just ran out of time to find the error.

## After the Test

After you get your test back, look it over. It is a great idea to correct the problems you missed. It is much easier to fill in the gaps of what you missed as you go along in the semester than to wait until you are studying for the final exam. Most final exams cover the material from all of that semester's tests. If you cannot correct the problems yourself or with the help of your study group, go see the instructor and ask for help. Remember, you will see this material again!

## Making the Most of Your Instructors

The instructor selects course material and decides what is important and how you will be evaluated. Few students take the time to get to know their teachers. Here are some reasons to get to know your teachers:

1. They are professionals and have experience in the field you are studying. Therefore, they have valuable perspectives.
2. Every student will need references (for scholarships, job applications, or school applications), and teachers can provide them.
3. Professors are in charge of the course and can help you focus your studying to be most effective.
4. They are the experts in the field and can answer the hard questions you do not understand.
5. They assign your grades, and at some point in your college career, you may be on the borderline between grades. If they know you and that you are working hard, they may be more likely to give you the higher grade.
6. They are likely to know employers and can provide leads for full-time positions or internships.
7. They may be aware of scholarship opportunities or other sources of money for which you could apply.

A school could send you the course materials at home and have you show up for one day to take your exams. Yet, students come to school to study. Why? One of the main reasons is for the interaction with the faculty, staff, and other students.

If so many reasons to know teachers exist, why do most students avoid this process? Many instructors do not encourage students to come see them or do not present a welcoming appearance. Teachers may somehow remind students of their parents, as people who seem to be out of touch with students. It may be that many students do not understand all of the advantages of getting to know their teachers.

When getting to know your teachers, remember they have spent their careers studying the material you are covering in your courses and should thoroughly enjoy it. Thus, telling your calculus teacher you find math disgusting would be insulting and suggest he or she has wasted his or her career.

Professors are teachers by choice, especially in engineering. Industry salaries are higher than academic salaries, so teachers teach because they want to and because they are skilled at it. If you are having a problem with an instructor's teaching style, approaching him or her with a problem-solving strategy such as the following can be constructive: "I am having trouble understanding the concepts in class. I find that if I can associate the concepts with applications or examples I understand them better. Can you help me with identifying applications?"

As a group, instructors love to talk, especially about themselves and their expertise. They have invaluable experience that you can benefit from. Ask them why they chose their particular field. Ask them what lessons they learned as students. You will get some valuable insights and get to know someone who can help you succeed in your class and possibly in your career.

Faculty members are busy and have many demands on their time (research, securing grants, writing, committee assignments, etc.) in addition to your class. These other demands are what keep the programs running. So, respect their time. It is okay to get to know them but do not keep popping in unless you have a reason. In addition, when you come to them with questions, show them the work you have been doing on your own; you only need to get clear of a particular hurdle. A busy faculty member will not mind helping a hard-working student understand a difficult to grasp idea or concept. However, showing up and asking how to do the homework can be interpreted as "Do my homework for me." This could create a negative impression of you with your instructor that would be difficult to overcome.

## Well-Rounded Equals Effective

Being an effective person goes beyond being a good student. Developing the habits that make you effective during high school and college will prepare you for success later. Part of being effective is functioning at full capacity. To do this, you must have your life in order. One analogy that is frequently used is seeing a person as a wheel. A wheel will not roll if it is flat on one side. Similarly, people cannot function optimally if they have a problem in their lives. Five key areas in life are graphically represented here in Figure 8.1. Maintaining and developing each area will make you effective as you can be.

### *Intellectual*

One of the main purposes of going to school is to expand the intellectual dimension of your life. Take full advantage of the opportunities at your learning institution. Besides engineering courses, schools will require general education courses (languages, humanities, or social sciences). These non-technical classes are a great way to broaden yourself. Many students find these classes a refreshing break from their engineering classes.

In addition to course work, schools have a wide range of activities to enhance your intellectual development, e.g., student organizations, seminars, and special events. Planning some non-engineering intellectual activities will help keep you motivated and feeling fresh for your engineering classes. Some students find that

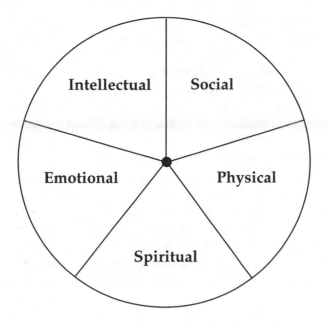

**Figure 8.1 Areas of wellness for students.**

reading a novel or two during the semester also provides a refreshing, intellectual break from engineering studies.

## Social

The school's social aspect is one that most students seem to master. Planning appropriate social activities will make your experience more fun. Enjoying school and being successful are strongly correlated. In addition, establishing a group of friends with which to socialize will provide a support structure to help you when times get tough. Friends can also provide a network that can be beneficial later in your professional life.

## Physical

All the studying and preparation is a waste of time if you are not physically able to perform when the tests or quizzes come. In addition, you can waste time if you are not able to study efficiently. Three physical areas are essential to becoming an effective student: fitness, sleep, and nutrition.

People, as well as students, are more productive when physically fit. Exercise will not only help you study better but will also allow you to live a longer, more productive life. Now is a great time to develop a fitness habit. Fitness activities can include aerobics, jogging, biking, walking, or participating in intramural sports. Every campus has people who can help you develop a fitness plan that is right for you.

Many students neglect sleep. The problem is that classes still come at the same time regardless of when you fall asleep. Sleep deprivation reduces a person's productivity, which has a direct impact on studying. Each person needs a different amount of sleep. Typically, most adults need to sleep six to eight hours per night. Schedule enough time for sleep or you will not be effective. Studies have shown that student performance drops by staying up late studying the night before a test. You cover more material the night before the test but are so tired at test time that you cannot retrieve the material. Evaluate how much sleep you need. Do you fall asleep in class? If so, you are not getting enough rest. Another test for sleep deprivation is to turn off your alarm one morning and see when you wake up. If you are on a schedule, most people will wake up at the same time without the alarm. If you sleep several hours past when the alarm would have gone off, you are possibly sleep deprived and need more rest.

The third physical area is diet. Our bodies are designed to work properly with the right input. While you do not have to become a nutritional fanatic, eating a sensible diet will enhance your ability to succeed in your academics.

The number one problem on most high school and college campuses is alcohol abuse. Every year, thousands of capable students fail due to alcohol abuse. If you choose to indulge, make honest assessments of the impact on your studying. Do not let excess alcohol consumption become the barrier to your success.

## Spiritual

Most would agree that humans are spiritual beings and function much better and more efficiently with a healthy spiritual component in their lives. Stephen Covey, in his book *The Seven Habits of Highly Effective People,* puts it this way:

> *The spiritual dimension is your core, your center, your commitment to your value system. It's a very private area of life and a supremely important one. It draws upon the sources that inspire and uplift you and tie you to the timeless truths of all humanity. And people do it very, very differently.*
>
> *I find renewal in daily prayerful meditation on the Scriptures because they represent my value system. As I read and meditate, I feel renewed, strengthened, centered and recommitted to serve.*

Search for the spiritual activities that keep you renewed. This may be through a small group study, attending worship services, meditating, or immersing yourself in music for a time of personal reflection. If you are not sure where to start, try something with which you are familiar. This will probably be the most comfortable for you. If it does not seem to be the right thing for you, branch out from there.

## Emotional

Bill Hybels, pastor of a suburban Chicago church, explained the need to attend to this dimension when he addressed a gathering in Detroit in 1996. He is an avid runner, eats well, and is physically sound. He has many casual friends and is socially active. He has written a number of books and articles and is sound intellectually. Yet, he reached an emotional crisis that prevented him from being effective in his job and with his family.

Hybels describes emotional reserves like a gas tank. We need to put reserves into our tank so we have the ability to take it out when we need it. There will be times in college and beyond when you need those reserves. The lesson Hybels shared is the need to watch the gas gauge and not to let the tank run empty. Students are not exempt from needing to pay attention to their emotional reserves.

Keep the emotional dimension in check by creating a support network. This network is what you use to help fill your emotional gas tank. Friends are an integral part of the network. Do you have a person or people to whom you can talk honestly about important issues in your life? This might be a close friend, a parent, a relative, an advisor, a teacher, or a counselor. Cultivate at least one friendship where you feel the freedom to share honestly.

The second aspect to maintaining emotional reserves is to schedule time that is emotionally neutral or energizing. Many students get themselves into many emotionally draining activities and do not schedule time for themselves. Eventually, this catches up with them and reduces their effectiveness.

Find advisors and counselors who will talk with you. Do not feel you have to have a severe psychological problem to simply talk with them. These people are trained to help you and will just listen if that is what you need. They are a great resource for maintaining your emotional dimension, thereby helping to keep your academic effectiveness high.

## Your Effective Use of Time

You may be wondering how to find time to maintain all these areas, e.g., studying for class, keeping a balanced lifestyle. What happened to all that fun time you have heard about?

The answer is effective time management. Good time management will help you succeed as a student and will serve you well throughout your entire career. Mastering time management will set you ahead of your peers. When engineers and managers are surveyed and asked why they do not take the time to manage their time, the most common response is inevitably that they do not have time. Students have the same feelings.

The key to making a schedule work is to make every minute count. Treat your education like a full-time job and work similar hours. This will not only help you manage your life during your academic years but will also provide excellent preparation for your life.

Making your day effective requires planning because you need to find suitable study times. You can use these times to do your individual studying or meet with a study partner or group. Deciding in advance what you will study and when is more effective. Deciding which subjects you will study and when is a waste of precious time and energy. Plan enough time for each subject so you can study between class sessions and not fall behind.

Another thing that scares students away from effective time management is that it sounds too rigid. One of the reasons for going to school is to have fun and meet people. How can you do these things if everything is scheduled? The answer is responsible flexibility. If you have your schedule planned out, you can decide when it is a good time to interrupt your work time to have some fun. One criterion for making this decision is recovery time. Does your schedule have enough non-work time that could be converted to work time before class deadlines?

Once you fall behind, catching up is hard. To avoid this, a weekly make-up time is an excellent idea. This is like an overtime period at work. Many professionals choose Saturday from 6 A.M. to 10 A.M. as a time to schedule this overtime work. This typically does not interfere with anything else in a normal schedule, still allowing a most of a full Saturday to enjoy. For some reason, 6 A.M. on Saturday is not a popular study time for students. However, an evening or two or a Saturday afternoon will also work. These extra time blocks will allow you to keep up, give you the flexibility to take time off with a friend for something fun but still maintain a disciplined weekly schedule.

## *Scheduling Work and Other Activities*

No matter how good you get at squeezing in study time, there are still only 24 hours in each day. As a result, there is a limit to what a person can do during any given period of time. Planning a schedule complete with classes, study time (including make-up time), eating, sleeping, and other activities is essential to deciding how much is too much. If you need to work while you are in school, schedule it into your week. If you join a student organization, plan those hours into your schedule as well. You need to see where it all fits if it does. Students commonly over-commit themselves to classes, work, and activities. Something has to give. It is easier to drop an activity than repeat a class.

## *Calendars*

It is important to have a daily and weekly calendar. Another critical item in effective time management is a long-range schedule. For students, this is a semester calendar. This will help you organize for the crunch times, like when all five of your classes have midterms within two days, or you have three semester-projects due on the same day at the end of the semester. Tests and projects tend to be bunched together, something you cannot control, because there are educationally sound times to schedule these things and most professors follow the same model. Given this fact, plan your work schedule, which you can control, so you spread it out in a manageable way.

## *Organizers: Paper or Electronic?*

Calendars and task lists can be kept in numerous forms, and you will need to determine which is most effective for you. There has been a proliferation of electronic tools, software, and gadgets designed to supplement or replace paper calendars and organizers. Among the most common electronic options are Personal Digital Assistants (PDAs), which can function like a personal organizer as well as a cellular phone, fax sender, and Internet portal. PDAs began with pen-based input but now incorporate handwriting recognition features as well as keyboard inputs. PDAs are also called palmtops, hand-held computers, and pocket computers. Most of the traditional time management companies, such as Franklin Covey, have created an electronic version of their products to be used with a PDA. Many students now own mobile phones. A great number of these products have date books built right into them. Software like Microsoft Outlook or ACT can be linked with a mobile phone date book and contact list though a manufacturer's software package and cable. Alarms can be set to remind students of appointments and due dates. If you use a PDA or mobile phone to organize, use it properly. Make sure any notification alarm or ringer is not auditory but rather set to vibrate. This can avoid embarrassing moments in the middle of a class when an alarm could interrupt the instructor or class.

The media you use to organize your schedule is your choice. For any system to be effective, it must fit your own needs and style. Our recommendation is that you explore the options and keep your decisions flexible.

## Organizing Tasks

A common method for organizing one's day is to make a To Do list each morning (or the evening before). This is a good way to ensure that everything needing attention gets it. Things not done on one day's list are transferred over to the next.

The one drawback to making a list is determining what to do first. You can approach the tasks in the order you write them down, but this ignores each task's relative importance. For instance, if you have a project due the following day for a class and it is listed last, will it be done? A popular planning tool is the Franklin Planner produced by the Franklin Covey Co. (Material here used courtesy of Franklin Covey.) They stress the importance of prioritizing your life's activities according to your values, personal mission statement, and goals. They suggest using a two-tiered approach to prioritizing activities. The first is to use A, B, and C to categorize items as follows:

A = Vital (needs attention today)
B = Important (should be taken care of soon)
C = Optional (no one will notice if it is not done today)

Keep your long-term and intermediate goals in mind when assigning a priority level to an item. Moreover, be sure to write simple statements for each task on your list, beginning each item with an action verb.

Once you have categorized the items, look at all the A priorities. Select the item to do first and give it a 1, so the item becomes A1. The next becomes A2, etc. Now you can attack the items on your list sequentially beginning with A1. Once finished with the A items, move on to B1 and work through the B items. Often, you will not get to the C priorities but you have already decided that it is okay if they are not done that day.

Table 8.1 shows a sample list for a typical student's day. If you are starting a time management routine, be sure to include making your list. This keeps you reminded of it and gives you something to check off right away!

You can learn many lessons by looking at how you schedule activities. One is that paying the phone bill is not necessarily a top priority, but it requires a little time to address. A helpful approach is to do the A items quickly and first. Then move on to the more time-consuming items. For example, you may want to plan a work-

## TABLE 8.1. Sample Prioritized List of Tasks

| Priority | Task |
|---|---|
| A3 | Finish chemistry lab (due tomorrow) |
| A5 | Read Math assignment |
| B3 | Do math homework problems |
| B4 | Outline English paper |
| A4 | Workout at Intramural Building |
| B2 | Debug computer program |
| A6 | Annotate Math notes |
| B1 | Annotate chemistry notes |
| A7 | Spiritual meditation |
| C1 | Call Stacy |
| C2 | Write parents |
| A2 | Mail phone bill |
| C3 | Shop for CD's |
| A1 | Plan Day |

out into the schedule, right after the chemistry lab. The workout can help motivate you to finish a difficult task to move to something more enjoyable.

Looking at this schedule, you may think that poor Stacy is never going to get called or the letter to the parents written. It often happens that the more fun things are, the more likely they end up as a C priority. However, it is okay to make relationship-building or fun activities a high priority. People need to stay balanced. Maybe today, Stacy is a C, but tomorrow that call might become an A priority.

Franklin Covey provides additional help for identifying priorities. It rates tasks by two criteria, urgency and importance, as shown in Table 8.2. Franklin Covey stresses that most people spend their time in Quadrant I and III. To achieve your

## TABLE 8.2. Franklin Covey's "Time Matrix" for Daily Tasks

|  | Urgent | Not Urgent |
|---|---|---|
| **Important** | I<br>• Crises<br>• Pressing deadlines<br>• Deadline-driven projects, meetings, preparations | • Preparation<br>• Prevention<br>• Values clarification<br>• Planning<br>• Relationship building<br>• Recreation<br>• Empowerment  II |
| **Not Important** | • Interruptions, some phone calls<br>• Some mail, some reports<br>• Some meetings<br>• Many pressing matters<br>• Many popular activities  III | • Trivia, busywork<br>• Some phone calls<br>• Time-wasters<br>• 'Escape' activities<br>• Irrelevant mail<br>• Excessive TV  IV |

(© 1999 Franklin Covey Co.)

long-term goals, Quadrant II is the most important. By spending time on activities that are in Quadrant II, you increase your capacity to accomplish what matters most to you. Urgency is usually imposed by external forces, whereas you assign importance to something.

You may well want to use a schedule that shows a whole week at a time and has your goals and priorities listed. A general example is shown in Table 8.3.

A weekly schedule allows you to see where each day fits into the bigger scheme. It also keeps your priorities in full view as you plan. Franklin Covey offers many versions of its Planner, including the Collegiate Edition. Contact them at 1-800-654-1776 or visit their website at www.franklincovey.com.

## Stay Effective

There are two final notes on time management that can help you stay effective. The first is to tackle the most important activities when you are at your best. We are all born with different biological clocks. Some people wake up first thing in the morning, bright-eyed and ready to work. These people are often less effective in the late evening. Other people would be dangerous if they ever had to operate heavy equipment first thing in the morning, especially before that important first cup of coffee. However, once the sun goes down, they can go on a tear and get a lot done. Know your own rhythms well enough to plan most of your studying when you are most alert. Students commonly leave studying until late in the evening,

**TABLE 8.3. Sample Generic Weekly Schedule and Planner**

| Weekly Schedule | | | Sunday | Monday | Tuesday | Wednesday | Thursday | Friday | Saturday |
|---|---|---|---|---|---|---|---|---|---|
| Musts | Goals | | Day's Priorities | | | | | | |
| Academics: | | | | Run laps | | Run laps | | Finish CS | Run laps |
| Math | Finish Prob | | | | | | | Finish Math | |
| Chem lab | Anal. data | | | | | | | Movie | |
| Comp Sci | Debug Prog | | | | | | | | |
| English | Essay | | | | | | | | |
| | | | Appointments/Commitments | | | | | | |
| | | 8:00 | | English | | English | | English | |
| | | 9:00 | Chapel Synagogue | Chem | | Chem | | Chem | |
| Hall Dance | | 10:00 | | | Chem Lab | | CS Lab | | |
| | | 11:00 | | CS | | CS | | | |
| | | 12:00 | | | | | | | |
| | | 1:00 | | | | | Chem Rec | | |
| | | 2:00 | | Calculus | Calculus | Calculus | | Calculus | |
| | | 3:00 | | Meet prof | | Engr 101 | | Engr 101 | |
| | | 4:00 | | | | | | | |
| | | 5:00 | | | | | | | |
| Physical | Run and Ride | Evening | | | | | | | |
| Mental | Read a novel | | | | | | | | |
| Spiritual | Devotions | | | | | | | | |
| Social | Hall dance | | | | | | | | |
| Emotional | Lunch w/Dave | | | | | | | | |

often because that is the only time they have. Know yourself well enough to plan your time most effectively.

Finally, pace yourself. You are entering a marathon, not a sprint. Plan your study time in manageable time blocks with built-in breaks. A general suggestion is to take a five-minute to ten-minute break for every hour spent studying. This will help keep you fresh and effective.

## Accountability

The single most important concept to being successful is to be accountable for your goals and intentions. The most effective way to do this is with an accountability partner. This is someone whom you trust and from whom you can comfortably receive honest feedback. Ideally, you will meet regularly with this person to share how things are going. Regularly means about once per week. Throughout a semester you can fall behind quickly, so weekly checks are a good idea. The person you choose can be a peer, a mentor, a parent, a relative, or anyone you respect and trust and with whom you feel comfortable.

Your accountability could also be to a group of people. Bill Hybels describes the change in his life when he entered into an accountability group with three other men. These men held each other accountable in all areas of their lives: professional, familial and spiritual. You can do the same as a student.

The first step is to give your accountability partners the standard to which you want to be held. An easy way to do this is to share your goals with them. Let them see and hear from you where you want to be in five years, in ten years, and beyond that. Let them also see the short-term goals you have set to help move you toward achieving those longer-term objectives. Then, on a weekly basis, share with them how you are doing in meeting your goals. You will find that it is much easier to stay on course with the help of at least one close, accountability partner. Success will come much more easily.

## Overcoming Challenges

At some point in our lives, we encounter adversity. As a student, adversity may take the form of a bad test or course grade or , a teacher who you feel is treating you unfairly, or any of a myriad other obstacles. Since you will encounter adversity and failure in your life, your personal success is not always dependent on how much you avoid failing, but rather on how well you handle it and learn from it. Were you able to move on and grow as a person? This is especially true as a student.

All students will run into a course or a test where they struggle. The key is to avoid focusing on the negative and continue to be productive. Every academic advisor has stories about good students who tripped up on a test or in one class and let that affect their performance in other classes.

## Focus on the Controllables

If something bad happens, figure out what, if anything, can be done to help rectify the situation. This may mean writing down a list of actions you can take. Do not focus on things you cannot change. If you receive a low test grade, you cannot change the grade. You may, however, be able to talk to the professor to get feedback. You can study the material you missed and be ready for the next one.

## Keep It in Perspective

Examine the real impact of the situation. What are the consequences of the situation? One class can affect your graduation GPA by about 0.05. This will not produce life-altering consequences. If you are unsure about the consequences, see your teacher, counselor, or advisor to help you put it into perspective.

## Move on to the Positive

If you hit a bump in the road, move beyond it quickly and remain effective in your studies. Examine the problem and learn from it. Perhaps you failed a test. This is a great opportunity to examine how you are studying and take corrective action to improve your effectiveness. You can affect your present and future but not your past.

## Remember, Everyone Fails

Again, keep things in perspective. If you run into difficulty, you are not alone. There are numerous examples of failures, e.g., Henry Ford neglected to put a reverse gear in his first automobile. Nevertheless, he recovered and went on to become successful.

Abraham Lincoln, likely our greatest president, provided a tremendous example of overcoming adversity. Here is a brief list of some of his accomplishments and setbacks:

1831—failed in business
1832—defeated for legislature
1833—failed in business again
1834—elected to legislature
1835—fiancé died
1838—defeated for Speaker
1840—defeated for Elector
1843—defeated for Congress
1846—elected to Congress
1848—defeated for Congress
1855—defeated for Senate
1856—defeated for Vice-President
1858—defeated for Senate
1860—elected President

Lincoln never stopped achieving just because he ran into a setback. You do not have to either.

## LEARNING ACTIVITIES

### Activity 8.1    Rating Consumer Products

Time:    2–5 hours

Everyday we go through an evaluation process when deciding which clothes to buy, which fast food to eat, or which computer we are going to purchase. People evaluate products differently. For instance, people choose differing over-the-counter medicines for a headache. Why does one family buy a name brand and another buy a cheaper, store brand with the same ingredients? When choosing pain relief medicine, did one family see the major brand marketed more on television or billboards and choose it because they had seen it before? Maybe the family buying the off brand has seen the advertising and actually compared the two brands, discovering the ingredients were the same.

In this assignment, students will develop rating scales and specific criteria, and then validate information provided for specific products. Students should review the list of products provided by the teacher to choose which features and companies to evaluate. The entire class should develop a rating scale, as a group, so there is a consistent standard for evaluating different products.

## REFERENCES

THE CENTER FOR APPLIED RESEARCH IN EDUCATION (1996) Hands-on math projects with real-life applications: Rating consumer products, pp. 126–130.

http://www.consumerreports.org/index.html Consumer Reports website

http://www.consumersunion.org/ This site provides informative and educational materials developed by Consumers Union's advocacy offices on various consumer issues.

http://www.cpsc.gov/indexmain.html CPSC, an independent Federal regulatory agency, helps keep American families safe by reducing the risk of injury or death from consumer products.

### Activity 8.2    Planning and Succeeding

Time:    2–4 hours

1. Writing goals as described is an important element in defining your goals.
   a. Write five goals to achieve by graduation.
   b. Write five goals to reach within ten years.
   c. Write ten goals to reach during your lifetime.

2. Write some specific goals:
   a. Set a grade goal for each class this semester.
   b. Calculate your expected semester GPA.
   c. Define a stretch goal for your semester GPA.

3. Identify a particular employer, and an entry-level position with that employer, which you would like to obtain following graduation. Find out the requirements for that position.

4. List the barriers you see to achieving your graduation goals.
   a. Categorize them as either those within your control or those beyond your control.
   b. Develop a plan to address those within your control.
   c. Develop strategies to cope with those beyond your control.

5. Identify one or more people with whom you might form a study group. Study together on a regular basis until the next test. Report on your experience.

6. Identify the type of environment in which you study the best. Describe it and include a sketch of the location.

7. Test your conclusion in item six with an experiment. Do one assignment in one environment (perhaps in your room, alone and totally quiet) and another in a different environment (perhaps in a more public place, with background noise). Report your findings.

8. At the halfway point in the semester, calculate your grades in each class. Compare these to the goals set in item four. Are you on track? If not, identify the changes you will make to reach these goals.

9. Define at least one long-term goal in each of the five wellness areas discussed in this chapter. Define short-term goals needed to help achieve these long-term goals.

10. List at least two needs you have in each of the five wellness areas. List at least two activities you can plan to help meet these needs.

11. In your weekly schedule, identify the times of each day when you study most efficiently. Are these the times you are actually scheduled to study? If not, explain why you have scheduled other items instead of studying at these times.

12. Make a To Do list. Prioritize the list using A, B, and C categories, and number the items in each category.

13. Assume your grades will be lower than expected at the end of the semester, and write out a strategy for recovering from these low grades.

14. List your long-term goals, including every area of your life. Fill out a schedule of what you actually did during the last seven days. Put each activity into categories corresponding to one of your long-term goals or an extra "not related" category. Grade yourself (using A, B, C, D, or F), indicating how well you did the past week working toward your long-term goals.

15. Explore the availability of electronic organizers and prepare a brief report on the options for students.

16. Compare and contrast electronic organizers versus paper ones. What are the benefits of each and which would you recommend to your classmates?

17. Interview two teachers and report what they use to organize and prioritize their time.

## Activity 8.3     Book Smart

Time:     2 hours

## CASE STUDY

The primary purpose of this case study is to develop measuring and averaging skills. Students will be required to find the average thickness of the paper used in three school textbooks. This answer will be compared to the actual paper thickness that is measured with a micrometer.

Your teacher will divide you up into teams of two and should assign you a Vernier caliper and a micrometer. Use the Vernier caliper first to total thickness of each books paper content, (measure to the nearest .001 inch). In order to find the thickness of a page, you must divide the number of pages by two; there are two sides to every page. Record each of your answers in separate columns for each book. Average your three books together to find a final average for your group. Turn in your results to your teacher or share them with the class. Calculate the class average for each books paper thickness; you should produce one final number for this step. Compare your group average to the class average. Which group was closer to the class average?

Now use a micrometer to measure the thickness of one page of each book. Multiply the thickness of the page by the number of pages in the book and divide by two. How does this number compare with the Vernier caliper average that you measured in the first part of this study? Average your group's micrometer-gauged pages together to get a final average thickness. Again, turn this in to the teacher or present the information to the class so you can make more comparisons.

Plot the resulting answers on a spreadsheet for each of the sections (Vernier measurement and micrometer measurement) and compare them to the average class measurements.

## RESOURCES

Teachers need to make sure that Vernier calipers and micrometers are available. DO NOT use dial or digital calipers since Vernier calipers force students to use math skills that they would not use otherwise. Calipers and micrometers are available from local hardware stores or national catalog supply companies.

## REFERENCES

Covey, S.R. *The Seven Habits of Highly Effective People.* New York: Simon and Schuster, 1989.

Douglass, S. and A. Janssen. *How to Get Better Grades and Have More Fun.* Bloomington, IN: Success Factors, 1997.

Maxwell, J. C. *The Winning Attitude: Your Key to Personal Success.* Nashville, TN: Thomas Nelson Publishers, 1993.

Landis, R. *Studying Engineering: A Road Map to a Rewarding Career.* Burbank, CA: Discovery Press, 1995.

Tobias, C.U. *The Way They Learn, How to Discover and Teach to Your Child's Strengths.* Colorado Springs, CO: Focus on the Family, 1994.

Lumsdaine, E. and M. Lumsdaine. *Creative Problem Solving: Thinking Skills for a Changing World,* 2nd ed. New York: McGraw-Hill, 1993.

Herrmann, N. *The Creative Brain.* Lake Lure, NC: Brain Books, 1988.

*Chapter 9*

# What Do Employers Expect?

"What are employers actually looking for?" is often a question that only people looking for jobs tend to think worth asking. Some common traits appeared after careful scrutiny of many job postings and employer surveys. This section will define some of the "soft" or "interpersonal skills" and important, non-technical characteristics desired by employers. These soft skills should be common sense and, at first glance, might appear trivial but do not be fooled. You need to pay close attention to these skills because if you neglect them it may be at the peril of your own job. One of the hardest parts of getting a new job is retaining the skills that helped initially obtain the position. Professionals will make every effort to keep up their skills through professional development, i.e., courses, seminars and many other venues that provide updates to their skills collection. A person would have a hard time learning these interpersonal or soft skills out of a book. Take time to practice these things within your academic, personal, and professional relationships and in any other appropriate setting.

The Wisconsin Governor's Work-Based Learning Board has defined several **skills** needed for success in the workplace:
- Communicates clearly with supervisor and others
- Acts professionally
- Learns effectively
- Manages self responsibly
- Plans for change (flexibility and adaptability)
- Plans for personal and professional growth
- Works productively
- Recognizes safe and unsafe work habits
- Demonstrates proper safety procedures
- Maintains a safe and healthy work environment

One of the most significant traits not listed above and not a large part of the research that appears below, punctuality, is one of the most critical and overlooked workplace issues. It is your responsibility to be at work on time and to have the materials needed to get the job done.

A successful engineering student will develop the ability to **network and communicate** with others. It has been said that more work can be done in eighteen holes of golf than a week of meetings. The point here is that people need to have the ability to talk to each other no matter what the setting. Competitors are doing everything they can to get business and if a personal relationship is established, it can bolster a company's competitive position. Argumentation, questioning techniques, negotiation methods, rhetorical abilities, discussion guidance, body language, presentation ability and strategies, and the ability to communicate with people from different cultures are all part of being a successful engineer.

Surprising as it may be, an engineer must posses the ability to think **creatively,** even illogically. You must entertain all conceivable ideas, even if they seem impossible. An overused cliché is that a person must be able to think "outside the box." What if that box is actually a sphere? Engineers must determine the best way to solve each unique problem. Here is a question for evaluation: seasoned engineers and fourth-graders sit down at a table and are given the same problem to solve. Which will come up with the more creative solution? The fourth-graders are more likely to because they have not been told they cannot do things yet. Seasoned engineers, having gone through schooling and formal training, have been told what they cannot do, and so have limited imaginations.

Like many professions, engineers must have a **positive attitude,** a love for the job, and a professional presence about them that allows others to work along side them. Frequently, people get complacent with their jobs and get into a rut emotionally. Get excited about projects, and get involved with the intricacies of what is at hand. A willingness to learn is always a positive trait to have.

Be prepared to ask, "What happens if _____?" or "Can we do this?" Take **educated risks** that make at least some sense, not risks that are long shots. If you take the long shot or educated risk, be ready to accept the consequences after failure. The experiences of the project team will bring these evaluations together. One person should not bear the brunt of the decisions or risks. Self-reflection is often involved before, during, and after the decision-making process. Being critical of a decision is a good way to analyze personal development as long as the analysis does not go to the extreme where every thing is critiqued and questioned.

Many firms demand that their employees have a firm understanding of business and finance coupled with an ability to explain what they are doing throughout a project to non-technical people. Profit motivates many employers and if the bottom line cannot be explained, they are less likely to pursue a project.

Life-long **learning** is necessary for all who excel in their own field; and engineering is no different. As the standards, practices, wants, and needs of a culture evolve, so must the engineers and the company. Some companies employ the practice of "that's how we've always done it." While this may be fine in certain arenas, certainly it is not the way to long-term success in engineering.

## Working with Others

In order to curtail possible conflicts between group members, one has to realize members may disagree. Everyone has different backgrounds, values, training and other variables that give the team significant advantages versus a team that is assembled of similar people. Conflict resolution can help navigate those tough times when group members have issues with each other. One must always consider the interests and suggestions of other people while working on a team. A team member needs to be able to step back from the situation and evaluate it positively. Many employers cite this trait as one of the most important they are looking for in an employee, particularly when that employee will be working as part of a team of people.

Employers want employees who can **cooperate** with others. People that work with each other do not necessarily have to be best friends, but close friendships sometimes develop. One of the key elements in employment that parents, teachers, or self-help books bring up to anyone seeking a career is to make sure you love your job. If people have more friendships at work, they will enjoy the time they spend there. This creates a positive environment that only increases productivity. Co-workers that do not get along with each other for personal reasons still have to figure out a way to get things done in the professional arena no matter how strong their feelings.

Employees must be able to **take criticism** or a critical comment or judgment. Employees do not have to like what they hear, but they must be receptive to other people's comments, evaluate the comments, and decide what to do next. Positive criticism serves to motivate and give appreciative feedback to an employee or co-worker.

Team members should possess the ability to sense other team members' tendencies. This does not mean mind reading, but a person should notice nonverbal signals and body language. Posture says a lot about how interested a person is with a given situation. If the shoulders are loose and forward in a hanging position and the person is leaning to one side it may appear that he or she is not interested and not ready to answer a question or engage in a conversation. Conversely, if that same person stands with head attentive, shoulders in a firm position and arms at the side rather than crossed, this means he or she is ready for a conversation.

Be **attentive** to others. Another good way to show that you are interested in what people say is to lean forward when they are speaking. We do this almost instinctively when we are interested, and may avoid it when we are disinterested. Eye contact is also a good indicator if someone is interested in a conversation. If people are not looking at the speaker, they are probably not interested in the conversation. If you have problems looking directly into people's eyes, pick a point on their forehead, not far above the eyes, and concentrate on that spot. Because it is so close to the speaker's eye, they will not realize you are not looking directly at them. This makes the listener more comfortable and conveys a sense of interest. So, be positive, lean forward, and look almost into the speaker's eyes. Something as simple as a smile can also take the edge off a conversation by comforting a listener or speaker. The smile, however, will send mixed signals if occurring at the same time as negative comments or thoughts.

Webster's defines **sarcasm** as a keen, reproachable expression; a satirical remark uttered with some degree of scorn or contempt. Sarcasm should never be used in a business or professional setting where people do not know each other. Even if it is just one new person in the group discussion, be professional. Sarcasm may have a place when two people are speaking who know each other well, and

wit can be a person's defining personality. However, sarcasm can also send unintended messages, so sarcasm should be used with caution.

Mental **flexibility** and openness are particularly important when a business
deals with people of different cultures or countries. Every culture has its own idea
of what things are acceptable in business situations. Each culture has its unique
values and protocols that dictate how to conduct business. A person must adapt
and improvise if necessary, which includes respecting a culture and its customs.
Whether or not a business is interacting with international enterprises or local
ones, be open to new ideas and thoughts. This returns to the philosophy of thinking illogically. If people shut off an unfamiliar or unpracticed idea, they may be
shutting off an important opportunity.

Analytical thinking is the ability to comprehend a situation by breaking it into its
component parts and identifying the underlying or complex issues. This process
systematically organizes and compares a problem's aspects and determines relationships to solve the problem in a sound, critical manner.

Written and oral **communication** skills are critical. A critical skill is the ability
to communicate ideas in written and oral form. Most employers require project
documentation and reports presented to others. This may take the form of an e-
mail to a supervisor, a technical memo, a presentation to engineers and technicians, or even a formal presentation to directors or executives. Weekly activity reports may also be part of a job. When writing these reports, memos, and
informational descriptions, be sure to use correct grammar and complete sentences and check all spelling. Spell Check on a computer may highlight spelling
mistakes, but it will not correct a misspelled word. When speaking in front of
people, be careful not to use comfort words or phrases, e.g., you know, uh, and
like.

**Networking** describes the ability to find opportunities, develop new contacts,
and work with people from many backgrounds. This is a great asset in reaching
short-term and long-term career objectives. Meet informally and formally with
many individuals, who are able to provide advice, lend support with projects, and
help get things done more efficiently. These same people can recommend new
assignments and write important letters of recommendation.

You will need a lot of **self-discipline** during the transition into college and the
workplace because you will need to take your organizational styles to a higher
level. You will be given a great amount of freedom and responsibility to succeed
(or fail). You must learn time management, punctuality, preparation, organization,
and specific protocols and procedures. As you acquire and develop these skills,
you will be able to apply them to a variety of situations, which will make you more
productive and successful.

Employers look for people who can interact with many people and groups,
e.g., engineers, technicians, scientists, production workers, labor union representatives, clerical staff, managers, directors, and CEOs. Each will help the employees learn as they experience the successes, challenges, frustrations, and
friendships that evolve from these interactions. Employers also look for people
that are involved in activities outside work. Do not hesitate to mention hobbies,

community involvement, church activities, and volunteer positions on a résumé or during an interview.

## Experience

**Supervisory** and management opportunities that require an employee to take responsibility for other individuals or groups can be key elements in successfully completing projects. These opportunities to manage the work of others can provide invaluable experience that can lead to promotions and or leadership positions within a company.

How do you get **experience** without a job, and how do you get a job without experience? This is the job seeker's first question. The National Commission used it in a nationwide publicity campaign for Cooperative Education. All engineering students have considered or will consider this question in their career. Career-related experience, commonly called experiential learning, has become an important supplement to a student's engineering education. Through the years and changing job markets, numerous graduate schools and their graduates' employers have continued to stress that they seek to fill their annual openings by selecting only those candidates who have done outstanding things to distinguish themselves. This also applies to high school students looking to get into college or take positions in business after graduation. In most cases, those students who have participated in some form of experiential learning have gained the types of skills and competencies sought by prospective employers and schools.

The educational experience has evolved considerably over the past century. For most college engineering students during the first half of the 20th century, college was completed in four years. An engineering education has always been a rigorous experience, thus most students chose to devote their time and efforts to experiences in the classroom, library, or laboratory. The conventional student population viewed those students who chose to **work while going to school** as non-traditional and working class people forced to work for pay as a means of affording their schooling. However, in the latter part of the 20th century, the costs of a college education soared. The rule, rather than the exception, is that most students are forced to work to pay for steadily increasing tuition, room and board, and ancillary expenses such as books, computers, and transportation. (Some countries fully subsidize college education with one of the benefits being that students can concentrate on their studies, with less need for extracurricular employment during school.)

For many students, the idea of combining their academic studies with a **career-related employment** experience is something new. The thought of stretching their college education to five or six years can seem an unnecessary burden. However, employers need employees with experience. Many employers expect today's employees to have significant work experience. Why? Because prospective employees with career-related work experience require less training and can produce results more quickly than the prospective employee without any.

Many forms of experiential learning are available to students. The most popular options include on-campus and off-campus jobs, summer work, volunteer experiences, academic internships, research assistantships, and perhaps the most popular and beneficial program for students, cooperative education. Each of these has advantages and disadvantages depending on a student's particular situation and background.

## MEYERS BRIGGS

The Myers Briggs model of personality is based on four preferences:

1. Where, primarily, do you direct your energy?
2. How do you prefer to process information?
3. How do you prefer to make decisions?
4. How do you prefer to organize your life?

The following information is provided and copyrighted by Team Technology, headed by Steve Myers, winner of the APT's International Award for Merit for the development of the MTR-i™ team roles, and author of "Influencing People Using Myers Briggs", the Management Team Roles—Indicator™, and the Ideal Team Profile Questionnaire™. For more information and additional publications visit the website: http://www.teamtechnology.co.uk.

As with all personality questionnaires, the results of any of these can be wrong. All questionnaires recognized by the psychological establishments have reliability and validity research which shows, how wrong, on average, they can be. The questionnaire can provide valuable information, but the real value of the Myers Briggs personality model is in deciding what type you are.

This article contains a description of each of the preferences, which you can use to help determine your Myers Briggs type. With the understanding this article provides, you will be able to read other pages that cover topics such as team development, personal growth, stress, etc. You may also like to read the Team Role descriptions (or download and print the free poster) to find out how you are directing your attention. Comparing your MBTI type (in-born preferences) with your *MTR-i™* team role (how you are behaving) can be valuable.

## 1. Where, primarily, do you direct your energy?

**To the outer world of activity, and spoken words**
**OR**
**To the inner world of thoughts and emotions**

If it is toward the outer world of activity or words, it is called Extroversion, denoted by the letter E. If it is toward the inner world of ideas, information, or thoughts, it is called Introversion, denoted by the letter I. Extro- is a prefix meaning *without* and intro- is a prefix meaning *within*.

During each day, you will spend time spontaneously doing or saying things in the external world, as well as retreating into the inner world of contemplation and thought. Even the clearest extrovert may desire, at day's end, to be left alone with his or her thoughts. Conversely, if an introvert has been working in isolation all day, he or she may feel the need to party in the evening to restore some balance.

You, like every other individual, need a particular balance of introversion and extroversion. However, laying aside special circumstances like those in the previous paragraph, extroversion or introversion will increase your energy level and the other will decrease it. The following table lists words and expressions often associated with extroversion and introversion:

| *Extroversion* | *Introversion* |
| --- | --- |
| Social | Private |
| Expressive | Quiet |
| Many | Few |
| Broad | Deep |
| Interaction | Concentration |
| Outward | Inward |
| Action before thought | Thought before action |

Which is your preference, i.e., what is your personality type? How are you behaving most of the time (i.e., what is your MTR-i™ team role)? Sometimes it can be difficult to tell. Every individual exhibits all of the above characteristics at some time or other, and one source of difficulty can be in distinguishing which behaviors are 'learned', or a response to current demands, and which reflect true preference. Distinguishing between the two is where comparing your MBTI® questionnaire and *MTR-i*™ questionnaire results can help.

## Preference, Role, or Learned Behavior

One feature differentiating E from I people is whether action or thought comes first. In situations that demand action, such as the sounding of a fire alarm, both types will act. Most people are trained to evacuate the building in an emergency or to take other appropriate action. So, the fire alarm results in most people doing something, and few people will sit and think. They will adopt an extroverted team role. But their underlying preference is still the same.

In situations that demand thought, such as solving a crossword puzzle, both types will think. Most crossword puzzles cannot be solved by taking action or by talking. Extroverts and introverts need to think first and to make some progress toward a solution. Their team roles are introverted, but their underlying preference remains the same.

Team roles, therefore, reflect how we respond. Finding a true, inner preference is more difficult because everyone adapts to some degree. However, the difference between people who have a preference for extroversion and introversion becomes more apparent when they have free choice. In these situations, the extrovert will act, and the introvert will think. However, few situations involve a free choice. Behavior (at work, for example) may be influenced by the following factors:

- the culture of the organization (some employers expect action-oriented behavior, others expect considered responses)
- your training or upbringing
- a range of environmental factors, such as whether the situation is a new or familiar one, whether receiving recognition or reward, and the effects of stress or illness. The need to restore balance may also be a factor (e.g., an extrovert may need some time alone after a busy week).

Nevertheless, your innate preferences will still influence your behavior as will those factors listed above. In a situation demanding action, an introvert may nevertheless bring a more thoughtful approach, or delay action. In a situation demanding thought, the extrovert may tend to talk the problem through, or act more quickly. The *MTR-i*™ team role you perform depends on a combination of the demands being placed on you. Isabel Briggs Myers believed your type is innate and stays the same throughout life.

Another indicator of your true preference may be the level of stress or enjoyment in a situation. Where your preferences coincide with the situation's demands, you may find it quite enjoyable. An extrovert may find the work frustrating or stressful if required to work in an introverted style, but find the work enjoyable or energizing if required to work in an extroverted style.

NOTE: In the sentence above about a demanding situation, will "you" find the situation or the demands more enjoyable?

## 2. How do you prefer to process information?

**In the form of known facts and familiar terms**
**OR**
**In the form of possibilities or new potential**

If it is in the form of facts or familiar terms, it is called Sensing, denoted by the letter S. If it is in the form of possibilities or new potential, it is called iNtuition, denoted by the letter N (N is used rather than I, to avoid confusion with Introversion). Sensing is used because information is taken in primarily by way of the senses. iNtuition is used because information is perceived primarily in an intuitive fashion.

Sensing tends to be interested in tangible reality, focusing on the present, and seeing what is, rather than what might be. At an extreme, Sensing can have its feet so on the ground that it misses out on possibilities for the future.

The preference for iNtuition gives a greater emphasis on insight and the future, focusing on what might be, rather than what is. At an extreme, iNtuition can focus so much on possibilities that it loses touch with reality.

Sensing tends to communicate in direct ways, and iNtuition prefers to communicate in creative ways. The following table shows words that are normally associated with each of these two preferences.

| Sensing | iNtuition |
|---|---|
| Facts | Possibilities |
| Experience | Novelty |
| Present | Future |
| Practicality | Aspiration |
| Enjoyment | Development |
| Realism | Idealism |
| Using | Changing |

## 3. How do you prefer to make decisions?

### On the basis of logic and objective considerations
### OR
### On the basis of personal values

If it is on the basis of logic and objective considerations, it is called Thinking, denoted by the letter T. If it is on the basis of personal values, it is called Feeling, denoted by the letter F. The following table lists words often associated with each of the two preferences.

| Thinking | Feeling |
|---|---|
| Analyzing | Sympathizing |
| Objective | Subjective |
| Logical | Personal |
| Criticism | Appreciation |
| Onlooker | Participant |
| Decides on principle | Decides using values |
| Long term view | Immediate view |

## 4. How do you prefer to organize your life?

### In a structured way, making decisions and knowing where you stand
### OR
### In a flexible way, discovering life as you go along

If it is in a structured way, making decisions and knowing where you stand, then it is called Judgment. If it is in a flexible way, discovering life as you go along, it is called Perception. (The reason for these terms is complicated. If you would like to know more, then read our page on the dynamic model after you have completed this page).

People whose preference is Judgment prefer, in their lifestyle, to make decisions. They prefer to make decisions about what to do, where to go, what to say, and so on. As a result of these decisions, their lifestyle appears organized. They prefer to make decisions in the world of actions and spoken words and appears organized.

People whose preference is Perception prefer, in their lifestyle, to learn or experience new things. They prefer to find out more, rather than making decisions, and are more comfortable when they keep their options open. As a result of this openness, they appear flexible. That is, they prefer to perceive new things in the world of actions and spoken words and appear flexible. The following table lists some words often associated with Judgment and Perception.

| *Judgment* | *Perception* |
|---|---|
| Close | Open |
| Decide | Explore |
| Structure | Meander |
| Organize | Inquire |
| Firmness | Flexibility |
| Control | Spontaneity |

## Working out your own preference

Everyone's personality reflects all aspects of the Myers Briggs model. You use Extroversion as well as Introversion, Sensing as well as iNtuition, Thinking as well as Feeling, and Judgment as well as Perception. You can perform any of the *MTR-i* team roles.

However, your type is a permanent influence in your personality, which influences your choice and which preference or team role to perform. The letters that represent your preferences are combined to produce your Myers Briggs Type, such as ENTJ. An ENTJ prefers Extroversion, iNtuition, Thinking, and Judgment. ENTJ types feel energized by having lots of things going on (E). They interpret events by seeing patterns or overviews (N). They make decisions on the basis of logic (T), and they organize life on a logical basis (J).

Look at the lists of words for each preference above, and think about your preference (not just the way you behave in, say, your work or social roles). List the letters in the four-letter form outlined above. If you are unclear about any of them, insert a question mark, e.g., IS?P.

If you have managed to put down four letters with no question marks, regard this as a provisional estimate of your type. You may find it useful to think about it again when you have finished reading this page. Take a look at the Team Roles, and work out which ones you enjoy most. Then use the table at the bottom of this page to see if you have a congruence between your team role and your preference. Getting to know your true preferences is a task that can last as little as a few hours, or several months, or even longer.

If you have included two question marks in your own type, that is all right. In some instances, you may find the following list helpful, as it suggests a likely answer to one of those question marks. These are only suggestions; you must conclude for yourself with which type you feel comfortable.

| *If your guess is:* | *then consider:* | *If your guess is:* | *then consider:* |
|---|---|---|---|
| ES?? or EN?? | ES?P or EN?P | IS?? or IN?? | IS?J or IN?J |
| ?S?P or ?N?P | ES?P or EN?P | ?S?J or ?N?J | IS?J or IN?J |
| E?T? or E?F? | E?TJ or E?FJ | I?T? or I?F? | I?TP or I?FP |
| ??TP or ??FP | I?TP or I?FP | ??TJ or ??FJ | E?TJ or E?FJ |

What is your Myers Briggs type? You have probably narrowed down your choice to a few types but perhaps have not yet settled on one. The final section on this page consists of some brief descriptions of the 16 types. Read the ones that you think apply to you, and see if you can narrow your choice. If at the end of this you are still unsure, many sources exist to help you decide, including websites that will test you free (search for Meyers Briggs test on the Internet) and a book listed on the Team Technology website at http://www.teamtechnology.co.uk.

## The Sixteen Types

This section contains a brief overview of the sixteen types that result from the Myers Briggs model. Everyone is an individual, but Myers Briggs highlights general themes or similarities between people. Reading this section may help you to consolidate your understanding of the preferences and help you identify your own personality type.

## ESTJ

ESTJs take their energy from the outside world of actions and spoken words. They prefer dealing with facts and the present and make logical decisions. Their lives are organized on a logical basis. They are practical and likely to implement trusted solutions to practical problems in a businesslike and impersonal manner. They prefer to ensure that the details have been taken care of rather than spend time considering concepts and strategies.

## INFP

INFPs take their energy from the inner world of thoughts and emotions. They prefer dealing with patterns and possibilities, especially for people, and prefer to decide on the basis of personal values. Their flexible lives follow new insights and possibilities as they arise. They are quiet and adaptable up to a point; when their values are violated the normally adaptable INFP can surprise people with their stance. They will seem interested in ideas, and they may sometimes make creative contributions. They have hidden warmth for people and a desire to see self and others grow and develop. They prefer to undertake work that has a meaningful purpose.

## ESFP

ESFPs take their energy from the outside world of actions and spoken words. They prefer dealing with facts, with which they usually take at face value. They prefer dealing with the present and with people, and probably derive much enjoyment out of friendships. Their live their flexible lives in the present, responding to things as they arise. They are impulsive and friendly, seek enjoyment out of life, and make new friends easily. They like taking part in solving urgent problems, such as firefighting or troubleshooting. They operate best in practical situations involving people.

## INTJ

INTJs take their energy from the inner world of thoughts and, maybe, emotions. They prefer dealing with patterns and possibilities for the future, and decide using impersonal analysis. Their lives are logically organized. They are strategists, identifying long-term goals and organizing life to meet them. They tend to be skeptical and critical, of themselves and others, with a keen sense of deficiencies in quality and competence. They often have a strong intellect yet are able to attend to details relevant to the strategy.

## ESFJ

ESFJs take their energy from the outer world of actions and spoken words. They prefer dealing with facts, and decide on the basis of personal values. They like dealing with people and organize life on a personal basis. They are warm, seeking to maintain harmonious relationships with colleagues and friends, who are an important part of their lives. They can find conflict and criticism difficult to handle. They have a strong sense of duty and loyalty, and are driven by a need to belong and be of service to people.

## INTP

INTPs take their energy from the inner world of thoughts and, maybe, emotions. They prefer dealing with patterns and possibilities and decide logically. Their lives are flexible, following new insights and possibilities as they arise. They are quiet and detached, and adaptable up to a point. Sometimes They may stop adapting and insist a clear principle is at stake. They are not interested in routine and will experiment or change things to see if they can be improved. They operate best when solving complex problems that require the application of intellect.

## ENFP

ENFPs take their energy from the outer world of actions and spoken words. They prefer dealing with patterns and possibilities, particularly for people, and decide on the basis of personal values. Their flexible lives follow new insights and possibilities as they arise. They are creative and insightful, often seeking new ideas that can be of benefit to people. They may neglect details and planning, but they enjoy work that involves experimentation and variety, working toward a general goal.

## ISTJ

ISTJs take their energy from the inner world of thoughts (and, maybe, emotions). They prefer dealing with facts, and making decisions after considering the various options. They organize their lives logically. They are quiet, serious and well prepared for most eventualities. They are a keen observer of life, developing a good understanding of situations, which is often not expressed. They have a strong sense of practical objectives and work efficiently to meet them.

## ESTP

ESTPs take their energy from the outer world of actions and spoken words. They prefer dealing with facts, which they usually view objectively, and they decide logically. Their flexible lives consist of a series of interesting activities. They are action-oriented problem solvers and prefer to work with practical organizational issues. They can be impulsive and like to take part in troubleshooting-type work. They sometimes neglect follow-through but work best when much going on needs organizing and solving.

## INFJ

INFJs takes their energy from the inner world of thoughts and emotions. They prefer dealing with patterns and possibilities, particularly for people, and decide using personal values. They organize their lives on a personal basis. They often have a private sense of purpose in life and work steadily to fulfill that goal. They demonstrate a quiet concern for people and are interested in helping them develop and grow. They are good at developing insight into people though their insights often remain unexpressed.

## ENFJ

ENFJs take their energy from the outer world of actions and spoken words. They prefer dealing with patterns and possibilities, particularly for people, and decide using personal values. They organize their lives on a personal basis and seek to develop and maintain stable relationships with those people they like. They are actively concerned with promoting personal growth in others. They are highly sociable, and expressive of feelings towards others, but can find conflict and criticism difficult, particularly if it might damage long-term relationships. They work best in situations involving people.

## ISTP

ISTPs take their energy from the inner world of thoughts (and, maybe, emotions). They prefer dealing with facts and decide logically. Their lives are flexible, and they demonstrate an interest in acquiring new information leading to a practical understanding of the way the world works. They are quiet and detached, and adaptable up to a point. They are often good at solving organizational problems. They are curious about how and why things work can seem impulsive, and sometimes produce surprising ideas or do something unpredictable.

## ENTJ

ENTJs take their energy from the outer world of actions and spoken words. They prefer dealing with patterns and possibilities, and decide after considering the consequences of the various courses of action. They organize their lives logically. They tend to control life and organize systems and people to meet task-oriented goals. They often take the role of executive or director and use a business-like and impersonal approach. They may appear intolerant of people who do not set high standards for themselves or do not seem to be good at what they do.

## ISFP

ISFPs take their energy from the inner world of thoughts and emotions. They prefer dealing with facts and people, and decide on the basis of personal values. They are adaptable up to a point, quiet, and friendly. They are interested in people, enjoying their company preferably on an individual basis or in small numbers. They take a caring and sensitive approach to helping others. They enjoy the present and dislike confrontation and conflict. They usually are a supportive team member.

## ENTP

ENTPs take their energy from the outer world of actions and spoken words. They prefer dealing with patterns and possibilities and decide logically. They are adaptable and focus on new ideas and interests as when they arise, particularly if they involve increasing their competence or skill. They are ingenious problem solvers, constantly test new ideas out, and can seem to enjoy a good argument. They are interested in instigating change and operate best in overcoming difficulties where the solution requires creative effort.

## ISFJ

ISFJs take their energy from the inner world of thoughts and emotions. They prefer dealing with facts and people and decide on the basis of personal values. They organize their lives on a personal basis and seek to enjoy relationships with people they like. They are quiet, serious observers of people and are conscientious and loyal. They prefer work that involves being of practical service to people. They are often concerned for and perceptive of how other people feel, and they dislike confrontation and conflict.

### *MTR-i* Team Roles and MBTI type

This table shows the relationship between *MTR*-i team roles and MBTI type. One measures preference, and the other measures your behavior. So, the results of your MBTI and *MTR-i* might differ.

| MTR-i(tm) ™ team role | MBTI® type |
|---|---|
| Coach | ESFJ/ENFJ |
| Crusader | ISFP/INFP |
| Explorer | ENTP/ENFP |
| Innovator | INTJ/INFJ |
| Sculptor | ESFP/ESTP |
| Curator | ISFJ/ISTJ |
| Conductor | ESTJ/ENTJ |
| Scientist | ISTP/INTP |

*Chapter 10*

# Systems and Optimization

A System is a means of achieving a desired result. It has input, process, output, and feedback loops. The system approach can also be considered a way of thinking. Whenever you are posed with a problem or a case study, simply break it down into smaller sub-systems to simplify the approach.

For example, an automobile has systems and sub-systems: mechanical systems (gears and pulleys); electrical systems (battery, wiring and computers in newer cars); and fluid systems. As the driver, you become the input part of the system when you feed the car information in the form of pressure on gas and brake pedals. You manipulate a wheel mounted to a column that controls direction of the vehicle. The automobile converts the input to process, obeying the driver's signals and converting it into output. Does the vehicle "want" to engage its four-wheel drive? Will the brakes "choose" to function correctly when the driver "miscalculates" and has to slam on them only ten feet from a tree or another vehicle? If one of these systems fails, the driver certainly wants to have feedback. In a vehicle, the gauges and displays give one form of feedback—from the fuel gauge to the large warning lights that tell you simply that something is wrong with your engine. We do not think of these "feedbacks" consciously until a crisis arises, like looking down at your speedometer after passing a police vehicle.

## INPUT

Inputs can vary but fall into several categories. They can be in many forms; however, we will discuss the top three. These are people, data, and time.

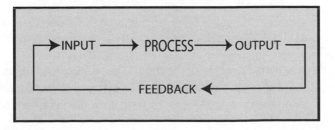

People have the ability to control their own environments and do so through input. They can design products and systems in order to safeguard a want or need of a single person or millions of people. People can make decisions without having data or information, data. These decisions are not always the best for the desired goal. Through ethical practices and sound decision making, people are always important in the systems process.

Data are everywhere in our lives, from the statement that customers receive about their bank account to the box scores in the Sunday paper, data are all around us. Data are the most critical inputs that an engineer or designer can have to attack a problem. For example, if you were to predict how many cars traveled on a road in a week's time, you would need to collect data. Civil engineers place counter boxes that have hoses attached to them that count the number of times vehicles passing over them every day. These boxes, placed in strategic places along travel routes, get the data that the engineers needs. These data can then be transferred and organized to tell a story. The story might allude to facts that might otherwise not be known. For instance, there may be a heightened period of travel during the middle of the day that could account for the increased number of accidents that otherwise could not be explained. The engineers may have to do some deeper investigation to find out why; however, with the data they at least know they have to look further. Was a major business convention in town that week? Did the local school only have half days? These are all questions that can be asked based on the data in order to obtain additional information needed to solve the problem.

Time could be a characteristic that fits well within information; however it is significant enough to be put into its own category. Time is always a factor in the process. If an engineers know that time is limited, they may make decisions that limits what can and cannot be done. Time can be an enemy or a friend. For example, if engineers knew they had three months in the summer to build a new roadway they could plan out what needed to be done each day and each hour. Because of weather, scheduling, and unforeseen setbacks, time becomes a major factor in what gets done on that project. Conversely, if the weather cooperated all summer, all the materials and machinery was available, and no major setbacks occurred, the engineer could actually finish the project early. In some contracts, bonuses are available for finishing a project early. These usually are a defined sum per day, week. or month ahead of a scheduled finish date. There is a precaution though. If safety, materials, and procedures are sacrificed in order to save time or finish early, the engineer and his or her team may create a final design that is flawed or wrong with immediate, future, or unforeseen problems.

## PROCESSES

All systems have actions associated with them. These steps can be completed in different ways and make up the processes in the systems loop. Many different processes can take place in this block of the loop, including problem solving, and production.

The philosophy associated with producing the world's goods and services has changed dramatically in recent years. As multinational corporations struggle to minimize costs while maximizing efficiency, the location of manufacturing facilities in foreign lands has increased significantly. Many firms have found distributing products from a facility close to their customer base more efficient and less costly. Others have found it more cost-effective to produce their products in places with cheaper labor, higher quality, and more favorable government policies. Therefore, in the realm of global production, firms can commonly have dozens of production facilities in different countries.

Typically, much of today's research, design, and development is done in central locations though actual production may occur in different parts of the world. Engineers may find that their work involves communication and input from colleagues around the globe. In addition to international telephone and e-mail communication, production problems, quality control issues, and meaningful design changes may necessitate frequent international travel.

The typical foreign-based manufacturing facility is usually staffed by foreign national employees. Almost all of the production staff and skilled and unskilled labor are hired from the particular region. The corporate staff is usually involved in the initial facility start-up, the employee hiring and training, and the monitoring of the manufacturing process. Often, corporations use local engineering talent for maximum efficiency, so many firms recruit international students attending U.S. colleges and universities to fill these roles. The overall objective is to provide the technical expertise to the local labor force so it will be able to run the operation. The number of actual U.S. employees permanently assigned to a particular facility will probably be small. Their role will be to manage the operation, deal with immediate issues, and serve as the liaison between the home office and the production facility.

A similar situation exists for foreign firms that have established manufacturing facilities in the United States. Most of the on-site corporate staff will be from the home country, but the objective is to hire and train as many U.S. employees as possible. Foreign firms will invest time and money to train these individuals to adapt to their particular business style. For many employees at these facilities, gaining a technical knowledge of the engineering process and a proper cultural perspective can be critical. Certain methods or procedures may actually arise from a particular cultural background, belief system, or age-old methodology. This can be quite an adjustment for U.S. workers hired by these firms. Often, newly trained technical employees may be recruited from American colleges and universities to take advantage of their solid technical education and a broader understanding of cultural issues.

Many of the ideas and concepts which have revolutionized the manufacturing industry have been a result of the globalization of world production processes. Engineers and managers are adopting some of the best manufacturing technologies developed in different countries. Today's world of manufacturing is often a blend of the best practices and procedures gained from the integration of systems and concepts from engineers around the world. Manufacturing is truly a global enterprise in the life of today's engineer.

There are three major parts of production. They are routing, scheduling, and controlling.

*Routing*—Production planners or industrial engineers have to determine the number of parts to be built and when those parts or products will be ready for consumers. These planners are responsible for routing the product through the facility. They determine the exact route that an item will travel though a plant when in production. This layout is done in advance of any parts being made and can resemble a maze of lines in a diagram that explains this in detail. For example, if you were to make a pizza in your kitchen this is the process that would take place:

1. Locate resources for the pizza, usually from the grocery store.
2. Arrange financing for the product.
3. Arrange for the transportation of the product to storage on site. This would involve driving to and from the grocery store.
4. Store the product until ready for use (usually in freezer).
5. Prepare machine for heating. (Turn on the stove.)
6. Remove product from storage.
7. Remove packaging from product.
8. Wait for machine to heat to temperature.
9. Place product in machine to be heated.
10. Wait 15 minutes.
11. Locate tools to remove product from machine heating it (usually hot pads).
12. Remove product from machine.
13. Stage product in a safe and sanitary place for cooling.
14. Locate tools to cut product into pieces.
15. Locate sanitary items needed to serve product.
16. Bring serving items and staged cut product to same area.
17. Separate product onto serving items.
18. Distribute product.

This is every step usually taken when making a frozen pizza. It is explained as if it were a procedure in a manufacturing or production plant, but it is an effective way to describe how complicated a simple process can be if done on a large scale.

*Scheduling*—Materials, workers facilities, and machinery are items that require planning in the scheduling process. In order to determine how many of a product will be made, production planners do several things. They forecast an estimated demand for the product; this may come in the form of a detailed market analysis or an evaluation of a competitive product. Customer orders also determine the minimum number of products to be made. Production planners might also have to allocate parts or products to be used by the business itself. For instance, if a company manufactured trash bags, it would obviously also need trash bags in the everyday course of its business. It would have to determine how many bags to add to manufacturing to supply its business with the bags.

Lead time is another thing that a production planner has to consider in scheduling. Lead time is the amount of time needed to get a particular part made, deliver a new machine to the plant, or have materials shipped to the production facility. In some startup businesses, lead time may also be an issue with personnel hiring.

Just in time (JIT) manufacturing is a system that reduces the inventory of raw materials, parts, or product components to the minimum. These materials are scheduled to arrive at the production facility as they are needed for use.

*Controlling*—Product level data reports are needed during production to assure a company that schedules from the plant are on time and that departments and workers are completing tasks on time. Changes in the way products are made usually come in small forms from an evaluation of the product level data. These reports not only catalog vital information for the current product, but are useful in the future development of other products.

Quality control is vital in the production of any consumer goods. Quality control involves inspecting materials, work products, and labor before and during production.

Materials have to be inspected when they arrive at the production facility. A quality product starts with quality materials. This doesn't necessarily mean high price. Quality materials can be defined as materials that meet the specification of a company for use in production. These materials may have to be tested for certain characteristics needed for the product. If the testing shows that the materials are not what are specified, the shipment of materials may be rejected. This may be as simple as a Rockwell C hardness test, or as complicated as a scanning electron microscope (SEM) test or a tensile test.

Work in progress checks are made frequently during production. When a material is processed, it has to be checked as it goes through the production process. For example, when first stamping the hoods for the Mini Cooper vehicle, plant personnel realized that the formed sheet metal was tearing where the hood was contoured. Engineers' initial discussion suggested the counters had to be removed so the tears could be eliminated. The designers of the vehicle and the contours in its hood felt this was unacceptable and that an alternative solution maintaining the hoods contours had to be found. After careful planning, the engineers provided specifications for the sheet metal to be thicker where the contours were being stamped to insure that the hoods would not tear. In the end, the solution worked and satisfied not only the engineers but the designers as well. Quality control was an important part of this process. If the tears had not been realized before the cars left the plants, it could have required an extensive, expensive recall of the vehicles to replace the hoods.

Work in progress checks produce three ratings on the product status. Rejects are parts that do not meet standards or cannot be repaired. Reworks are items that fail to meet standards but can be repaired. Accepted parts meet company standards for that point in the process. Final products have to meet these three company standards initially set forth in the design process. If these standards have certain minimums (e.g., must support at least 10 pounds), then the final products will have to meet the minimum standards to be allowed to go to the consumer. These minimum standards may be checked using machines, jigs, and fixtures, or even the human eye for visual checks. If a product has to meet a minimum standard and is to be tested for that standard in its final form, it has to be destroyed. This process is called destructive testing. It may be that the part is put though multiple uses or split in half to check its internal features for flaws. These destructively tested parts are not sent to the customer but are discarded or documented as a part of a group of parts that have passed a test. Many times inspectors will label a part "inspected by" with a stamp or a sticker to assure the customer that it has been through the process.

## OUTPUT

An output can be a myriad of things. It might be goods or services, products, structures, messages, or even people moved from one place to another. These outputs are direct results of or anticipated outcomes from the design. If a company wanted to provide homes for families in an undeveloped area of a state, it would go through many different processes in order to complete that goal. In doing so, it would produce a final product and unwanted outputs such as scrap wood, metal, and other wastes associated with its product. Pollution is another large unwanted output. Pollution can come in many forms. Chemical by-products are a large problem for companies that make film for cameras. They have to spend millions of dollars each year to dispose of these chemicals properly. Noise pollution is generated by construction companies building new homes for people. While they may not have to pay because they pollute neighborhoods with noise, they must follow city rules in regard to the hours they may operate their equipment and work on a site. Air and water pollution are significant problems for our country. Many companies that produce unwanted outputs must also spend money on systems to clean the air and the water they will put back into the environment. (For more information refer to Chapter 13, Technology/Society Interaction and Ethics.)

## FEEDBACK

Feedback generates data that allow the user, consumer product, system, or company to make informed decisions about what comes next. This could be used by a system to regulate itself at some set time interval. For example, if the temperature within the cooling system of a passenger vehicle reaches a certain temperature, a mechanical part (thermostat) is opened in order to allow fluid to flow to other parts of the engine. If this feedback information is ignored, or the mechanical part does not respond to the feedback, the entire system may be jeopardized and could fail catastrophically. Similarly, in that same vehicle, feedback gauges allow the operator to monitor the vehicles characteristics. These same monitoring actions can be done automatically by complex designs that can adjust levels based on feedback from a system. Nuclear power plants, space agency vehicles, monitoring systems, and interactive homes are examples of essential feedback.

## OPTIMIZATION

Optimization is doing the most with the least. A crude but distinct example would be cutting a 2" by 2" square out of as piece of paper. Would cutting it out of the center be the best solution? Or might a better solution be to use the edges of the paper in a corner to eliminate two unnecessary cuts and have a large portion of the paper uninterrupted for future use?

Optimization can save companies millions of dollars if done correctly. For example, an engineer took a position with a respected company that built landfill tractors. These tractors and their parts were cut out of one-inch thick steel by a computer numerically controlled plasma cutter. The company had laid out the parts on large sheets of steel with no apparent order. This engineer's first assignment on the job was to cut waste in the production. The first place he looked to optimize was in the layout of the cuts from the one-inch thick steel. By going through and organizing how the parts were laid out and cut, he learned that he could save over 25 percent of the steel that would otherwise be wasted. This 25 percent that went to recycling was now being used to make additional parts for the tractor and, therefore, saved money. Think of it this way. If that 25 percent savings represented a monetary savings of $1,000 and the engineer's labor was $500 for the work, the company would have saved five hundred dollars. However, if the company made on an average 50,000 tractors annually, the savings would total 25 million dollars that year alone.

Most design problems we face have no definitive answer. Most solutions are a series of compromises, which in turn allow the product to function well. An engineer or technician must assess all the parts of a problem and optimize each before the job is resolved. Webster's defines optimization as "The procedure or procedures used to make a system or design as effective or functional as possible, especially the mathematical techniques involved."

Optimization moves the innovation process further. It favors gaining understanding by abstract thinking. But rather than diverge, an individual with this thinking style prefers to converge. This results in converting abstract ideas and alternatives into practical solutions and plans. Individuals rely on mentally testing ideas rather than on testing things. People who favor the optimization style prefer to create optimal solutions to a few well-defined problems or issues. They prefer to focus on specific problems and sort through large amounts of information to pinpoint what's wrong in a given situation. They are usually confident in their ability to make a sound logical evaluation and to select the best option or solution to a problem. They often lack patience with ambiguity and dislike dreaming about additional ideas, points of view, or relations among problems. They believe they know the problem.

## Activity 10.1   System loops

Time:      1–2 class periods, 4 days at home collecting data.

In this case study, student teams of three will examine bicycles to determine what systems they have. The teams will also create input, process, output, and feedback loops for each system in order to explain it to the class. A presentation will be required of at least one of the systems of a bicycle. Teams will present detailed information to the class though electronic presentations, physical and graphical models of the specific system, and written text of the system. The instructor of the class will assign the team a specific system and the following are the minimum requirements of the presentation:

1.  Define the input, process, output, and feedback loops specifically.
2.  Replicate the system to scale (determined by the team) with the use of mechanical design software or through orthographic hand drawings.
3.  Replicate the teams' recommended modifications to the system to scale (determined by the team) with the use of mechanical design software or

**This CNC (Computer Numerical Control) Router has many systems within it. How many can you name?**

through orthographic hand drawings. This can be an overlay for a hand drawing, a layer in a CAD drawing, or a pop on part for a scale model.

4. Build a scale model of the system assigned to your team.
5. Write a report on the system and describe the specific uses, advantages and disadvantages of the loops, and what your team would design to improve the system.

## CASE STUDY

The philosophy of a bicycle has been around since 1817 when Baron Karl Drais von Sauerbronn invented a walking machine (the Laufmaschine) that helped him get around better. This was not pedal powered; he simply walked along side of it and pushed it for power. In 1865, the first bicycle known as the Velocipede or Boneshaker used pedals to power it.

In the 21$^{st}$ century, you can now build your own bike online through websites like the one that TREK offers. Composite materials, special frames, and complicated aerodynamics are a few of the systems involved in bicycles today. The future of this industry holds many advancements and innovations designed into prototype bicycles every day. Many of these developments may become part of the bicycles sold in stores in the coming years.

## RESOURCES

http://www.trekbikes.com Trek is a privately held corporation that still resides in Waterloo, the small town in Southeastern Wisconsin where, in 1976, five employees started making hand-built bicycle frames in an old wooden barn. Nearly a quarter of a century later, Trek is the world leader in bicycle products and accessories, with 1,500 employees worldwide.

## Activity 10.2   Material Systems Dissection

Time:     3–4 class periods

## CASE STUDY

Materials are usually things that society tends to pick off the shelf or purchase based on a known supply. Kevlar was not a viable solution to many manufacturers in the past because it did not exist. Scientist's and engineers created it based on a need. Materials can be created based on parameters defined for a project. In our culture, a many small appliances are found around the home. These appliances are part of what technology has done to improve our lives. The materials used in these appliances are usually selected based upon availability, ability to serve the purpose, and economics. As you carefully dismantle the appliance, attempt to determine the purpose of each part and why the material that was used to make it was chosen.

## Part A

1. Carefully dismantle the appliance using the tools needed to remove the appliance's casing and inner parts. In this process, attempt to keep each of the parts intact. In your journal, note the type of appliance dismantled, your observations, and your impressions when dismantling it.
2. Categorize the disassembled parts, and place them into containers or bags labeled metals, ceramics, polymers, and composites, such as in Figure 15. Do this to the best of your ability.

## Part B

1. Obtain a piece of poster board. Use the parts from your appliance to make a poster about your appliance. Use your creativity. The poster could include parts, historical developments, the appliance as you envision it in the future, an alternative to this appliance, etc.

## Clean-Up Instructions

1. The parts that are junk need to be placed in the area designated by your instructor. Recycle parts as needed or as your community prescribes.
2. Tools must be returned to their proper places.

## RESOURCES

ENERGY CONCEPTS, INC. (1999) *Materials Science Technology: Solids.* Pg. 2.23–2.25

http://www.science-ed.pnl.gov/mst.stm This is the Department of Energy's Materials Science Curriculum website. The site gives a glimpse into the full document with information and selected activities in PDF format.

http://www.energy-concepts-inc.com This national vendor sells the reorganized Materials Science Curriculum, so it is more teacher friendly. This curriculum covers many topics beyond the DOE curriculum.

*The New Way Things Work*

David Macaulay, Neil Ardley ISBN: 0395938473; Publisher: Houghton Mifflin Company, September 1998.

## Activity 10.3    Texas Traffic

Time:    5–10 class periods

## CASE STUDY

The Houston TranStar Automatic Vehicle Identification (AVI) traffic monitoring system collects real-time information showing current travel conditions on Houston area freeways and high occupancy vehicle (HOV) lanes. This information is provided to personnel within the Houston TranStar Center for detecting freeway congestion. This travel information is provided to the public through media reports and is displayed on selected roadside electronic message signs and the Houston TranStar website (http://traffic.tamu.edu/).

The system uses Automatic Vehicle Identification (AVI) technology developed by the Amtech Systems Division of TransCore to collect the real-time traffic information. Houston was the first city to apply AVI technology for monitoring traffic conditions.

The AVI system operates through the use of AVI antennas and readers, which are installed on structures along Houston freeways. The AVI antennas and readers monitor the passage of vehicles equipped with transponder tags. The transponder tags are powered by a small battery, which enables them to reflect signals transmitted from the antennas/readers.

The system uses vehicles equipped with transponder tags as vehicle probes. The main source of vehicle probes is commuters using the EZ-Tag automatic toll collection system installed by the Harris County Toll Road Authority (HCTRA). Transponder tag readers are placed at 1- to 5-mile intervals along freeways and HOV lanes. Each reader senses probe vehicles as they pass a reader station and transmits the time and location of the probes to a central computer over a telephone line. As the probe vehicles pass through successive AVI readers, software calculates average travel times and speeds for a roadway segment. The averages are made available to software, which provides the data for the Houston TranStar website.

In this case study, students will access the TransStar website (or your local website that has proper information) to study information and habits of drivers as

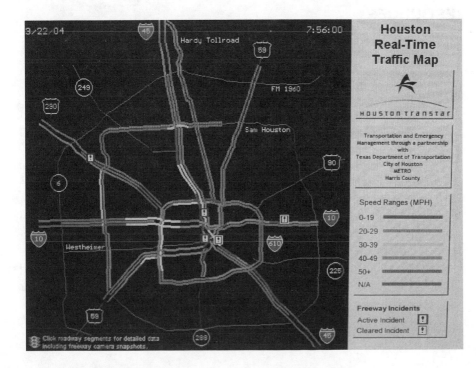

well as trouble spots within a system. Breaking into groups of three, students should study the habits and mistakes of drivers over a five-day period. The student team is required to collect and log major accident areas, major delay areas and areas that it determines may be a future area of concern. Eleven major arteries flow into and out of the Houston area. Each team will be assigned a specific road for data research.

## ASSESSMENT

Students will be required to present their spreadsheets and the charts derived from the sheets to the entire class. They may choose to create large posters or PowerPoint presentations, or they may draw the chart on a chalkboard or whiteboard for the class. The presentation should include maps, charts, and pictures that explain their specific researched road as well as the main accident spots, slowdowns, and predicted future problem spots. The following questions should also be answered in written format by each member of the group:

Identify the input, process, output, and feedback parts of this system.
What are the pros and cons of this technology?
How would you use it if it were available in your hometown?
Could this technology provide a positive impact on your community? Why?

## RESOURCES

http://traffic.tamu.edu/ This is the Houston TranStar website where the data can be downloaded and traffic can be observed. It also has links explaining how the technology works and other topics to make the lesson more comprehensive.

http://www.dot.gov/about_dot.html This page gives you information about the Department's of Transportation's organization, key officials, mission, and other important information related to managing the department and accomplishing its mission.

*Asphalt Nation: How the Automobile Took Over America, and How We Can Take It Back,* Jane Holtz Kay

ISBN: 0520216202; Publisher: University of California Press (November 1998)

This book is an excellent reference for alternative transportation methods and a bit of history on the automobile.

http://www.microsoft.com/office/excel/default.asp This is Microsoft's Excel homepage. It has tips and tricks, usage procedures, and the top ten frequently asked questions. If the site changes, search Microsoft's website, or search your school's software.

*Chapter 11*

# Materials

## Introduction

Engineers in materials science designed the materials in a bulletproof vest used by police in all fifty states. The warm winter coat that you wear is lightweight and waterproof because of materials science research. Long ago, humans realized the need to find something that would help them achieve a goal or a desire in their lives. From a rock tied to a stick to make a hammer, to a ceramic cooled in a superconductivity application, humans have always relied on materials. Since the dawn of history, materials have played a significant role in the field of engineering and education. Dr. Thomas Stobe, a professor at the University of Washington in Seattle says, "Materials can effectively be used to generate and capture students' interest in science, mathematics, engineering and technology."

The military and space industries have defined specific service requirements for their systems that led to the development of advanced composites and ceramics. For instance, military troops in the field use armor, which is too heavy. Troops are carrying up to 100 pounds of equipment each. This creates issues and inefficiencies in movement and combat. Scientists have developed ceramics to reduce the weight carried by troops and vehicles, while being more effective than steel. This solution decreases the amount of fuel to be transported by the military as some of their vehicles get only 1/2 mile to the gallon. This new ceramic armor is expected to perform much better in close combat and city combat, is stronger than the old steel plates, and will provide better protection, even against stronger armor-piercing, steel bullets.

This new armor not only provides solutions for personal armor but will be used on tanks and amphibious vehicles as well. The ceramics can withstand 140,000 pounds of pressure per square inch and is ideal for taking a hit from a large shell and remain strong.

Interesting points about advanced materials:
1. Markets for advanced ceramics grew from less than $2 billion in 1987 to over $20 billion in the year 2000.
2. Materials consume 30 and 50 percent of manufactured goods cost.

The engineering component of our educational system has embraced materials science and has taught it in a way so students can gain the needed knowledge, no matter what their needs might be. Teachers explain the characteristics of materials and the background needed to understand why a material does what it does. Why, for instance, is an alloy stronger than a native metal, instead of just picking the alloy off a shelf because the teacher says its stronger? Is it even the right metal for the application? Now, students have all the information available to them to make decisions about what materials are even appropriate for their need.

According to a U.S. Office of Technology Assessment report, a key to remaining competitive with other countries and companies is to train more scientists and technologists with a broad background in advanced materials.

## HISTORY

We can trace materials science to the Babylonians, who were makers of ceramic building materials (bricks). With the development of primitive furnaces, clays could also be used to make durable products.

An example of materials use is imprinted clay tablets produced in 2200 B.C. that stated "every workman taught his own trade to his children and these in turn would instruct theirs."

The history of human civilization and social development is strongly intertwined with the pervasive role of materials—namely, substances accessible to mankind that can be processed to exhibit the desired properties for making things. This connection between the human race and its materials has expanded enormously over the ages. Billions of tons of raw materials are taken annually from nature to be manipulated into innumerable products and systems. We have come to recognize materials as one of the primary resources of humankind, ranking with living space, food, energy, information, and even manpower.

Materials were so important throughout the ages that the time periods were named after the dominantly-used material, e.g., the Stone Age, the Bronze Age, and the Iron Age. Below are some historical developments made possible after specific materials were developed:

| | |
|---|---|
| 8000 B.C. | Hammered Copper |
| 7000 B.C. | Clay Pottery |
| 6000 B.C. | Silk Production |
| 5000 B.C. | Glass Making |
| 4000 B.C. | Smelted Copper |
| 4000 B.C. | Bronze Age Begins |
| 1000 B.C. | Iron Age Begins |
| 500 B.C. | Cast Iron |
| 300 B.C. | Glass Blowing |
| 105 A.D. | Paper |
| 600–900 A.D. | Porcelain |
| 1540 A.D. | First Foundries |
| 1774 A.D. | Crude Steel |

| | |
|---|---|
| 1789 A.D. | Discovery of Titanium |
| 1800 A.D. | Battery |
| 1824 A.D. | Portland Cement |
| 1850 A.D. | Reinforced Concrete |
| 1856 A.D. | Bessemer Steel-making Process |
| 1870 A.D. | Celluloid Production |
| 1871 A.D. | Periodic Table |
| 1884 A.D. | Nitrocellulose |
| 1886 A.D. | Electrolytic Reduction of Aluminum |
| 1891 A.D. | Silicon Carbide |
| 1907 A.D. | First Totally Synthetic Polymer |
| 1923 A.D. | Tungsten Carbide |
| 1930 A.D. | Fiberglass |
| 1937 A.D. | Nylon |
| 1947 A.D. | Germanium Transistor |
| 1950s A.D. | Silicon Photovoltaic Cells and Transistors |
| 1958 A.D. | Ruby Laser |
| 1959 A.D. | Integrated Circuit |
| 1966 A.D. | Fiber Optics |
| 1986 A.D. | High Temperature Superconductors |

Data courtesy of the Department of Energy and Energy Concepts, Inc.

## NEW MATERIALS

New materials are developed based on a need. For instance, if a fashion designer wanted to make polyester feel like silk and make it breathable and comfortable, could that material be created? As a designer, you do not have to rely on what is on the shelf to design dreams.

Engineers can design without worrying if a material is available for their application. A materials scientist can create a material based on a specific need. Of course, the client would have to define what they want do to create the material to satisfy that need. The following characteristics help develop them:

1. **Strength (stiffness)** measures the ability to resist the effects of forces such as tension, compression or torsion.
2. **Ductility** measures how well the material can be shaped without fracturing. Vehicle body parts are typically made out of ductile sheet metal that can be stamped into shapes.
3. **Brittleness** is a physical property indicating when the material will break while undergoing small deformations. For instance, a rose that has been dipped in liquid nitrogen becomes brittle.
4. **Hardness** is the ability to resist indentation and wear. A Rockwell hardness test is the most typical test that forms indentations compared to the relative hardness of materials. Several other hardness tests are the Brinell hard-

ness test, a micro-hardness test (a more precise hardness test), and a Durometer hardness test (for plastics and elastomers).

5. **Elasticity** is the ability to return to the original shape after being deformed by a stress or after removing a load.
6. **Electrical conductivity** is the measure of the ability to conduct electricity or electrons.
7. **Thermal conductivity** is the measure of the ability to conduct heat.

"Materials can effectively generate and capture students' interest in science, mathematics, engineering, and technology" says Dr. Thomas Stobe, a professor at the University of Washington in Seattle.

## CLASSIFYING MATERIALS

### Metals

Webster's defines a metal as "Any of a category of electropositive elements that usually have a shiny surface, are generally good conductors of heat and electricity, and can be melted or fused, hammered into thin sheets, or drawn into wires. Typical metals form salts with nonmetals, basic oxides with oxygen, and alloys with one another. An alloy of two or more metallic elements."

Humans have used metals for thousands of years. The earliest metals used were "native" metals, found in their pure form in nature. Examples are copper, gold, silver, and meteoric iron. These metals were used for everything from ornamentation to specific tools.

Some of the mechanical properties associated with metals include strength, toughness, malleability, and ductility. Most metals are opaque, lustrous, dense, are good conductors of heat and electricity, and have high melting points. Metals can be classified as ferrous or non-ferrous. Ferrous metals contain significant iron content (50% or more). These materials, like wrought iron and steel, attract magnetic materials. Non-ferrous metals contain less than 50% iron and do not attract magnetic materials. These metals also have a higher corrosion resistance than ferrous metals. Examples of these are aluminum, zinc, lead, titanium, copper, tin, and magnesium.

Iron and steel are the first and second most commonly used metal, respectively. Aluminum, the third most commonly used metal, is lightweight and can be stronger than steel. For example, the space shuttle program chose an aluminum alloy for the chassis of the orbiter because of its strength to weight ratio. As most know, the aluminum of the space shuttle is skinned with protective ceramics and composites to keep the extreme temperatures out. Because of a breach of the orbiter's protective layer, high temperature gasses entered the vehicle and melted the aluminum superstructure, creating a catastrophic and fatal event.

At the tip of the Washington Monument is a 100-ounce aluminum pyramid. Originally placed as part of the structure's lightning protection apparatus, the aluminum pyramid was so rare in the late 19th century, that is was actually displayed before the public at Tiffany's jewelers.

Colonel Thomas Lincoln Casey, of the Army Corps of Engineers, was in charge of the Washington Monument. As the construction neared completion, Casey sent a request to William Frishmuth, a man whose foundry had done the plating work for the monument. Casey asked if Frishmuth could make a metal pyramid that could serve as the monument's lightning rod. At the time, copper, bronze, or brass, plated with platinum, were the preferred materials.

The capping ceremony of December 6, 1884, and the formal dedication of the monument on February 21, 1885, were given front-page publicity in the nation's newspapers and the aluminum point or apex was correctly described. People who had never before even heard about aluminum now knew what it was.

Important dates in history for metals:

| | |
|---|---|
| 8000 B.C. | Hammered Copper |
| 4000 B.C. | Smelted Copper |
| 4000 B.C. | Bronze Age Begins |
| 1000 B.C. | Iron Age Begins |
| 500 B.C. | Cast Iron |
| 1540 A.D. | First Foundries |
| 1774 A.D. | Crude Steel |
| 1789 A.D. | Discovery of Titanium |
| 1856 A.D. | Bessemer Steel-making Process |
| 1886 A.D. | Electrolytic Reduction of Aluminum |

## Ceramics

The word ceramic is derived from the Greek word, *keramos,* meaning burned material. Ceramics are a family of important materials used for thousands of years. Evidence exists that ceramic products were in use by 6500 B.C. Others claim that ceramic materials were used in Japan as early as 13,000 years ago. Glazed pottery was being used in Egypt by 4000 B.C., and glass beads appeared by 2500 B.C.

Early applications of ceramics were as building materials and for containers. While still being used for these functions, they are used for video tape components, automobile engine parts, bathroom fixtures, electronic parts, catalytic converters, the space shuttle, and as nuclear fuel.

Ceramics can be crystalline solids with properties related to the covalent or ionic bonds that hold them together. They can also be non-crystalline in nature like glass. Metal oxides are used to color glass. These oxides, like cobalt, appear with a different color than the final colored glass. For example, cobalt oxide in its powdered form is black. When mixed and heated correctly, this oxide helps create a blue glass. Glass is usually considered a separate part of ceramics because it lacks crystalline organization, which means the glass does not have an orderly atomic structure.

Historical dates in the development of glass:

| | |
|---|---|
| 4000 B.C. | Glass Glazes |
| 2500 B.C. | Glass Beads |

| 1500 B.C. | First Glass Objects (other than beads) |
| 200 B.C. | Glass Blow Pipe; First Flat Glass |
| 100 A.D. | First Colorless Glass |
| 1200s | First Quality Glass Mirrors |
| 1676 | Lead Crystal Glass Developed |
| 1688 | First Plate Glass |
| 1820s | Mechanical Glass Press |
| 1900 | Development of Tank Furnaces |
| 1912 | Borosilicate Glass Developed |
| 1914 | First Automated Window Glass |
| 1926 | Ribbon Machine for Making Light Bulbs |
| 1940s | Glass Cookware |
| 1957 | Glass-Ceramics |
| 1960s | Float Process for Plate Glass/Optical Wave Guides |
| 1980s | Use of Modern Optical Fibers |

Ceramics can be classified as clay products, refractories, and abrasives, and glasses. Refractories are used in high temperature applications and are usually made of clay. They also line the walls, ceilings, and floors of furnaces. These ceramics keep the metal exteriors of furnaces from melting along with the contents places in the furnace. Abrasives are usually extremely hard, pure, ceramic compounds or mixtures and are used to make sandpaper and grinding wheels.

## Polymers

The term polymer is formed from the Greek words *poly* meaning many, and *mer* meaning parts. The words polymer and plastic are often used interchangeably; however, this is incorrect. Polymer is a term that is used to define an entire group of materials, of which plastic is a member.

Plastics are synthetic polymers. These synthetic polymers can be classified as either thermoplastic polymers or thermosetting polymers. A thermoplastic polymer softens when heated and re-solidifies when cooled. It is the primary type of polymer used for recycling. A thermosetting polymer is a type of polymer that, when heated, sets into a solid form permanently and cannot be softened by heat.

Other types of polymers include natural materials like wood, leather, cotton, wool, silk, and rubber. Additional natural polymers like proteins, enzymes, starches, and cellulose are processed by plants and animals. These materials exhibit many of the same types of characteristics as plastics as they are molecular materials.

Polymers, as a group, are not strong. However, they are good electrical insulators, e.g., used as electrical extension cord housing. They also have low melting temperatures.

Polyethylene Terephthalate (PETE): Most of the recycled PETE is from beverage containers. Recyclers are able to make 99% pure, granulated PETE that sells at approximately half the cost of new PETE. Approximately 50% is used as fiberfill for jackets and strapping. What remains is used to make liquid soap bottles, surfboards, paint brushes, fuzz on tennis balls, and more soft drink bottles.

| Recycle Code | Abbreviation and Chemical Name | Types of Uses and Examples |
|---|---|---|
| 1 | Polyethylene Tereph- thalate (PETE) | Clear, 2 2-liter beverage bottles |
| 2 | High High-density Polyethylene (HDPE) | Milk jugs, detergent bottles, some water bottles |
| 3 | Polyvinyl Chloride or Vinyl (PVC or V) | Saran wrap, plastic drain pipe, shower curtains, some water bottles |
| 4 | Low- density Polyeth- ylene (LDPE) | Plastic bags, garment bags, coffee can lids |
| 5 | Polypropylene (PP) | Aerosol can tops, rigid bottle caps, candy wrappers, bottoms of bottles |
| 6 | Polystyrene (PS) | Hard clear plastic cups, foam cups, eating utensils, deli food containers, some packing popcorn |
| 7 | Other | Biodegradable, Ssome packing popcorn |

High-density Polyethylene (HDPE): The process for recycling HDPE is well-developed. Recycled HDPE is used for drainage pipes, flower pots, plastic lumber, trash cans, automotive mud flaps, kitchen drain boards, beverage bottle crates, pallets, signs, stadium seats, recycling bins, traffic barrier cones, golf bag liners, and toys.

Polyvinyl Chloride or Vinyl (PVC or V): Recycled PVC has been used in drainage pipes, pipe fittings, floor tiles, bottles, doormats, hoses, and mud flaps. It is generally not burned in incinerators because it releases hazardous fumes such as dioxins and furans.

Low-density Polyethylene (LDPE): LDPE is burned in incinerator-powered generators to produce electrical energy. It is also recycled into items in which color is not critical such as garbage can liners, grocery bags, paint buckets, fast food trays, lawn mower wheels, and automobile battery parts.

Polypropylene (PP): Recycled polypropylene is used in license plate holders, desktop accessories, hanging files, food service trays, flower pots, and trash cans.

Polystyrene (PS): One of the most challenging polymers to recycle is polystyrene, the material from which Styrofoam cups and packing material are made. Although some methods for recycling are in place, chemists are still looking for more effective ways to recycle the huge amounts of waste polystyrene.

Important dates for polymers:

| | |
|---|---|
| 1830–1839 | Christian Schonbein produced cellulose nitrate by dissolving cellulose in nitric acid. Charles Goodyear patented vulcanization |
| 1860–1869 | First commercial production of celluloid to produce photographic film and dentures. |
| 1880–1889 | First synthetic fibers formed from rayon. Cellulose nitrate extruded and carbonized to form lightbulb filaments. |
| 1900–1909 | Cellophane invented. |

|            | Leo Bakeland developed the first truly synthetic plastic called bakelite. |
|------------|--------------------------------------------------------------------------|
| 1910–1919  | Manufacture of rayon stockings for women.                                |
| 1920–1929  | The understanding of the chemistry of polymers begins.                   |
| 1930–1939  | Nylon is created. Polyethylene, PS, and Teflon are produced.             |
| 1940–1949  | Epoxies are introduced.                                                   |
| 1950–1959  | Polyurethanes enter the market.                                           |
|            | PVC becomes the main ingredient for phonograph records.                  |
|            | Reinforced plastic is used to make the Corvette body.                    |
| 1960–1969  | Synthetic turf used in the Houston Astrodome.                            |
|            | Permanent press fabrics introduced.                                      |
| 1970–1979  | Kevlar and Gore-Tex are developed.                                       |

## Composites

Are you a person who enjoys outdoor activities? Maybe you have used a skateboard in the summer or snowboard in the winter. Whether or not you enjoy these types of activities, composites are ubiquitous in recreational equipment.

A common definition of composites is a combination of two or more constituent materials bonded together in an effort to provide better properties than those of the individual materials. Components of composites produced in 1988 were valued at 2 billion dollars. By the year 2000, this value was estimated at 20 billion dollars. Composites are used extensively in the International Space Station and make up over 10,000 pounds of each space shuttle.

Composites are different than the combinations found in metals, polymers, and ceramics. Alloys, copolymers, and glass ceramics are all combinations that occur on the molecular or microscopic level. Composites are almost always identified without using any special instruments. All composites consist of a reinforcement, a matrix, and a boundary in between. The reinforcement is the part of the composite that provides strength. This typically comes in the shape of a fiber, whisker, or particulate. The matrix is the glue that holds it all together. A common example is concrete; sand, and stone as its reinforcement and cement as its matrix.

Composite materials and their use date back thousands of years. The ancient Israelites and Egyptians added straw to their bricks to help hold them together. Three thousand years ago, the Incas used plant fibers to strengthen their pottery. The Colosseum (Coliseum) and other ancient Romans structures were held together with cement that was made of slacked lime and pozzolana. Slacked lime is made by heating lime and then crumbling it by adding water. Pozzolana is a volcanic ash from Mount Vesuvius. This combination, called hydraulic cement, may represent the first cement that hardens while submerged in water. In 1824, Joseph Aspdin, who was a bricklayer in Leeds, England, combined finely pulverized lime, silica, alumina, and iron to create Portland cement. Portland refers to the Isle of Portland just off the British coast. The cement that Aspdin created resembled the color of stone quarried on the Isle of Portland.

Unidirectional and bidirectional carbon fiber, Kevlar, and plain-weave fiberglass reinforced composites are a few of the materials use in composite lay ups.

Lay ups are composed of consecutive layers of fabric, resin, and sometimes a core material. Applications of these composite types range from high-end mountain bikes to furniture to bullet-proof vests. They can be found in the open wheel racing series like Formula One (F1), CART and the Indy Racing League. Also, such composites are often used in Department of Defense applications where high strength and low weight are desired.

These fabric materials can be laid up by hand, forming them into a mold and then painting them on the matrix of resin that is typically a type of epoxy. They can also come from the factory with the epoxy matrix already impregnated into them (pre-impregnated or "pre-preg"). This requires that the fabric be refrigerated or stored at a specific temperature. Pre-preg fabrics are more expensive than fabrics without the matrix but are easier to use as the mess is reduced. When the matrix is mixed by hand, it is combined from two different parts, a resin and a hardener. An epoxy matrix, when mixed, has a specific time that it can spend in the mixing container. A chemical reaction is taking place between the two ingredients and that chemical reaction can speed up when it is in larger volumes. This characteristic is called "pot life" or how long it can be in the container before it starts its hardening process. This matrix also has a prescribed work time based on the amount of hardener used. This is the time available to work with the materials placing and forming them into their mold or application. The start of the hardening process is called "going off". When the matrix "goes off," little time remains to work with it.

Effective lay ups involve the following process:

1. The fabric is cut to the appropriate size.
2. The bag, peel ply, perforated plastic, and bleeder are cut to the appropriate sizes.
3. The mold is prepared with gel coat, mold release and/or wax.
4. The correct amount of resin and hardener are used.
5. Pot life is not compromised.
6. Material is laid up within appropriate work time.
7. No sections of the lay up are "starved" (without the correct amount of matrix or, more specifically, the resin or epoxy)
8. No sections are over filled with matrix
9. There is a good seal on the vacuum bag (12–15 psi)
10. The peel ply is able to be removed with no folds or creases
11. The mold is released from the fabric
12. Clean up of work area is completed promptly to insure work and tool surfaces have no remaining fabric or resin

## INDUSTRY SUPPORT

ASM International (ASMI) is the society for materials engineers and scientists. ASMI is a worldwide network dedicated to advancing industry, technology, and applications of metals and materials. ASMI has said that the United States needs engineers and technicians to move us forward in the development and implementation of the newest research in materials, polymers, and composites. It has also stated that a solid materials science curriculum will enable our students to excel in this fast-paced, rapidly changing environment.

The American Ceramic Society (ACerS) is dedicated to the dissemination of scientific, commercial, and educational information about ceramic materials and industries. ACerS has more than 10,000 members in 80 countries. Mark Glasper, the director of communications for ACerS, says of ACerS that it is "an excellent tool for teachers searching for a program to augment their curriculum. Materials will be an important part of, and the focus of, the future."

Louis Ludke, President and CEO of the National Composites Center in Kettering, Ohio, says that composites are the next generation of materials. A composite is a blending of two or more distinct materials resulting in a new material with properties superior to the properties of the individual materials. " As new composite materials are developed, a newly trained workforce will be required to work with them. The materials science curriculum is an excellent tool to introduce students to the principles and concepts required to pursue a career in advanced structural composites."

Without the interest and support of industry and businesses, our programs would exist within only our own four walls. Now, because of the significant interest of business from all over the world, our students get to experience firsthand what only a select few would do in colleges and engineering careers.

## THE UNIVERSITY CONNECTION

Many universities have had materials science departments within their colleges for decades. Historically, they have focused mainly on metallurgy. However, that trend is leveling off. Each university offers courses ranging from traditional metallurgy to high technology superconductors and composites.

The Department of Materials Science and Engineering at the Massachusetts Institute of Technology (MIT) traces its history back to the founding of MIT in 1865. Over the years, the department's central function has been providing students with the opportunity to conduct independent and creative research at the forefront of materials science and engineering. Today, the department is the largest of its kind in the United States and is the leader in many areas of materials education and research. The department shares in creating the tradition of excellence for which MIT is known.

The Department of Materials Science and Engineering at the University of Wisconsin-Madison is a dynamic community of outstanding faculty, dedicated educators, and talented students. Graduate students enrolling in the department pursue advanced degrees in Metallurgical Engineering, with M.S. and Ph.D. degrees offered. Graduate research in Metallurgical Engineering covers a full range of cutting-edge technologies with an emphasis on engineering. These include metals, ceramics, semiconductors, superconductors, thin films, and the next-generation super alloys and composites. Active research programs also exist in ferrous and non-ferrous metals. The ceramics unit includes an experiment created in partnership with the University of Wisconsin-Madison's College of Engi-

# Properties of Elements

| Element | Symbol | Atomic Number | Atomic Mass | Crystal Structure (pm = $10^{-12}$ m) | Atomic Radius | Ionic Radius* |
|---|---|---|---|---|---|---|
| Actinium | Ac | 89 | 227.03 | FCC | | 118 |
| Aluminum | Al | 13 | 26.98 | FCC | 125 | 54 |
| Americium | Am | 95 | 243.06 | HCP | | |
| Antimony | Sb | 51 | 121.76 | RHB | 141 | 5+ 60 |
| Argon | Ar | 18 | 39.95 | FCC | | |
| Arsenic | As | 33 | 74.92 | RHB | 121 | 5+ 46 |
| Astatine | At | 85 | 209.99 | | 145 | 7+ 62 |
| Barium | Ba | 56 | 137.33 | BCC | 198 | 135 |
| Berkelium | Bk | 97 | 247.07 | | | |
| Beryllium | Be | 4 | 9.01 | HCP | 89 | 45 |
| Bismuth | Bi | 83 | 208.98 | RHB | 146 | 5+ 76 |
| Boron | B | 5 | 10.81 | RHB | 81 | 3+ 23 |
| Bromine | Br | 35 | 79.90 | ORH | 114 | 196 |
| Cadmium | Cd | 48 | 112.41 | HCP | 141 | 97 |
| Calcium | Ca | 20 | 40.08 | FCC | 174 | 100 |
| Californium | Cf | 98 | 251.08 | | | |
| Carbon | C | 6 | 12.01 | HCP | 77 | 4+ 16 |
| Cerium | Ce | 58 | 140.12 | FCC | 182 | 3+ 118 |
| Cesium | Cs | 55 | 132.91 | BCC | 235 | 167 |
| Chlorine | Cl | 17 | 35.45 | ORH | 99 | 181 |
| Chromium | Cr | 24 | 52.00 | BCC | 117 | 6+ 52 |
| Cobalt | Co | 27 | 58.93 | HCP | 116 | 72 |
| Copper | Cu | 29 | 63.55 | FCC | 117 | 1+ 96 |
| Curium | Cm | 96 | 247 | | | |
| Dysprosium | Dy | 66 | 162.50 | HCP | 159 | 91 |
| Einsteinium | Es | 99 | 254 | | | |
| Erbium | Er | 68 | 167.26 | HCP | 157 | |
| Europium | Eu | 63 | 151.97 | BCC | 185 | 3+ 95 |
| Fermium | Fm | 100 | 257.10 | | | |
| Fluorine | F | 9 | 19.00 | CUB | 64 | 136 |
| Francium | Fr | 87 | 223.02 | BCC | | |
| Gadolinium | Gd | 64 | 157.25 | HCP | 161 | 3+ 94 |
| Gallium | Ga | 31 | 69.72 | ORH | 125 | 3+ 62 |
| Germanium | Ge | 32 | 72.61 | CUB | 122 | 4+ 53 |
| Gold | Au | 79 | 196.97 | FCC | 134 | 137 |
| Hafnium | Hf | 72 | 178.49 | HCP | 144 | 4+ 84 |
| Helium | He | 2 | 4.00 | HCP | 120 | |
| Holmium | Ho | 67 | 164.93 | HCP | 158 | |
| Hydrogen | H | 1 | 1.01 | HCP | 30 | 1- 154 |
| Indium | In | 49 | 114.82 | TET | 150 | 3+ 81 |
| Iodine | I | 53 | 126.90 | ORH | 133 | 216 |
| Iridium | Ir | 77 | 192.22 | FCC | 127 | 4+ 68 |
| Iron | Fe | 26 | 55.85 | BCC | 117 | 2+ 75 |
| Krypton | Kr | 36 | 83.80 | FCC | 196 | |
| Lanthanum | La | 57 | 138.91 | HCP | 169 | 3+ 102 |
| Lead | Pb | 82 | 207.2 | FCC | 154 | 4+ 84 |
| Lithium | Li | 3 | 6.94 | BCC | 123 | 68 |
| Lutetium | Lu | 71 | 174.97 | HCP | 156 | 85 |
| Magnesium | Mg | 12 | 24.31 | HCP | 136 | 72 |
| Manganese | Mn | 25 | 54.94 | CUB | 117 | 80 |
| Mercury | Hg | 80 | 200.59 | RHB | 144 | 2+ 110 |
| Molybdenum | Mo | 42 | 95.94 | BCC | 129 | 6+ 62 |
| Neodymium | Nd | 60 | 144.24 | HCP | 164 | 3+ 100 |
| Neon | Ne | 10 | 20.18 | FCC | 141 | |
| Neptunium | Np | 93 | 237.05 | ORH | | |

# Properties of Elements (*continued*)

| Element | Symbol | Atomic Number | Atomic Mass | Crystal Structure (pm = $10^{-12}$ m) | Atomic Radius (pm = $10^{-12}$ m) | Ionic Radius* |
|---|---|---|---|---|---|---|
| Nickel | Ni | 28 | 56.69 | FCC | 115 | 70 |
| Niobium | Nb | 41 | 92.91 | BCC | 134 | 5+ 69 |
| Nitrogen | N | 7 | 14.01 | HCP | 70 | 171 |
| Osmium | Os | 76 | 190.2 | HCP | 126 | 4+ 88 |
| Oxygen | O | 8 | 16.00 | CUB | 66 | 140 |
| Palladium | Pd | 46 | 106.42 | FCC | 128 | 90 |
| Phosphorus | P | 15 | 30.97 | MCL | 110 | 5+ 35 |
| Platinum | Pt | 78 | 195.08 | FCC | 130 | 2+ 80 |
| Plutonium | Pu | 94 | 244.06 | MCL | | |
| Polonium | Po | 84 | 208.98 | MCL | 146 | 6+ 67 |
| Potassium | K | 19 | 39.10 | BCC | 203 | 138 |
| Praseodymium | Pr | 59 | 140.91 | HCP | 165 | 3+ 101 |
| Promethium | Pm | 61 | 145 | HCP | 163 | 3+ 98 |
| Protactinium | Pa | 91 | 231.04 | ORH | | |
| Radium | Ra | 88 | 226.03 | BCC | | |
| Radon | Rn | 86 | 222.02 | FCC | 222 | |
| Rhenium | Re | 75 | 186.21 | HCP | 128 | 4+ 72 |
| Rhodium | Rh | 45 | 102.91 | FCC | 125 | 3+ 75 |
| Rubidium | Rb | 37 | 85.47 | BCC | 216 | 152 |
| Ruthenium | Ru | 44 | 101.07 | HCP | 124 | 4+ 67 |
| Samarium | Sm | 62 | 150.36 | RHB | 162 | 3+ 96 |
| Scandium | Sc | 21 | 44.96 | FCC | 144 | 81 |
| Selenium | Se | 34 | 78.96 | HCP | 117 | 198 |
| Silicon | Si | 14 | 28.09 | CUB | 117 | 4+ 40 |
| Silver | Ag | 47 | 107.87 | FCC | 134 | 126 |
| Sodium | Na | 11 | 22.99 | BCC | 157 | 102 |
| Strontium | Sr | 38 | 87.62 | FCC | 191 | 118 |
| Sulfur | S | 16 | 32.07 | ORH | 104 | 184 |
| Tantalum | Ta | 73 | 180.95 | BCC | 134 | 5+ 68 |
| Technetium | Tc | 43 | 98.91 | HCP | 130 | 7+ 98 |
| Tellurium | Te | 52 | 127.60 | HCP | 137 | 221 |
| Terbium | Tb | 65 | 158.93 | HCP | 159 | 3+ 92 |
| Thallium | Tl | 81 | 204.38 | HCP | 155 | 3+ 89 |
| Thorium | Th | 90 | 232.04 | FCC | 165 | |
| Thulium | Tm | 69 | 168.93 | HCP | 156 | 3+ 87 |
| Tin | Sn | 50 | 118.70 | TET | 140 | 4+ 69 |
| Titanium | Ti | 22 | 47.88 | HCP | 132 | 4+ 64 |
| Tungsten | W | 74 | 183.85 | BCC | 130 | 6+ 68 |
| Uranium | U | 92 | 238.03 | ORH | 142 | |
| Vanadium | V | 23 | 50.94 | BCC | 122 | 5+ 59 |
| Xenon | Xe | 54 | 131.29 | FCC | 215 | |
| Ytterbium | Yb | 70 | 173.04 | FCC | 174 | 3+ 86 |
| Yttrium | Y | 39 | 88.91 | HCP | 162 | 93 |
| Zinc | Zn | 30 | 65.39 | HCP | 125 | 74 |
| Zirconium | Zr | 40 | 91.22 | HCP | 145 | 4+ 87 |

| Crystals | *Ionic Radius |
|---|---|
| BCC = Body-centered cubic | CUB = Cubic Includes includes the |
| FCC = Face-centered cubic | charge |
| ORH = Orthorhombic | on the ion whose size |
| HCP = Hexagonal close-packed | MCL = Monoclinic is listed |
| TET = Tetragonal | RHB = Rhombohedral |

NOTE: This chart does not include all elements. The ones not included are either uncommon, non-reactive, or there are disagreements on as to how to determine their sizes.

## Typical Properties of Materials

| | Specific Gravity | Tensile Strength (MPa) | Young's Modulus (GPa) | Flexural Strength (MPa) |
|---|---|---|---|---|
| **Metals** | | | | |
| Aluminum 6061 | 2.70 | 124 | 69 | |
| Copper | 8.91 | 170 | 117 | |
| Steel 1020 | 7.86 | 750 | 207 | |
| Solder (50-50) | | 42 | | |
| Titanium Ti-6 Al-4 V | 4.42 | 960 | 114 | |
| **Ceramics** | | | | |
| Alumina ($Al_2O_3$) | 3.8 | 172 | 379 | 330–550 |
| Borosilicate Glass | 3.04 | | 60 | 69 |
| Silicon Carbide (SiC) | 3.17 | 165 | 430 | 170–500 |
| Soda-Lime Glass | 2.47 | | 69 | 107 |
| **Polymers** | | | | |
| Epoxy | 1.11 | 69 | 6.9 | |
| Nylon 66 | 1.15 | 83 | 2.8 | |
| Polyester (thermoplastic) | 1.1 | 46 | 0.585 | |
| Polyester (thermoset) | 1.1 | 28 | 6.9 | |
| High-density Polyethylene (HDPE) | 0.96 | 28 | 0.830 | |
| Low-density Polyethylene (LDPE) | 0.93 | 14 | 0.170 | |
| Polypropylene | 0.90 | 34 | 1.4 | |
| Polystyrene (PS) | 1.05 | 48 | 3.1 | |
| Polyvinyl Chloride (PVC) | 1.32 | 41 | 2.8 | |
| **Composites** | | | | |
| Boron (70% fiber in epoxy) | 2.17 | 1800 | 23 | |
| Carbon (67% fiber in epoxy) | 1.64 | 1200 | 221 | |
| Concrete | 2.3 | 27 | 27 | |
| E-Glass (73% fiber in epoxy) | 2.15 | 1600 | 56 | |
| Kevlar (82% fiber in epoxy) | 1.43 | 1500 | 86 | |
| Douglas Fir Wood | 0.49 | 85 | 11 | |

neering. The following is an overview of the superconductivity experiment used in class.

In the zero-resistance experiment, students measure the loss of resistance in a superconductor by monitoring the voltage due to a constant current passing through the superconductor as it is cooled in a liquid nitrogen bath. For comparison, students will also measure the resistance of a pure metal foil, which will not drop to zero but will decrease upon cooling. The superconductor in this experiment is a tape composed of a $Bi_2Sr_2Ca_2Cu_3O_{14}$ ceramic core and a silver sheath. The bismuth oxide ceramic is the super conducting material which loses resistance at about 110 K, well above the temperature of a liquid nitrogen bath, 77 K. This experiment was created in partnership with the University of Wisconsin-Madison's College of Engineering.

## REFERENCES

Gomez, Alan G. *Materials Science: What is it and why is it important?* Tech Directions, 10629351, Mar. 2002, Vol. 61, Issue 8

Cohen, Morris. *Materials Science and Engineering—A Perspective.* Massachusetts Institute of Technology http://dmse.mit.edu/matsci/

Materials Science Technology, Energy Concepts Inc, Mundelein, IL, 1997.

Materials Science and Technology Project (MST). Pacific Northwest National Laboratories: http://science-ed.pnl.gov/mstdata.stm

Helsel, L and Liu, P. *Industrial Materials.* Goodheart-Willcox Publisher, Tinley Park, IL, 2001.

ASM International Michael J. Kenney, Ph.D. 9639 Kinsman Road Materials Park, OH 44073-0002 USA

American Ceramic Society Mark A. Glasper, Director of Communications, P.O. Box 6316 Westerville, OH 43086-4700

National Composite Center Louis A. Ludke, President & CEO 2000 Composite Drive Ketterling, OH 45420

Massachusetts Institute of Technology's Department of Materials science and Engineering: http://dmse.mit.edu/

The University of Wisconsin's Department of Materials Science and Engineering: http://www.engr.wisc.edu/mse/

## LEARNING ACTIVITIES

## Activity 11.1    Alloying Copper and Tin

Time:      1–2 hours

## CASE STUDY

Many elements exist in the world and beyond. Most of the elements on earth are metallic. In reality, few of these metals occur freely in nature. Copper, gold, silver, and platinum are the only elements that are occasionally found as pure "native" metals. Historically, most of these metals were used for jewelry and other decorative applications. Because of its availability, copper was used for tools and weapons before being replaced with bronze. Copper and bronze were the first metals used by humankind.

In mid-1982, the rising costs of copper changed the composition of pennies from an alloy that was 95% copper and 5% zinc to a copper-clad zinc coin. In this activity, you will experiment with both types of pennies to see how they compare. First, students will plate zinc on to a penny to form a "silver" penny. Next, you will provide the "silver" penny with enough energy using a burner to diffuse the copper and zinc atoms, turning the penny to "gold".

## ASSESSMENT

In this experiment, students will have several steps to complete. Students will place four of the six pennies into the solution until they turn completely "silver". At this point they are taken out of the solution and one of each penny is heated until they turn "gold". These are then compared to the pennies that were not put into solution (the control group). Grades may be given for successful completion of each step in the experiment. Summary questions, lab report entries, and traditional test items associated with the experiment will provide the teacher with measures regarding student understanding.

## RESOURCES

ENERGY CONCEPTS, INC. (1999) Materials Science Technology: Metals. Pg. 3.1–3.15

http://www.science-ed.pnl.gov/mst.stm This is the web site for the Department of Energy's Materials Science Curriculum. The site also gives a glimpse into the full document with information and selected activities in .PDF format.

http://www.energy-concepts-inc.com This is a national vendor that sells the reorganized Materials Science Curriculum so that it is more "teacher friendly". This curriculum covers many topics beyond the DOE curriculum.

## Activity 11.2   Recycling Display

Time:      1–3 hours

## CASE STUDY

There is a misconception that many plastics are non-recyclable. Many people do not know how to dispose of plastics properly after using them. However, public service announcements and programs have been educating people to recycle plastics. In the late 1990's, studies found that plastics account for 12–14% of landfills, the same percentage as in the 1970's, although the use of plastics has increased significantly. People and industry are recycling more so there is little or no increase in the landfill data. Many communities like Madison, Wisconsin require people to separate their trash so it can be easily recycled. However, some major cities have not started recycling in part due to high start up costs or lack of suppliers for the waste materials.

In this activity, you will be creating a display that shows several different categories associated with each recycling code. Included in the display should be: the recycling symbol and number, the type of plastic, the code or abbreviation (e.g. HDPE), the structural formulas, and a sample of the plastic as well as its uses.

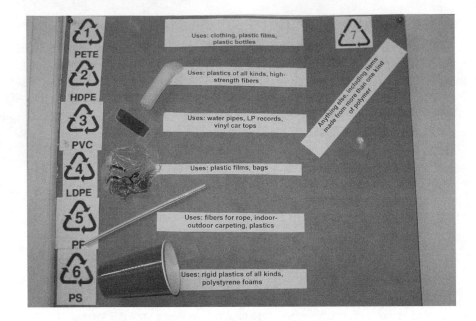

## ASSESSMENT

Teachers and their students should gather samples of materials that are recyclable and non-recyclable. These products can be found at home, in packaging, as well as at local recycling centers. The samples should be laid out and numbered for the students to observe and categorize as recyclable or not. Use the assignment as the basis for discussing why particular plastics are or are not recyclable. The teacher may provide a reading on recyclable materials and with questions pertaining to the reading. The student display should be based on the information read and the requirements of the activity: recycling symbol and number (the type of plastic, the code or abbreviation (e.g. HDPE), the structural formulas, a sample of the plastic, and its uses. Encourage students to be creative in their presentation of this information. Consider posting the displays in prominent areas around the school and in the community.

## RESOURCES

ENERGY CONCEPTS, INC. (1999) Materials Science Technology: Polymers. Pg. 4.3–4.6

http://ice.chem.wisc.edu/materials/links.html Various Materials Science links.

http://www.engr.wisc.edu/mse/newsletter/2000_winter/highschool.html An example of a high school materials science course with connections to the university environment.

# Chapter 12

# Technology and Society

## Introduction

In addition to technical expertise and professionalism, engineers are also expected by society and their profession to maintain high standards of ethical conduct in their professional lives. This chapter will cover the most important issues of ethics insofar as they are relevant to careers in the various engineering disciplines. A brief concluding section will also summarize the various legal guidelines and requirements with which engineers are expected to be familiar.

## SUMMARY

To begin, here is a summary of the basic areas and kinds of questions to be covered:

1. Since engineering ethics is a branch of ethics, we start with a clarification of the nature of ethics, and how ethical values and concerns differ from other kinds of values and concerns.
2. Next, the specific concerns of engineering ethics will be examined and distinguished from other concerns about values which engineers may also have. Such non-ethical concerns include personal preferences, social or political values, and more technical professional values related to becoming and remaining an excellent engineer from a purely engineering or scientific point of view.
3. After the preliminary clarifications and distinctions above, we will examine the basic issues, concepts, and special topics of engineering ethics. Most of these are also included in or assumed by the codes of ethics; provided by various professional engineering societies. It will be useful to examine one of these codes in more detail, namely the National Council of Examiners for Engineering and Surveying (NCEES) Model Rules of Professional Conduct. The study of this code will give engineers a good understanding of the kind of real-world requirements, which professional engineering societies uphold for their members. We will supplement the discussion with some model an-

swers to typical ethics questions, which provide a useful test of familiarity with and understanding of the ethical issues covered in such codes.

## GENERAL ETHICS

### *Ethical standards exist independently of any particular group of experts.*

Generally, ethics is concerned with standards, rules, or guidelines for morally or socially approved conduct such as being honest or trustworthy or acting in the best interest of a society.

Not all standards or values are ethical standards though. For example, personal preferences and values such as individual choices of food or clothing are not important enough to qualify as ethical values or choices. Ethical standards apply only to conduct which has some significant effect on the people's lives.

Thus, if an engineer used a substandard grade of steel in the construction of a bridge, he or she would definitely have violated ethical standards because of the safety hazards potential. On the other hand, using an inferior brand of ketchup on one's fries does not violate any ethical standards, since at worst it would only violate personal, non-ethical standards of what tastes good.

Ethics and ethical standards should also be distinguished from matters of legality and legal standards. Roughly speaking, the distinction is that legal standards are defined in legal documents by some properly appointed legal body, and those documents and legal experts determine what the law is and who should obey it. Ethical standards exist independently of any particular group of experts (being accessible to all through the exercise of their own thinking capacities). A codification or written form of the standards merely describes or summarizes what those preexisting ethical standards are rather than (as in the case of the law) defining their very nature.

### *Even legal standards must give way to ethical standards in the case of conflicts.*

For example, if there were a legal requirement that any load-bearing beam in a bridge must be able to bear five times the average real-world stress on that beam, then an engineer could simply calculate whether the beam met the legal standard or not. However, an ethical standard for what would be an adequate safety factor would be a different matter, and could not be settled by calculation or appeal to a rulebook since an ethically concerned engineer could convincingly argue that conventional or routine safety factors are ethically inadequate and unacceptable. Such a person cannot be appeased merely by appeal to some rulebook or to current legal standards for such matters.

In ethics, its standards are always more important than any other standards. Thus, in the contrast between personal and ethical standards discussed above, if one's personal standards conflict with ethical standards, then one must suppress his or her personal standards to resolve the conflict in favor of the ethical standards. For example, if one personally values finding the highest paying job but it turns out that the job in question involves some ethically wrong conduct, then his or her ethical duty would be not to accept that job but to take a lower paying, more ethical job.

Even legal standards must give way to ethical standards in the case of conflicts. For example, suppose that a law governing motor vehicles did not require a vehicle recall when engineering defects were found in its construction. If a good case could be made that this was an ethically unacceptable situation, then the law would have to be changed to conform to the ethical standard. So, here too, other kinds of values or standards must be suppressed or adjusted in the case of a conflict with ethical standards, which are always of overriding importance.

## JUSTICE

Intuitively, it is ethically wrong to treat people in unjust or unfair ways. People should be given what they deserve, and all people should be treated equally. Still, reducing justice to a series of rules is difficult, and justice may even conflict with a utilitarian approach since what is best for most may be unfair to minorities.

The U.S. Declaration of Independence talks about the right to life, liberty, and the pursuit of happiness. As with justice, moral rights are not easily reducible to rules, and they may conflict with the greatest good for most people.

## ETHICS DISCUSSION QUIZ 1

1. A person's behavior is always ethical when he or she:
   A. Does what is best for oneself
   B. Has good intentions no matter how things turn out
   C. Does what is best for everyone
   D. Does what is most profitable

2. Which of the following ensures that behavior is ethical?
   A Following the law
   B. Acting in the best interest of a society
   C. Following non-legal standards for socially approved conduct
      a. All of the above
      b. II and III only
      c. None of the above
      d. I only

## ENGINEERING ETHICS

An ethics statement from the Engineers Council for Professional Development says, "Engineers shall hold paramount the safety, health, and welfare of the public in the performance of their professional duties." Engineers and technicians must uphold and advance the integrity, honor, and dignity of the engineering profession.

Engineering ethics is the study of moral issues and decisions confronting individuals and organizations involved in engineering and is the study of related questions about moral conduct, character, ideals, and relationships of peoples and organizations involved in technological development (Martin and Schinzinger, *Ethics in Engineering*).

Much of engineering ethics is an applied or more specific form of what could be called "general ethics," that is, ethical standards which apply to any human activity or occupation. For example, ethical duties of honesty, dealing fairly with other people, obeying the relevant laws of one's country or state, and so on apply in any situation. Thus, some ethical obligations of engineers, of this general kind, are also found in other occupations and activities.

## A conflict between personal standards and engineering ethical standards must be resolved in favor of supporting the relevant engineering ethics standards

Other kinds of standards found in engineering ethics are more specific but not unique to engineers, such as the principles governing the ethical activities of any professionals in their contacts with clients or customers. These are an important part of engineering ethics, and they show up in the large overlap in the contents of many ethics codes derived from different professions (e.g., engineering, law, medicine, academia).

Finally, some standards apply primarily to the engineering profession, such as those dealing with the proper, ethical manner of approval of designs or plans by managers which require them to have professional engineering qualifications and expertise. These standards are what make engineering ethics distinctive relative to other kinds of applied ethics, but they are only one small part of the whole of engineering ethics.

Now to the question of how ethical standards for engineers differ from other standards of value which engineers might use. The aim is to clarify which choices an engineer might make are relevant to engineering ethics as opposed to other choices, which are based on other values or standards.

Personal choices or values are irrelevant to engineering ethics. For example, an engineer who submits a bid to a potential client on a project might decide to factor in a higher profit rate for him/herself than is usual for such projects.

However, as long as an open bidding process exists (where others are free to submit possibly lower bids) and as long as the personal pricing scheme does not affect the quality of the work, then the pricing equation remains a purely personal or economic decision with no ethical implications.

On the other hand, if an engineer were to decide to maximize his/her profit by submitting a low bid and then secretly compromising the materials quality so as to achieve personal profit, that would be a violation of engineering ethics standards, which require the use of high-quality materials and construction methods. In this case, a conflict between personal standards (maximum profit) and engineering ethical standards (honoring the contract and using high-quality materials and methods) would have to be resolved in favor of supporting engineering ethics standards.

## *Engineering ethics codes include prohibitions on unethical behavior while off the job as well.*

These contrasts between personal versus ethical values apply to corporate or economic values versus ethical values. The above examples of legitimate versus ethically illegitimate maximization of profits would apply even if the motivation were the interest of a corporation or economic group. A corporation or engineering firm would be wrong to seek profit at the risk of lower engineering product or service quality. An engineer's duty is to uphold engineering ethics standards even if his/her job is at risk, such as when doing the ethically right action conflicts with the non-ethical interests of a boss or corporate structure.

We should also discuss the relation of engineering ethics issues to legal issues. Although legal and ethical values are distinguishable kinds of values, one still has an ethical obligation to obey the prevailing laws in a given jurisdiction. People voluntarily entering the engineering profession agree to abide by ethics codes which explicitly or implicitly require compliance with existing laws. Here, too. good engineering ethics standards will be in close harmony with existing legal codes governing engineering.

Another area of close harmony between different engineering ethics values is in technical or scientific values. Any scientific community dealing with pure sciences or applied sciences, such as engineering, will have technical values concerning the proper and correct ways to construct theories, carry out calculations, and make statistically satisfactory estimates where exact results are not possible. These actions by themselves are not ethical values (good science is not identical to good ethics). But engineers do have an ethical obligation to use good scientific methods at all times; so, there is a close correlation between practicing good science and being an ethical engineer.

One other topic should be mentioned here. Most engineering ethics concerns an engineer's behavior while engaged in professional engineering activities. But be aware that engineering ethics codes also include prohibitions on unethical behavior while off the job as well if those activities would affect public perceptions of one's professional integrity or status. This would include activities such as gambling which might tend to bring the profession of engineering into disrepute, deceptive or inappropriate advertising of services, or any other activities suggesting a lack of integrity or trustworthiness.

## ETHICS DISCUSSION QUIZ 2

1. Engineers should follow their professional ethics code because:
   A. It helps them avoid legal problems, such as getting sued
   B. It provides a clear definition of what the public has a right to expect from responsible engineers

     C. It raises the image of the profession and, hence, gets engineers more pay

     D. The public will trust engineers more once they know engineers have an ethics code

2. Engineers should act ethically because:
   A. If they do not, they risk getting demoted or fired
   B. The boss wants them to
   C. It feels good
   D. That is the way responsible engineers behave

3. The first and foremost obligation of registered professional engineers is to:
   A. The public welfare
   B. Their employer
   C. The government
   D. The engineering profession

4. Registered professional engineers should undertake services for clients only when:
   A. They really need the fees
   B. Their own bid is the lowest one
   C. They are fully technically competent to carry out the services
   D. Carrying out the services wouldn't involve excessive time or effort

## THE BASIC ISSUES, CONCEPTS, AND SPECIAL TOPICS OF ENGINEERING ETHICS

We will use the NCEES Model Rules of Professional Conduct as the basis for our discussion. Its study will help you understand the typical ethical requirements imposed by professional engineering societies.

## THE PREAMBLE

The Preamble (the Introductory section) to the Rules is important in that it describes NCEES's purpose, which is to safeguard life, health, and property, to promote the public welfare, and to maintain a high standard of integrity and practice among engineers. We will paraphrase the code's statements for clarification.

This part of the code makes its connection explicit with ethical concerns to which all ethical persons should adhere. We will examine each of the items covered in turn in this statement of the purpose.

*Engineers will be honest and trustworthy and maintain high standards of professional conduct and scientific expertise.*

**1. Normative ethical theories universally agree that to cause harm to people is ethically wrong.** This is why the Preamble mentions the purpose of "safeguarding" life, health and property. Ethical persons, including engineers, should avoid doing anything which would damage or adversely affect other people. More positively, one should also take measures which will safeguard or preserve people from future harm.

For engineers, the more positive measures would include such things as building devices with extra "fail-safe" features included which make harmful consequences of their use as unlikely as possible. Overall, then, ethical persons take great care not to cause harm to others, and they take whatever extra steps are necessary to minimize risks of potential harm to others as well.

**2. The code's second purpose is to ". . . promote the public welfare."** Standard normative ethical theories support this rule. Engineers have a duty or obligation not only to act in a harmless or safe way but also to take active, professional steps that result in definite benefits and improved conditions for the general public.

For example, as an engineer planning a new highway this rule would require one not only to plan and build it safely but to do such things as choose the shortest feasible route between its endpoints or to choose that route which would permit the most efficient road construction techniques. The engineer, therefore, maximizes the highway's utility and minimizes its cost for the general public.

**3. The code's last purpose in the Preamble, is ". . . to maintain a high standard of integrity and practice among engineers."** Thus, the code ensures engineers' honesty and trustworthiness, and maintains high standards of professional conduct and scientific expertise.

The Preamble explains why engineers should adhere to the code. Adherence is ethical and emphasizes the vital practical benefits to society which can only be achieved by engineers committed to following the code at all times.

## OTHER PREAMBLE ISSUES

*"Engineering registration is a privilege and not a right. This privilege demands that engineers responsibly represent themselves before the public in a truthful and objective manner."*

A privilege is a socially earned license to do something granted by society under certain conditions. For example, a driver's license requires the passing of a written and a driving test. Similarly, an engineering registration must be earned by recipients (they do not automatically have a right to that status) and the granting of it requires them to adhere to ethical standards.

*"Engineers must compete fairly with others and avoid all conflicts of interest while faithfully serving the legitimate needs and interests of their employers and clients."*

This encapsulates a range of ethical requirements, which are more fully covered in the code's body. The various issues it raises will be discussed along with those requirements below.

## THE ENGINEER'S OBLIGATION TO SOCIETY

When developing a new technology, engineers and technicians must consider its impact on society. When considering the new technology, two types of decisions must be made.

## RISK BENEFIT ANALYSIS

The definition of risk is "the perceived extent of possible loss". Does the risk of building the product outweigh the negative societal impact? Risk analysis helps assess the risks. When a company uses a risk analysis, it should be cautious in order to remain focused on the plan's original goals and intents. This cautiousness will produce strategies used to control risks and keep the study cost-effective. What one person defines as minimal risk will be different from another person's definition.

*Risk = probability of event × cost of event*

The code's first group of rules addresses the engineer's obligation to society. As before, we will paraphrase the code's statements so as to clarify their meaning:

1. While performing services, the engineer's foremost responsibility is to the public welfare.

This rule is also featured in the Preamble, as a duty to promote the public welfare. The idea of responsibility to the public welfare includes the idea of safeguarding the public from harm:

2. Engineers shall approve only those designs that safeguard the life, health, welfare, and property of the public while conforming to accepted engineering standards.

Designs and materials may be unacceptable because of wider issues about the public interest. These two rules together imply a broader context of responsibility for engineers than those arising from any one task or project. Designs and materials that seem perfectly adequate and ethically acceptable within the bounds of a given project may nevertheless be unacceptable because of wider issues about the public interest.

For example, until recently a refrigeration engineer could have specified Freon (a chlorinated fluorocarbon, or CFC, product) as the prime refrigerating agent for use in a product, and could have defended it as an efficient, inexpensive refrigerant with no risks to the appliance purchaser. However, these chemicals pose significant risks to the public at large because of the long-term damage to the environment when they leak out. Rules one and two tell engineers they must keep such wider, possibly longer-term issues in mind on every project.

3. If an engineer's professional judgment is overruled resulting in danger to the life, health, welfare, or property of the public, the engineer shall notify his/her employer or client and any appropriate authority.

## *The duty is to be forthcoming about all pertinent or relevant information in reports.*

This important rule, which may place the engineer in a difficult position if his or her employer or client is contributing to the problem states the engineer's duty is clear: ". . . any authority that may be appropriate" must be notified, even if the employer/client tries to prevent it. (Cases of this kind are popularly referred to as whistleblowing.)

4. Engineers shall be objective and truthful in professional reports, statements, or testimonies and shall provide all pertinent supporting information relating to such reports, statements, or testimonies.

5. Engineers shall not express a professional opinion publicly unless it is based upon knowledge of the facts and a competent evaluation of the subject matter.

Rules four and five together implement the general ethical requirement that a person ought to tell the truth in the specific context of an engineer's professional duties. Note that the duty as mentioned in rule four is not only to be truthful in what is said but also to be forthcoming about all pertinent or relevant report information.

Rule four also mentions being objective, which adds the element of being unbiased and basing beliefs and reports only on objective, verifiable fact or theory.

Rule five enlarges on the idea of being objective in reports. Others should be able to rely upon a professional opinion, and they can do this only if they know all of the relevant facts and are completely competent to evaluate the matter.

6. Engineers shall not express a professional opinion on subject matters for which they are motivated or paid, unless they explicitly identify the parties on whose behalf they are expressing the opinion and reveal the parties' interest in the matters.

Rule six expresses what is sometimes called the duty of full disclosure. Even if engineers honestly seek to be truthful and objective (see rules four and five), doubts might be raised about an engineers' motivation or objectivity unless they reveal on whose behalf they are expressing an opinion, and the interests that such persons have in the case. This rule is also related to the issue of conflicts of interest. (See rules six and eight.)

7. Engineers shall not enter business ventures or permit their names or their firm's names to be used by any person or firm which is engaging in dishonest, fraudulent, or illegal business practice.

This "clean hands" rule (shake hands only with those whose hands are as ethically clean as your own) shows that to be completely ethical in your own or your company's practices is not sufficient. Engineers must ensure others do not profit from their own good name with unethical activities. Rule seven, clearly related to rule one, concerns the public welfare in that engineers must promote this rule in their external dealings just as much as in their own activities.

8. Engineers who have knowledge of a possible violation of any of the rules listed in this and the following two parts shall provide pertinent information and assist the state board in reaching a final determination of the possible violation.

Rule eight generalizes rule three (a duty of disclosure when an engineer's professional judgment is overruled). In terms of the public welfare, each profession must regulate itself in this way to minimize or eliminate future rules infringements. Strict adherence to this rule will lead to wider appreciation and respect for the engineering profession because of its willingness to "clean its own house."

## THE ENGINEER'S OBLIGATION TO EMPLOYERS AND CLIENTS

The second decision, a cost benefit analysis, asks if the cost justifies the product? Cost benefit analysis is easy to use and is used in many fields. This analysis helps businesses and organizations decide what direction they should move. When a company chooses to use this technique, it will add the benefits of a course of action and subtract the costs associated with it. Costs can be a one-time event or can be spread out over time. Benefits are usually associated with long-term measurements. Time is an effect of the analysis and can be built into the analysis by using a payback period. (The time it takes for the benefits of a change to repay the costs is a payback period.) Remember to determine a specific time period to look for a payback period. For example if a company was to purchase computer system for $30,000 and if the payback period was 5 years, the system would cost $6,000 for those first 5 years and then would be paid for. The question the company would needs to ask is if the computer system will be current 5 years from now?

The code's second group of rules addresses the engineer's obligation to employers and clients.
1. Engineers shall not undertake technical assignments for which they are not qualified.

2. Engineers shall approve or seal only those plans or designs that deal with subjects in which they are competent and which have been prepared under their direct control and supervision.

*An important kind of trust for a professional engineer is to be trustworthy, honest, not reveal confidential information.*

Rules one and two require an engineer to be professionally competent, in undertaking technical assignments and in approving plans or designs. Rule two requires a double kind of knowledge, i.e., technical competence in the approved matters and direct control and supervision over their preparation. Engineers can only assure their legitimate and warranted approval in this way.

3. Engineers may coordinate an entire project provided that each design component is signed or sealed by the engineer responsible for that design component.

Rule three, in effect, invokes rule two. As long as each component of a project is satisfactorily approved, per rule two, then an engineer may coordinate an entire project. Rule three underlines rule two's importance as project managers have to rely on the approvals' validity for each project part.

4. Engineers shall not reveal professional information without the employer's or client's prior consent except as authorized or required by law.

This confidentiality requirement is the other side of an engineer's duties (See rules four and five of the first Obligations to Society section). Just as engineers must not lie or misinform, they must also restrict to whom they reveal professionally relevant information. Being trusted not to reveal confidential information is an important kind of trust.

Confidentiality is a central factor in assuring employers and clients that a person's professional service for them is for them alone. The clients and/or employers must know their engineers will not reveal any private information without their full consent. Rule four is related to rules five through eight below, in that any revealing information to others would probably create conflicts of interest and problems inherent in "serving more than one master".

5. Engineers shall not solicit or accept direct or indirect considerations, financial or otherwise, from contractors, their agents, or other parties while performing work for employers or clients.

Rule five is the first of four rules dealing with conflicts of interest in which an engineer has a primary professional interest or group of interests, to carry out a project for an employer or client. The conflict of interest arises when other factors activate non-professional interests, which then conflict with the primary interests.

In rule five, soliciting or accepting such things as gifts, hospitality, or suggestions of future job offers activates non-professionally related, personal interests (e.g., for additional pay, career advancement) which would be in conflict with an engineer's primary professional interests and duties.

Pay special attention to this rule and other conflict of interests; you must follow these rules. People sometimes think that breaking these rules is ethically harmless; they believe that if they have a strong enough character, they will not be pro-

fessionally influenced in a detrimental way and, hence, accepting such inducements cannot do any harm.

### Even the appearance of a conflict of interest can create serious ethical problems.

The appearance of a conflict of interest (however careful one is to avoid under-
mining his or her professional interest) can create serious ethical problems. This
conflict could cause a potential loss of trust from an employer or client. Just as
clients need to know that the engineers will keep their information confidential (as
in rule four), they also need to know that the engineer is single-mindedly working
with only their interests at heart.

Any doubts raised because of the appearance of a conflict of interest could
damage the client/engineer professional relationship.

Further rules are necessary to deal with conflicts of interest because the appearance or possibility of a conflict of interest may be unavoidable no matter how
ethically careful everyone is. However, a powerful method is available to minimize
any ethically bad effects. This method of full disclosure of potential conflicts to all
interested parties is addressed in the following two rules.

6. Engineers shall disclose to their employers or clients potential conflicts of
   interest or any other circumstances that could influence or appear to influence their professional judgment or their service quality.

7. An engineer shall not accept financial or other compensation from more
   than one party for services rendered on one project unless the details are
   fully disclosed and agreed to by all parties.

Rules six and seven address issues of full disclosure and keeping all parties
informed as to areas of potential conflict or potentially undue external influences.
The basic idea behind full disclosure is that maintains trust and confidence between all parties as detailed below.

First, if engineer A informs party B about a potential conflict or influence, then A
has been honest with B and has maintained or reinforced B's trust in A. Further-
more, if B is not further concerned about the matter, then the potential conflict or
potentially bad influence on A's professional conduct has been completely defused.

Suppose, on the other hand, that B is initially concerned about the issue after
A revealed it to B. Even so, the problem is already lessened because A has revealed the area of concern. Things would be much worse if B later discovered the
problem. If A had not fully disclosed the issue, the trust between A and B would
have been destroyed.

### When a governmental body is involved, another interested party is not directly represented in negotiations, namely the electorate.

Now that B knows about the area of concern and knows that A has fully cooperated in disclosing the problem, both can proceed to work out mutually acceptable

ways of minimizing or disposing of the problem. Even if the full disclosure leads to an initial problem which needs to be resolved, the problem does not break down the trust between A and B. In fact it may even reinforce trust, in that A's willingness to disclose a conflict/influence and negotiate with B is evidence of A's professional honesty and sincerity.

8.  To avoid conflicts of interest, engineers shall not solicit or accept a professional contract from a governmental body on which a principal or officer of their firm serves as a member. An engineer who is a principal or employee of a private firm and who serves as a member of a governmental body shall not participate in decisions relating to the professional services solicited or provided by the firm to the governmental body.

Rule eight deals with a special case of potential conflicts of interest, namely, when one of the interested parties is a governmental body. In such a case, a somewhat stricter rule is required than for the involvement of non-governmental agencies.

In non-governmental cases, it is ethically sufficient to disclose potential conflicts of interests to all parties and then to negotiate with the other parties as to how to deal with the potential conflicts. For example, if one's engineering firm has an official who was also on the board of directors of a bank, accepting a professional contract from that bank would be ethically permissible as long as all parties are fully informed about the official's joint appointment prior to the agreement and as long as they agree the joint appointment is not an impediment to their signing a contract.

The reason for rule eight is as follows: In the case of agreements among private persons or businesses, they are the only parties having a legitimate interest in the negotiations. Whatever they decide among themselves (assuming that no other ethical rules or laws are being broken) is acceptable.

On the other hand, when a governmental body is involved, another interested party is not directly represented in negotiations, namely the electorate or citizens of the jurisdiction covered by that governmental body. The governmental body must act in ways which respect the interests and concerns of the electorate. In such a case, ensuring that full disclosure of the potential conflicts of interest to all of the citizens of the electorate is impossible. Therefore, rule eight completely prohibits this kind of conflict of interest. Only then can the public trust in the engineering profession and in governmental bodies.

## AN ENGINEER'S OBLIGATIONS TO OTHER ENGINEERS

The code's third group of rules addresses an engineer's obligations to other engineers. These rules are more precise specifications of rules already introduced when applied to the specific context of obligations to other engineering professionals. The first rule also covers obligations to potential employers (whether or not they are engineers) when one is seeking employment.

1. Engineers shall not misrepresent or permit misrepresentation of their or any of their associate's academic or professional qualifications. They shall not misrepresent their level of responsibility or the complexity of prior assignments. Pertinent facts relating to employers, employees, associates, joint ventures, or past accomplishment shall not be misrepresented when soliciting employment or business.

*This duty not to misrepresent applies to issues of prior levels of responsibility and to the previous assignments' complexity.*

This rule one is an application or more specific form of rules four and five of the first section (concerning an engineer's obligation to society), and it requires objectivity and truthfulness in all professional reports, statements and opinions.

Rule one requires a person not to misrepresent their own qualifications, or those of associates, which is an important kind of truthfulness and objectivity. The rule further specifies that this duty not to misrepresent applies also to issues of prior responsibility levels and to the previous assignments' complexity as well. The rule concludes with a statement about any and all pertinent facts in one's history, and it requires that they, too, should not be misrepresented.

Rule one may seem unnecessary since it applies general ethical principles already in the code to special cases and circumstances centering on issues of employment and qualifications. However, as with issues of conflicts of interest, special temptations exist, which are best addressed by spelling out what is ethically required in such situations.

For instance, even well-meaning, honest engineers might be tempted to make their qualifications seem more impressive to a potential employer. They might rationalize that they are the best people for the job, and therefore they are doing the employer a favor by making their qualifications seem more impressive. That employer would then hire them instead of less-qualified people whose paper qualifications might misleadingly look as good.

Many variants on this approach would involve rationalization (inventing dubious reasons for what one wants to believe) and wishful thinking instead of objectivity, rationality, and truthfulness. So, rule one serves a useful function in explicitly stating requirements which some might be tempted to ignore or overlook.

2. Engineers shall not directly or indirectly give, solicit, or receive any gift or commission, or other valuable consideration, in order to obtain work, and shall not make a contribution to any political body with the intent of influencing the award of a contract by a governmental body.

Rule two continues the prohibitions against conflicts of interest which were found in rules five through eight in the previous section (rules covering the engineer's obligations to employers and clients). Those previous rules mainly covered cases where an engineer was already employed, while rule two here specifically applies to attempts to obtain future work, including the award of a contract by a governmental body.

This emphasizes that attempting to influence someone else (a potential employer, for instance) is as wrong as it would be to allow others to influence oneself. Thus the rule underlines that it is as ethically unacceptable to cause conflicts of interest in others as it is to allow oneself to be trapped in improper conflicts of interest.

*It is ethically unacceptable to cause conflicts of interest in others as it is to allow oneself to be enmeshed in improper conflicts of interest.*

That the second part of rule two, ". . . and shall not make a contribution to any political body with the intent of influencing the award of a contract by a governmental body," specifically mentions the intent of the person making the contribution. Contributions are not forbidden, only contributions with the wrong intent.

This part of the rule could be difficult to apply or enforce in practice because establishing what an engineer's actual intent was in making a political contribution may be difficult. The freedom to make political contributions to organizations of one's own choice is generally viewed as an ethical right which should be limited as little as possible. So, this part of the rule depends on and appeals to the ethical conscience of the engineers, who must judge their own intentions in such cases and avoid such self-interested contributions in the manner prohibited by the rule.

Note also that the first part of rule two, "Engineers shall not directly or indirectly give, solicit, or receive any gift or commission, or other valuable consideration, in order to obtain work . . . ," mentions the reason or intention behind giving or receiving gifts in the phrase "in order to obtain work." However, in practice it is much easier to judge when gift giving is ethically unacceptable than when political contributions are unacceptable. This is because there are more behavioral and social tests for suspicious inducements to obtain work than there are for suspicious political support. So policing and regulating infringements of this first part of rule two is easier than for the second part.

3.  Engineers shall not attempt to injure, maliciously or falsely, directly or indirectly, the professional reputations, prospects, practice or employment of other engineers, nor indiscriminately criticize the work of other engineers.

Rule three is somewhat unclear, so some discussion will bring out its ethically legitimate core.

First, how should the subordinate phrases "maliciously or falsely" and "directly or indirectly" be interpreted? On one possible interpretation, Rule three says outright that engineers should never attempt to injure in any way or for any reason the professional reputations of other engineers. On this interpretation those phrases just give examples of possible modes of injury which are prohibited, leaving unmentioned any other possible modes of injury which are nevertheless also assumed to be prohibited.

However, another interpretation is possible, according to which it is only certain kinds of injury which are prohibited by rule three, namely those spelled out by those same phrases interpreted so that the "directly or indirectly" part modifies the "maliciously or falsely" part. On this interpretation, rule three prohibits only malicious or false attempts (whether carried out directly or indirectly) to injure the reputations of other engineers.

This second interpretation would ethically permit attempts to injure the reputations of other engineers, as long as the attempts were carried out in a non-malicious and honest, truthful way (and presumably with the public welfare in mind as well).

Some support for this second interpretation comes from rule three's final clause, "nor indiscriminately criticize the work of other engineers." In this case, not all criticism of other engineers' work is being prohibited, only indiscriminate criticism. Thus, the last section of rule three outlaws criticism which is overemotional, biased, not well reasoned or factually accurate, but it does not prohibit well-reasoned, careful, accurate, factually based criticisms of other engineers.

*If criticism of other engineers ever becomes necessary, it should be done cautiously and objectively, and with all due respect for the professional status of the person being criticized.*

In support of the second interpretation, not all attempts to injure reputations are being prohibited, only those which are malicious, false, indiscriminate or otherwise ethically questionable.

However, something can be said in favor of the first interpretation as well (which involves a blanket condemnation of all attempts to injure reputations). Those defending it might say it is arguable on general ethical grounds that any attempt to injure someone's reputation must be viewed as going too far and therefore unethical. Even if one is convinced that someone else's work is poor, dishonest, at most one has a duty to point out the problems and shortcomings in the work. It is a big leap from criticizing an engineer's actions, on the one hand, to condemning the engineer in a way designed to injure his or her reputation, on the other hand.

One should criticize an engineer only out of a disinterested desire for the truth and not with the aim of injuring someone's reputation. others must judge whether the highlighted truth will diminish the reputation of the engineer.

Fortunately, deciding between these different interpretations of rule three is unnecessary. What is important is sensitivity to the ethical issues involved in each interpretation. And for the purposes of conforming to rule three, both sides can agree that if criticism of other engineers ever becomes necessary, it should be done cautiously and objectively, and respect for the person's professional status.

## ETHICS DISCUSSION QUIZ 3

1. With respect to the Moral Rules of Professional Conduct for engineers:
    A. The rules are a bad thing because they encourage engineers to spy on and betray their colleagues.

B. The rules are a useful legal defense in court when engineers can demonstrate that they obeyed the rules.

C. The rules enhance the image of the profession and hence its economic benefits to its members.

D. The rules are important in providing a summary of what the public has a right to expect from responsible engineers.

2. The Model Rules of Professional Conduct require registered engineers to conform to all but one of the following rules. Which rule is not required?
   A. Do not charge excessive fees
   B. Do not compete unfairly with others
   C. Perform services only in the areas of their competence
   D. Avoid conflicts of interest

3. You are a quality control engineer, supervising the completion of a product whose specification includes using only U.S.-made parts. However, at a late stage, you notice that one of your subcontractors has supplied you with a part having foreign-made bolts in it; these are not noticeable and would function identically to U.S.-made bolts. Your customer urgently needs delivery of the finished product. What should you do?
   A. Say nothing and deliver the product with the foreign bolts included, hoping the customer will not notice them.
   B. Find (or, if necessary, invent) some roughly equivalent violation of the contract or specifications for which the customer (rather than your company) is responsible. Tell them you will ignore the violation if they will ignore your company's violation.
   C. Tell the customer about the problem. Let the customer decide what to do next.
   D. Put all your efforts into finding legal loopholes in the original specifications, or in the way they were negotiated, to avoid your company's appearing to have violated the specifications.

4. You are the engineer on a building project which is behind schedule and urgently needed by the clients. Your boss wants you to certify some roofing construction as properly completed even though you know some questionable installation techniques were used. What should you do?
   A. Certify it, and negotiate a raise from your boss as your price for doing so.
   B. Refuse to certify it.
   C. Tell the clients about the problem, saying that you will certify it if they want you to.
   D. Certify it, but keep a close watch on the project in case any problems develop.

5. You are an engineer and a manager at an aerospace company with an important government contract supplying parts for a space shuttle. As an engineer, you know that a projected launch will face unknown risks because

the equipment for which you are responsible will be operating outside its tested range of behaviors. However, since you are also a manager, you know how important it is to your company that the launch be carried out promptly. What should you do?

A. Allow your judgment as a manager to override your judgment as an engineer and permit the launching.

B. Toss a coin to decide since one's engineering and managerial roles are equally important, and neither should take precedence over the other.

C. Abstain from voting in any group decision on the matter since as a manager and an engineer, you have a conflict of interest.

D. Allow your judgment as an engineer to override your judgment as a manager and not permit the launching

6. Your company buys large quantities of parts from various suppliers in a competitive market sector. As a professional engineer, you often make critical decisions on which supplier should be used for which parts. A new supplier is eager to get your company's business. Not only that, you find it is eager to provide you with many benefits, e.g., free meals at high-class restaurants, free vacation weekends for (supposedly) business meetings and demonstrations, and other more confidential and expensive gifts. What should you do?

A. Do not accept any of the gifts that go beyond legitimate business entertaining, even if your company would allow you to accept such gifts.

B. Report all the gifts, etc., to your company, and let it decide whether or not you should accept them.

C. Accept the gifts without telling your company because you know that your professional judgment about the supplier will not be biased by the gifts.

D. Tell other potential suppliers about the gifts, and ask them to provide you with similar benefits so you will not be biased for any particular supplier.

## ENGINEERING ETHICS AND LEGAL ISSUES

This section will provide an introduction to the various ways in which the law regulates and affects the engineering profession. We will discuss the impact of laws and regulations on various engineering ethics issues.

Engineers, like any citizens, are expected to obey society's legal rules and regulations. Here, we will concentrate on those laws and legal concepts with special relevance to engineers and the engineering professions. Every engineer needs to understand these legal matters.

## CONTRACT LAW

A contract is a mutual agreement between two or more parties to engage in a transaction which provides benefits to each of them. Here is a breakdown of this concept, each of which is required for a contract's valid existence. For simplicity, we will concentrate on two-party contracts.

1. Mutual consent. Each of the contractual parties must consent to the contract and agree to be bound by its terms.

*Without evidence of definite benefits for each party, it would often be impossible to decide whether or not each party had fulfilled its side of the agreement.*

2. Offer and acceptance. One of the parties must make an offer(e.g., to manufacture some equipment, carry out some services), and the other party must accept it.

3. Consideration. Each party must provide something of value to the other party.

The first two conditions, mutual consent and offer/acceptance, are needed because discussing a possible contract in provisional terms establishes no legally binding agreement. A contract exists only when each party has established that the other consents to it, and that one is offering something which the other has decided to accept. In other words, a legally enforceable agreement requires a definite promise by each party to do something specific.

The issue of consideration is also important but in more complex ways. The idea behind it is that a legally enforceable contract cannot exist unless each side stands to benefit from the contract. After the contract is fulfilled, each has received some benefit which they did not have prior to being fulfilled.

Without such evidence of definite benefits for each party, deciding if each party had fulfilled its side of the agreement might be impossible. For example, if an engineer were to hire relatives to do some work and paid them for doing it but did not specify exactly what the work would be, then the courts would likely rule that no valid contract existed and so the payments to his/her relatives were illegal or invalid.

## A contract does not have to be in writing to be valid.

Another kind of example where lack of consideration raises legal and ethical questions is as follows: If an engineering firm provided certain services to a client but never asks for nor receives any payment for the services, a strong suspicion would be that the services provided were some kind of bribe or illegal payback for other hidden services or benefits rather than part of a valid contract between the parties.

Some cases involving the issue of consideration are less clear-cut. For example, if an engineer decides to undertake a difficult and complex project and does so at a rate below standard prices, could such a contract be held invalid because of the inadequate reward provided to the engineer? Generally, the courts have held that the contractual parties must decide for themselves what constitutes adequate consideration or reward. So, the law will not provide protection to engineers against any possibly self-destructive intentions or other errors of judgment in which they may become involved.

On the other hand, issues about specific amounts of consideration or reward are central in settling contractual disputes between parties. (See the section on breach of contracts below.) A contract does not have to be in writing to be valid. For example, an engineer could make a contract over the phone with a client. A written contract is desirable for clarity, and it provides documentation should any questions about the terms and conditions of the contract arise. Engineers should prepare specific documentation on any agreements they make as soon as possible after the agreements are finalized.

## BREACH OF CONTRACT

One of the main values of having a legally enforceable system of contracts is case of any disputes between the parties, the contract provides an independent check or test of the validity of the party's claims in a dispute. For example, if a client orders some parts from an engineering firm but is dissatisfied with them and wants replacements or a refund, then the issue can be resolved by checking the contract to see if the parts conform to what is in the contract. If the engineering firm has carried out its part of the contract by supplying parts as specified, then the client has no case against it. Each party to a contract must decide ahead of time whether it wants what is in the contract. Each party cannot blame the other party to the contract even if unhappy about the agreement's outcome .

*For an actual breach of contract to occur, an actual violation of the terms of the contract must occur.*

If specified items were not supplied, or were supplied but were of substandard quality, or if items were not supplied until long after a deadline had expired, then the other party could claim a breach of contract and that it is entitled to compensation or contract termination.

Those suing for contract breach will want to obtain damages or compensation to recover the value they would have obtained under the contract had it not been violated. In other words, the other party is being required to provide an equivalent value which it had previously offered under the contract provisions.

Material breaches concern a contract's vital elements, the presence of which would justify termination and a suit for damages. Less severe deviations from a contract (such as a contractor not cleaning up a site at the end of a day's work or being a few days late in delivering a product) would be handled differently. Immaterial breaches will require the offending party only to make reparations or to accept a reduced fee for their work.

*We should "read between the lines" in terms of the intent of those documents as understood by those who formulated them.*

Often the distinction between a material versus immaterial breach of a contract will depend on issues about consideration (which is the value being received by each contract party).

For example, suppose engineering firm A fails to supply client B with certain parts X as specified in a contract, but supplies substitute parts Y instead. If the contract specified parts X, then a contract breach has occurred. Whether the breach is material or immaterial will depend on whether the substitute parts Y are similar enough in value and quality to parts X so client B will accept them as a substitute for parts X. Intuitively, if B is as well off (or nearly so) with parts Y as with parts X, then the contract breach would be immaterial. However, if parts Y are substantially worse than parts X, the contract breach would be material requiring stronger legal action.

This material/immaterial distinction has ethical implications. Any party to a contract unable to fulfill the contract provisions is under an ethical and legal imperative to do everything possible to provide an equivalent value to the other party.

## THE LETTER VERSUS THE SPIRIT OF THE LAW

This important distinction comes about because laws or contracts do not have specific terms and are open to varying degrees and types of interpretation in problematic cases.

How should such interpretations, or attempts to read between the lines, be carried out? A standard view is to do so in terms of the intent or desired interpretations of those documents as understood by those who formulated them.

In this way ethical standards enter as an important element in understanding the spirit of laws and contracts. Lawmakers or contracting parties wrote their rules or agreements with ethical standards in mind as to what would constitute the best and most ethically acceptable laws or contracts, given the purposes they wanted to achieve. Basic issues of engineering ethics are relevant even in cases when dealing with legal documents, insofar as such documents leave any room for disputes about proper interpretation.

## CASE DISCUSSIONS

As with any form of ethics, engineering ethics demands more than an impersonal, spectator point of view from us. Ethical problems are not simply some other person's problem to be handled with impartial advice or citing of ethical rules. They are, instead, a vital problem which must be resolved in livable terms.

To encourage this perspective, many of the cases are presented in the second person, i.e., as problems for you rather than some abstract person. In answering them, maintain this perspective, and explain how you would deal with each situation.

### Case One

Your company urgently needs a new manager in another division, and you have been invited to apply for the position. However, one of the job requirements is competence in the use of sophisticated computer-aided design (CAD) software.

You do not have that competence, but you are enrolled in a CAD course, so you will have that competence but not before the application deadline.

The application form for the job requires you to state if you have the necessary CAD software competence. Is it ethically permissible to claim that you have (in order to get the job) even though strictly speaking this is not true?

Would your answer be any different if your current boss informally (off the record) advised you to misrepresent your qualifications, his or her reason being that you are the best qualified person for the new managerial position?

## Case Two

You are a supervisor for a complex design project, and are aware of the NCEES code rule according to which engineers may coordinate an entire project provided that each design component is signed or sealed by the engineer responsible for that design component.

You are overworked and have got into the habit of, in effect, rubber-stamping the design decisions of your subordinate engineers. You do not test or check their work and routinely approve their plans (after they have signed or sealed them) with only a superficial examination. You justify this behavior because the NCEES rule says nothing about the supervisory responsibilities of someone in your position. Besides, you trust the engineers to do a good job and see no need to check up on them. Is the supervisor's position here ethically justifiable?

## Case Three

Newly hired as a production engineer, you find a potential problem on the shop floor: Workers are ignoring some government-mandated safety regulations governing the presses and stamping machines.

The workers override safety features such as guards designed to make it impossible to insert a hand or arm into a machine. They rig "convenience" controls, so they can operate a machine while close to it instead of using approved safe switches, etc., which requires more movement or operational steps. Their reason (or excuse) is if they followed the safety features strictly, production would be difficult, tiring, and inefficient. They feel that their shortcuts provide safe operation with improved efficiency and worker satisfaction.

Should you immediately insist on full compliance with all the safety regulations, or do the workers have enough of a case so that you would be tempted to ignore the safety violations? If you were tempted to ignore the violations, how would you justify these actions to your boss?

How much weight should you give to the worker's clear preference for not following the regulations? Ethically, can you relax safety standards if those to whom they apply want relax them?

## Case Four

Your automotive company is expanding into off-road and all-terrain vehicles (ATVs). As a design engineer, you are considering two alternative design concepts for a new vehicle, one having three wheels and another with four.

Engineering research has shown that three-wheeled vehicles are considerably less stable and safe overall than four-wheeled ones. Nevertheless, both designs would satisfy the existing safety standards for the sale of each vehicle class, so either design could be chosen without any legal problems.

However, market research has shown that a three-wheeled version would be easier to design and produce and would sell more vehicles at higher profit margins than would a four-wheeled version.

As an engineer, can you ethically recommend that your company should produce the three-wheeled version in spite of its greater potential safety risks? Do engineers have an ethical obligation to recommend the safest possible design for any new product?

## Case Five

Your company has supplied prefabricated wall sections, which you designed, to construction companies. One day, you have a new idea how these might be fabricated more cheaply using composites of recycled waste materials.

Pilot runs for the new fabrication technique are successful, so it is decided to switch to the new technique on all future production runs for the prefabricated sections. But you enter managerial debates about how, or even if, to inform the customers about the fabrication changes.

The supply contracts, written with specifications in functional terms so load-bearing capacities and longevity, etc., of the wall sections, were specified but no specific materials or fabrication techniques were identified. Thus, it would be possible to make the changeover without any customer contract violation.

Since the new fabrication method will save money, does your company have an ethical obligation to inform its customers and perhaps renegotiate supply at a reduced cost, so its customers share in the technique's benefits?

More specifically, do you have any special duty, as a professional engineer and designer, to advocate that you should inform customers of the new technique and the cost savings?

## Case Six

Your company manufactures security systems. These have raised few ethical problems since your products were confined to traditional forms of security, e.g., using armed guards, locks, and reinforced alloys.

However, as a design engineer you realize you could provide more comprehensive security packages using modern technology. These packages could include extensive video and audio surveillance equipment along with biometric monitoring devices of employees or personnel seeking entry to secure areas and would use highly personal data such as a person's fingerprints, or retinal or voice patterns.

A literature search reveals ethical concerns about the collection and use of personal data. For example, this high-tech surveillance could become spying, carried out without the employees' knowledge and violating their privacy. The data could be sold or used outside legitimate workplace by unscrupulous customers.

Your boss wants you to include as much of this advanced technology as possible because customers like these new features and are willing to pay well for them.

However, you are concerned about the ethical issues in making these new technologies available. As an engineer, do you have any ethical responsibility to exclude ethically questionable technologies in products which you design and sell, or should you include them in forms difficult to misuse? Is the misuse of such technologies an ethical problem for your customers and not for you?

## LEARNING ACTIVITIES

### Activity 12.1    Gilbane Gold

Time:     2 hours

## CASE STUDY

This case was originally prepared by the National Institute for Engineering Ethics of the National Society of Professional Engineers. It is a fictional but highly plausible case suggested by actual situations. Engineers will identify with the junior environmental engineer, David Jackson, easily who is caught between his desire to be a good employee and his sense of obligation as an engineer to protect the public's health, safety, and welfare.

Although this case's primary ethical issue is whistle blowing, secondary ethical issues include the engineers' obligations with respect to environmental issues, management problems having to do with honesty and trust between business and its host community, the issue of a community's fairness toward local manufacturing plants, the problems raised for individuals and groups by the necessity for action in the face of inconclusive scientific evidence, and the relationship of law and morality.

The case takes place in the imaginary town of Gilbane. The sludge from the Gilbane sewage plant has been used for many years as a fertilizer and is sold under the name "Gilbane Gold." The revenue from the sale of Gilbane Gold enables the city to supplement its tax revenues, saving a family of four approximately

**The National Institute for Engineering Ethics (NIEE) is an official component of the Murdough Center for Engineering Professionalism in the College of Engineering at Texas Tech University. NIEE was initially created by the National Society of Professional Engineers (NSPE), in 1988, and became an independent organization in 1995.**

$300 per year in taxes. In order to protect this source of income, the town placed severe restrictions on the discharge of heavy metals into the sewage, so the sewage would be safe for use by farmers as fertilizer. The restrictions were ten times more stringent than federal regulations.

Before implementing these regulations, Gilbane had aggressively marketed itself as a city with a good business climate, offering tax abatements to industries that chose to move there. After several high-tech firms moved to the area, the more stringent regulations were enacted. Z CORP was one of the companies that moved to Gilbane. Its Gilbane plant manufactures computer components, but the plant's manufacturing process creates substantial quantities of toxic materials, primarily heavy metals. Z CORP monitors its waste discharge monthly.

Two facts about the regulations affect the case's resolution. First, plants in Gilbane are responsible for supplying test data to the city. The data must be signed by an engineer, who attests to its accuracy. The law governing effluents is flawed, however, for it only regulates effluent discharge in terms of the amount of toxic material for a given volume of discharge but not in terms of the total contaminant quantity. So, a plant can always operate within Gilbane standards by simply increasing the volume of discharge.

Second, a newer and more sensitive (but more expensive) test for heavy metals has been developed since the city enacted its standards. The newer test is not required by Gilbane, and the city does not use it. Z CORP employees have access to the test, and it shows that the plant has apparently been slightly exceeding the allowable emissions on a number of occasions. This produces a problem for Z CORP. If it discloses the new test's results, the city might take legal action against it. If it does not disclose the results, some of its own employees may believe it is exhibiting bad faith.

The plant's junior environmental engineer, David Jackson, is a new employee. He has replaced a consultant who believes he was released because of his warnings about the discharge of toxic materials. David is concerned about Z CORP's heavy metals discharge, and his concern is intensified when he learns that Z CORP has signed a contract that will result in a five-fold increase in the discharge of heavy metals. David finally decides to blow the whistle on the plant's discharge levels by talking to the local TV newscaster.

## RESOURCES

www.nspe.org The National Society of Professional Engineers (NSPE) is the only engineering society that represents individual engineering professionals and licensed engineers (PEs) across all disciplines. Founded in 1934, NSPE strengthens the engineering profession by promoting engineering licensure and ethics, enhancing the engineer image, advocating and protecting PEs' legal rights at the national and state levels, publishing news of the profession, providing continuing education opportunities, and more. NSPE serves some 60,000 members and the public through 53 state and territorial societies and more than 500 chapters.

www.niee.org The National Institute for Engineering Ethics (NIEE) gives a discount to educational institutions for the Gilbane Gold video. They charge $92 plus $8 shipping and handling, which includes a study guide for the video. To purchase the Gilbane Gold material, contact NIEE:

National Institute for Engineering Ethics
Box 41023
Lubbock, TX 79409-1023
Email: ethics@niee.org
Ph: 806-742-NIEE (6433)
Fax: 806-742-0444

## Activity 12.2     GPS Implants

Time:      7–10 hours

## CASE STUDY

George Orwell created "Big Brother" in his book, *1984.* We are and will continue to be a technological society that strives on knowing everything. Many global citizens do not like the thought of someone or something always watching their every move. In extreme circumstances, of course, someone might want relatives or authorities to find them. Imagine your grandfather suffering from Alzheimer's. Should your family make the choice to fit him with an implant to guarantee his safety should he need medical treatment and no one was around? These devices can record information like home address and medical information that would be beneficial when scanned by hospital personnel. Would this kind of technology help parents keep track of their toddlers and young children? Could it prevent children from being kidnapped? Individual families have to make these decisions for this procedure.

With technology's advance, newer devices will always become available; microchips have been designed for affordable implementation. Due to their small size, microchips can be placed in animals, humans, wearable objects, vehicles, etc. Their uses vary from Global Positioning System (GPS) devices to units carrying medical information. UPS and Federal Express use similar GPS devices to help track shipping information.

The difficulty with many of these devices is the loss of privacy. Many devices may be used for identification and informational purposes. Global citizens are reluctant to carry devices capable of providing their personal information to governments and private industry.

Your team of three has been charged with deciding where to implement the uses of these microchips. Your task is to research microchips and possible locations for microchip use. Finally, the students must propose an application for this technology. The device can be used in the public or private sector but must have a significant market that does not exist. The design must include the following:

**Implanted 100 pin array for directly linking the human nervous system to a computer**

1. Research report
   a. The primary use
   b. Listings of the pros and cons
   c. Any ethical dilemmas that need resolution
   d. Directions for use
   e. A cost benefit analysis
2. A scale model of the device
3. An electronic (PowerPoint) presentation
4. A computer-aided drawing of the microchip AND the object it is planted in
   a. Dimensions of the device to be included

## RESOURCES

http://www.kevinwarwick.com/ In 1998, Kevin Warwick shocked the international scientific community by having a silicon chip transponder surgically implanted in his left arm. A series of further implant experiments have taken place in which Kevin's nervous system was linked to a computer.

http://www.digitalangel.net/ The exclusive licensee under U.S. patents for Personal Tracking and Recovery Systems and In-building Location Systems. Marketer of the first consumer safety and location system that combines GPS location and biosensor technologies as well as provides phone and e-mail alerts to subscribers. The worldwide leader in pet recovery, with approximately one

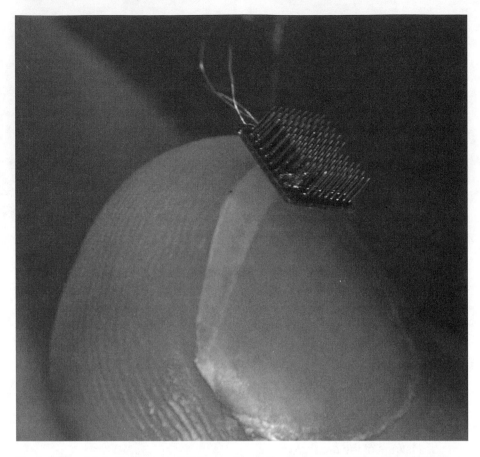

**Kevin Warwick is Professor of Cybernetics at the University of Reading, UK where he carries out research in artificial intelligence, control and robotics. In 1998 he shocked the international scientific community by having a silicon chip transponder surgically implanted in his left arm. A series of further implant experiments have taken place in which Kevin's nervous system was linked to a computer.**

million pets in the United States and over ten million in Europe protected by patented microchips

*Introduction to GPS: The Global Positioning System* Ahmed El-Rabbany An introduction to the Global Positioning System which accomplishes its task without the need for advanced mathematics. Emphasizing GPS applications, the book describes its signal structure, the types of measurement currently in use, and errors affecting these measurements. Recommendations for minimizing errors and integrating GPS technology with other systems are also included. Artech House, Incorporated, 1996

## Activity 12.3   Green Map Making

Time:     7–10 hours

## CASE STUDY

Green Maps give us a fresh perspective of our cities and towns. They help us locate eco-resources such as bike lanes, farmers markets, and wildlife habitats, along with cultural sites that make our hometowns special. Green Maps encourage us to discover new ways to experience the local urban ecology. Every city has wonderful places where you can get involved creating a more natural and enjoyable lifestyle. Find out ways to put the places and things you like about the local environment on the map.

Each mapmaking team of three people should start by choosing an area to explore. This area should be one that is familiar or one designated by the class. Draw or study a map showing landmarks and crossroads of the built environment, including streets, parks, and gardens. Discuss the target audience: neighbors, city residents, tourists, students, planners and policymakers. Discuss the general outline and goals for your Green Map. Consider where your map will be displayed and distributed when completed. Your introduction should include issues of concern to youth regarding the community's environmental health and cultural resources.

Scout around your area and find places described by the Icons (download from the website or use the worksheet from the teacher). While you are in the field, use notebooks or index cards to keep track of the category, name, and location of each green site. The first identifiable areas should be green businesses, eco-smart transportation, recycling reuse sites, special cultural and historic places, natural areas, habitats, and gardens. What other sites are important? Are toxic hot spots, blight sites or issues important for you to raise others' awareness? Put these on your Green Map, too.

Each team should discuss the green sites you've found and decide which of the Icons describe them best; you may use more than one, if necessary. Compare your sites to the descriptions on the Icon sheet. In addition, you can collect notes on noises

Getting Around Without a Car, Charted by
**Recycle-A-Bicycle's**
2001 Summer Youth Program

and smells, as well as share general impressions about the health and physical beauty of area's environment.

Write a short survey to help gather more information from local residents, shopkeepers, and other community members about wildlife, significant organizations, public transportation, cherished cultural sites, and other green places. Do different age groups suggest different kinds of sites?

All members of your map team should compare their lists of green sites they've discovered. Share your combined list with your class and community. Have you missed any categories? Must you do more research? Should you include more recreational, cultural or historical places? Are you ready to compile a final list?

If you have not already done so, draft your map by tracing a base map or drawing a diagram of your area that fits the size of your paper. You can work free-hand, trace an aerial photo or copy an existing map. Be sure to credit the source of the base map, which may be a road map, a lot survey plan, a city planning map, etc.

Use a copy machine or tracing paper and create a "working map". Use colors, the icons, and design elements to enhance the map and reproduction. You can select colors for the Icons, but make sure they stand out clearly. Plot the Icons in the appropriate places on your map. To identify each site by name, you can do one of the following:

1. Put the site's name right on the map next to the Icon
2. Number each Icon on the map, then put a numerical list on the side
3. Number the map, then put Icon and site name in a list on the side. Include the site's contact information and coordinates.

Please include the following:

- An arrow pointing North, the scale of the map, and the date of creation.
- A list of your mapmaking team members and the sources for your information and base map.
- The Green Map System's copyright for the Icons (Copyright GreenMap System, Inc. all rights reserved). You can also add the logo.
- A title block or logo for your Green Map, as well as your own copyright (if desired) and your contact information.
- Leave a white border (at least 3/8 inch) all the way around the map.

When you have placed everything on your working map and you are happy with its appearance, copy or trace over it to create a clean finished Green Map. If it is to be printed, special preparation will be needed. A computer-based map can also be created, using graphic design, desktop publishing, or geographic information systems (GIS) software.

Once the Green Map is nearing completion, you should consider strategies for sharing your findings with your school community, nearby schools, community groups, and elected officials. Send or deliver Green Maps, or your web-based map's address, to friends, relatives, the green businesses (and other sites) on your map, and to newspapers or the media. Include a press release or other background information about your project.

## RESOURCES

http://www.greenmap.org The Green Map System is a globally connected, locally adaptable eco-cultural program for community sustainability. Green Maps (both printed and online) utilize Green Map Icons to chart the sites of environmental significance around the world. Visit the website to see youth-made Green Maps and more resources for educators.
PO Box 249, NYC 10002
(212)-674-1631
*Introduction to Environmental Engineering.* MacKenzie L. Davis, David A. Cornwell, David A. Corwell New York: McGraw-Hill Companies, 1997.

---

## Activity 12.4    Recorded Music: Then, Now, and the Future

Time:      4–7 hours

## CASE STUDY

The history of recorded music has seen incredible technological advancements since Emile Berliner and Thomas Edison simultaneously invented their recording devices (1877). Mr. Berliner developed and patented the cylindrical and disc phonograph system. Mr. Edison invented the phonograph. In order to record with a phonograph, an indentation on a moving strip of paraffin coated paper tape was made by means of a diaphragm with an attached needle. This mechanism eventually lead to a continuously grooved, revolving metal cylinder wrapped in tin foil.

Since the 1950s, we have seen the marvels of 78s, 45s LPs, 8-track tapes, cassette tapes, Digital Audio Tape (DAT), Compact Discs, Mini-Discs, and MP3s. The future of recorded music will continue to change in the years to come. The demands of consumers will keep the music industry an integral part of our culture. The research and development divisions of major corporations will introduce and invent better technologies for consumers to listen to this music wherever they are. From inside our personal vehicles to our work establishments and back home, technology is all around to introduce sound into the air.

In this case study, your team will study the effects of technology on society. The interaction between technology and humans is obvious in recorded music. Some "experts" think digital music eliminates some of the artist's original contents. Other

 **Shawn Fanning wrote the code to help his friends find more MP3s online.**

experts believe that the developments in recorded music has increased our ability to listen to music in more convenient ways, with less space used for storage and equipment. Survey at least 25 adults (18 or older), asking several prescribed and team-developed questions. One of the objectives of the case study is to define what technology adults use to listen to music. Your additional team-developed questions should help create a hypothesis of what technology could be developed in the future.

Your team should develop a mission statement and a series of mock-ups or models of systems and products with drawings and data for consumers. The prescribed questions in the survey are as follows.

a. Age and gender
b. What medium do you currently use to playback music?
c. How many different mediums have you used in your lifetime?
d. What is your biggest complaint about the current medium you use?
e. Were there any good things about the previous mediums?
f. What do you think would improve the way music is stored and replayed?

## RESOURCES

http://memory.loc.gov/ammem/berlhtml/berlhome.html Emile Berliner and the Birth of the Recording Industry home page
http://www.jonesencyclo.com/ Audio Recording: History and Development
www.napster.com Roxio, Napster, the cat's logo, the Napster logo and the burning CD logo are either trademarks or registered trademarks of Roxio, Inc. in the U.S. and/or other countries.
*A Century of Recorded Music: Listening to Musical History.* Timothy Day

## Activity 2.5   Political Contributions Ethics Debate

Time:      1–3 hours

## CASE STUDY

Engineer A is the principal in a small consulting engineering firm. Approximately 50% of the work performed by Engineer A's firm is for the county in which the firm is located. The value of the firm's work is approximately $150,000 per year. Engineer A is requested to make a $5,000 political contribution, the maximum amount allowed by law, to help pay the cost of the media campaign of the county chairman.

After subsequent thought, Engineer A makes a $2,000 contribution to the campaign of the chairman, whom Engineer A has known for many years through mutual public service activities as well as activities on behalf of the political party. The county board chairman serves in a part-time capacity and receives $9,000 per year for his services. Other members of the board receive $8,000 per year for their services. As required under the laws of his state, Engineer A reports the campaign contributions to the state board of elections and correctly certifies that the contributions do not exceed the limits set by state law.

These contributions and the contributions of other firms in the county are reported by members of the local media who appear to suggest that Engineer A and other firms have contributed to the campaign in anticipation of receiving work from the county. Engineer A continues to perform work for the county after making the political contribution.

## Group Debate Question

Is it unethical for Engineer A to continue to perform work for the county after making the $2,000 contribution to the county board chairman's campaign?

## RESOURCES

*NSPE Case 73-6.* For many years, the engineering profession has been grappling with the ethical issues involved with political contributions by individuals to state and local candidates. Political contributions were the subject of a keynote address by the National Society of Professional Engineers (NSPE) at a recent national meeting and continue to be examined by a special task force charged with developing a political contributions policy.

### References: NSPE Code of Ethics

*Section II.3.a.* "Engineers shall be objective and truthful in professional reports, statements, or testimony. They shall include all relevant and pertinent information in such reports, statements, or testimony."

*Section II.5.b.* "Engineers shall not offer, give, solicit, or receive, either directly or indirectly, any political contribution in an amount intended to influence the award of a contract by the public authority, or which may be reasonably construed by the public as having the effect or intent to influence the award of a contract. They shall not offer any gift or other valuable consideration in order to secure work. They shall not pay a commission, percentage, or brokerage fee in order to secure work except to a bona fide employee or to bona fide established commercial or marketing agencies retained by them."

*Section III.1.f.* "Engineers shall avoid any act tending to promote their own interest at the expense of the dignity and integrity of the profession."

*Chapter 13*

# Concurrent Engineering and Teamwork

## Introduction

More than ever, engineering schools are requiring their students to work in teams. These teams take various forms: collaborative study groups; laboratory groups; design groups as part of individual classes; and design groups participating in extracurricular competitions, such as the solar car, super mileage vehicles, human powered vehicles, SAE formula cars, concrete canoes, bridge designs, Rube Goldberg projects, and many others. This team emphasis in engineering schools mirrors a management philosophy, which capitalizes on the use of teams that has swept through the corporate world in the last few decades.

Engineering employers, in fact, are the principal drivers for the use of teams in engineering schools. They seek engineering students who have team skills as well as technical skills. The purpose of this chapter is to communicate a team vision, to show why teams and collaboration are so important to organizational and technical success, and to give practical advice how to organize and function as a team.

If you study this chapter, work through the exercises, and examine some of the references, you should have most of the tools you need to be a successful team member and team leader while in engineering school. Apply the principles in this chapter to your team experiences in school and learn from and document each. By so doing, you will have a portfolio of team experiences that will be a valuable part of your résumé. Above all, you will have the satisfaction that comes from being a part of great teams, and great teams can be fun.

## WHY DO CORPORATIONS FOCUS ON TEAMS?

A full-page advertisement in *USA Today* (1991) elucidated the highly competitive climate in industry and what Chrysler (now DaimlerChrysler), for example, was doing about it. The use of interdisciplinary teams was an important part of the solution:

> "No more piece-by-piece, step-by-step production. Now it's teams.
> Teams of product and manufacturing engineers, designers, planners,

*financing and marketing people-together from the start. . . . It's how we built the Dodge Viper from dream to showroom in three years, a record for U.S. car makers. From now on, all our cars and trucks will be higher quality, built at lower cost and delivered to the market faster. That's what competition is all about."*

The greater the problem, the greater the need for the use of teams. Individuals acting alone can solve simple problems, but tough problems require teams. This section lists and then discusses the reasons many organizations have embraced the use of teams. Why is the use of teams so popular today?

- Engineers are asked to solve complex problems.
- More factors must be considered in design than ever before.
- Teams provide for greater understanding through the power of collaboration.
- Many corporations are international in scope, with design and manufacturing engineering operations spread across the globe.
- Since Time to Market is extremely important to competitive advantage, concurrent engineering, inherently a team activity, is widely employed.
- Corporations are increasingly using project management principles.

## QUICK PROFILES OF GREAT TEAMS

We understand more today than ever before of the importance of and power of creative collaboration. One recent book on the subject is *Organizing Genius: The Secrets of Creative Collaboration,* by Warren Bennis and Patricia Ward Biederman (1997). Their first chapter is titled "The End of the Great Man," in which they argue that great groups can accomplish more than talented individuals acting alone, stating that "None of us is as smart as all of us." They cite a survey by Korn-Ferry (the world's largest executive search firm) and The Economist in which respondents were asked who will have the most influence on global organizations in the next ten years. Sixty-one percent of the respondents answered "teams of leaders," whereas 14 percent said "one leader." Individual chapters of their book focus on great groups. A chapter on the Manhattan Project discusses the development of the atomic bomb during the Second World War by a team of great scientists. One chapter discusses Lockheed's Skunk Works, which developed the U-2 spy plane, the SR-71 Blackbird, and the F117-A Stealth fighter-bomber used in the Persian Gulf War. Another example is the development of the graphical user interface at Xerox's PARC (Palo Alto Research Center) and its adoption into the Macintosh computer by Steve Jobs and Steve Wozniak. Steve Jobs' goal was to lead his team to create something not just great, but "insanely great," to "put a dent in the universe."

Another recent book on collaboration is *Shared Minds: The New Technologies of Collaboration,* by Michael Schrage (1990). He defines collaboration as "the process of shared creation: two or more individuals with complementary skills interacting to create a shared understanding that none had previously possessed or could have come to on their own." He cites examples of collaboration in science (Watson and Crick; Heisenberg, Bohr, Fermi, Pauli, and Schrodinger; Einstein and Marcel Grossman), music (Gilbert and Sullivan; Rodgers and Hart; McCartney and Lennon), art (Monet and Renoir; Van Gogh and Gauguin; Picasso and Braque), and literature (editor Maxwell Perkins and writers F. Scott Fitzgerald and Thomas Wolfe; Ezra Pound and T. S. Eliot).

James Watson and Stanley Crick, winners of the Nobel Prize for discovering the double helix structure of DNA credit their collaboration for their success: "Both of us admit we could not have done it without the other. We were interested in what the answer is rather than doing it ourselves." Crick wrote that "our . . . advantage was that we had evolved . . . fruitful methods of collaboration." Sir Isaac Newton, father of classical physics, in a letter to his contemporary (and rival) Robert Hooke, recognizing his debt to fellow scientists, wrote: "If I have seen further (than you and Descartes) it is by standing on the shoulders of Giants." Modern physics also found its genesis in collaboration. Heisenberg, Bohr, Fermi, Pauli, and Schrodinger, all Nobel laureates for their roles in founding quantum physics, worked, played and vacationed together as they developed their revolutionary new ideas together.

## THE INCREASING COMPLEXITY OF PROJECTS

Engineers are asked to solve increasingly complex problems. The complexity of mechanical devices has grown rapidly over the past two hundred years. David Ullman, in his text *The Mechanical Design Process* (1997), gives the following examples. In the early 1800s, a musket had 51 parts. The Civil War era Springfield rifle had 140 parts. The bicycle, first developed in the late 1800s, has over two hundred parts. An automobile has tens of thousands of parts. The Boeing 747 aircraft, with over five million components, required over 10,000 person-years of design time. Thousands of designers worked over a three-year period on the project. Most modern design problems involve not only many individual parts, but also many subsystems—mechanical, electrical, controls, thermal, and others—each requiring specialists acting in teams.

Engineering designers must consider more factors than ever before, including initial price, life cycle costs, performance, aesthetics, overall quality, ergonomics, reliability, maintainability, manufacturability, environmental factors, safety, liability, and world market acceptance. Satisfying these criteria requires the collective teamwork of design and manufacturing engineering, marketing, procurement, business, and other personnel. The typical new engineering student, being interested primarily in technical things, tends to believe that engineers work isolated from other factors, always seeking the best technical solution. This is not true. Engineering involves solving difficult problems and finding technical solutions while considering numerous constraints. Engineers make things happen; they are doers. Engineers often stand astride boundaries in organizations. They are technical experts who also understand the economics and dynamics of the business enterprise.

## THE INTERNATIONAL FACTOR

Many corporations are international in scope, with global design and manufacturing engineering operations. These operations require teams to work together who may never physically meet. Instead, they meet and share data via electronic means. For example, design engineers in the United States commonly collaborate on a design with Japanese engineers for a product to be manufactured in China. Manufacturing personnel from the Chinese plant may also contribute to the project. As another example, software engineers in the United States, the United Kingdom, and India may collaborate on a software project around the clock and around the world. At any time during a 24-hour period, programmers in some part of the world will be working on the product.

## THE NEED FOR SPEED

Concurrent engineering is widely employed in order to achieve better designs and bring products to market more quickly. Time to Market is the total time needed to plan, prototype, and procure materials and to create marketing strategies, devise tooling, begin production, and bring a new product to market. In the traditional company, each of these steps is done serially, one at a time. It is a relay race, with only one person running at a time, passing the baton to the next runner. The design engineer "tosses a design over the wall" to a manufacturing engineer who may not have seen it yet. Then it is tossed over the wall to the procurement personnel, and so on. Concurrent engineering, on the other hand, is a parallel operation. Like rugby, all of the players run together, tossing the ball back and forth. Marketing, manufacturing, and procurement personnel are involved from the beginning of the design phase. The use of teams, as well as new technologies such as CAD/CAM, rapid prototyping, shared data, and advanced communications have radically changed the process of engineering. By using teams and CAD/CAM technologies, in the face of fierce international competition, U.S. auto makers have cut the time needed to bring a new car to market from approximately five years to fewer than three years.

Speed—the timely delivery of products to the marketplace—is critical for companies to be profitable. Jack Welch, General Electric's widely respected former CEO (see his biographical sketch in Chapter 6), has said "we have seen what wins in our marketplaces around the globe: speed, speed, and more speed." Xilinx Corporation has stated that its research shows that a six-month delay in getting to market reduces a product's profitability by a third over its life cycle. A Business Week article stated: "Reduce product development time to one third, and you will triple profits and triple growth."

Developing new products is a key to growth and profits. During a recent year, Rubbermaid Corporation introduced 400 new products—more than one per calendar day! At least one third of Rubbermaid's $2 billion in annual sales come from products which have been in existence fewer than five years. 3M Corporation CEO L.D. DiSimone, facing flat revenues and stiff competition, adopted a business focus of generating 30 percent of revenues from products fewer than four years old. The creative thought processes, the innovation needed to create a steady stream of new ideas, and the expertise to manufacture and bring these to market quickly are all functions that teams perform well.It is important for engineering students to understand the importance of speed (without compromising quality). Delays have many negative consequences. Engineering professors who supervise teams should provide incentives to students for early completion of team projects or at least penalties for failure to meet deadlines. Student teams should recognize the importance of speed, push themselves and their teams toward early completion of projects, and document their successes on their résumés and/or portfolios.

## PROJECT MANAGEMENT USES TEAMWORK

Project management is widely practiced in industries and government labs. Many engineers discover on their first job that they will frequently work in project-oriented teams. Unfortunately, most engineering students never take a project management class. Project management principles were developed in the defense industry in the 1950s and 1960s as a way to manage department of defense contracts. The typical corporation is organized by vertical divisions or lines, where individuals are clustered by job functions. For example, a corporation may have separate research, manufacturing, engineering, human resources, product development, marketing, and procurement divisions. Project management is a way of organizing individuals not by function but by products or projects. Selected individuals from research, engineering, manufacturing, human resources, procurement, and other divisions are gathered together as a team for a particular purpose, for example, solving a corporate problem, developing a new product, or meeting a crucial deadline. A project management approach, therefore, is inherently a cross-functional team approach and an excellent way to solve problems. Successful project management, however, is not an easy task.

The project managers' job is to complete a project, on time, within budget, with the personnel that they are given. Unfortunately, they are rarely able to hand pick their personnel, and their supervisors and/or customers typically set the project completion time and budget. In other words, project managers are never given all the people, time, and money they need. These tough constraints are often mirrored in student design teams. Their teacher picks the team members, they do not have much money or resources to work with, and their teacher determines the time frame. These conditions are uncomfortable for students, but they prepare students for the engineering world.

Project managers use tools that any team leader would benefit from learning how to use. Project managers plan the work requirements and schedules and direct the resource use (people, money, materials, and equipment). One popular project management tool that student teams should learn to use is the Gantt chart. (A simple Gantt chart is pictured in Fig. 13.1 below.) A Gantt chart organizes and clearly presents information on task division and sequencing. Tasks are divided into subtasks. The shading conveys the sequence in which the tasks

| Team A-9 ("Semi-Conscious Objectors"): Gantt Chart for Object Relocator: Fall 2000 | | | | | | | | | | |
|---|---|---|---|---|---|---|---|---|---|---|
| Task | Leader(s) | Oct 11-17 | Oct 18-25 | Oct26-31 | Nov1-7 | Nov8-14 | Nov15-21 | Nov22-28 | 11/29 -12/5 | Dec6-12 |
| Finalize Design | Team | ▓ | ▓ | | | | | | | |
| Procure Materials | Joe | | ▓ | | | | | | | |
| Pick-Up Module | Jaime | | | ▓ | | | | | | |
| Conveyor Assembly | George | | | ▓ | | | | | | |
| Working prototype | Team | | | | ▓ | | | | | |
| Fine Tuning Machine | Jaime/Team | | | | | ▓ | ▓ | | | |
| Instruction Manual | George | | | | | ▓ | | | | |
| Professor Tests Projects | Team | | | | | | ▓ | | | |
| Report Rough Draft | Sarah | | | | | ▓ | | | | |
| Finalize CAD Drawings | Ahmad | | | | | ▓ | | | | |
| Presentation | Ahmad/Joe | | | | | | | | ▓ | |
| Final Report Due | Team | | | | | | | | | ▓ |
| Judges test best two projects per class | Team | | | | | | | | | ▓ |

**Figure 13.1 A Sample Gantt Chart.**

should be completed and how long a task is expected to take. In our version, the person(s) responsible for each task is listed.

Project managers monitor projects by tracking and comparing progress to predicted outcomes. They use project management software such as Microsoft Project for planning and monitoring purposes. Project managers carefully guard the scope of the project from excessive mission creep (a never-ending expansion of purpose). A project is successful if it is completed on time, within budget, at the desired quality, and with effective utilization of resources. For more information on project management, consult any text on Project Management and read the publication titled "PMBOK: Project Management Body of Knowledge" at http://www.pmi.org, the website of the Project Management Institute.

## WHAT MAKES A SUCCESSFUL TEAM?

Have you ever been a member of a great team? It's a great experience, is not it? On the other hand, have you ever been part of an unsuccessful team? It was a disappointing experience, wasn't it? How do you measure or evaluate a team's success or failure? What specific factors make a team successful? What factors make a team a failure?

First, it is important to define what a team is and is not. A team is not the same as a group. The term "group" implies little more than several individuals in some proximity to one another. Team, on the other hand, implies two or more persons who work together to achieve a common purpose. The two main elements of this definition are purpose and working together.

All teams have a purpose and a personality. A team's purpose is its task, the reason the team was formed. Its personality is the collective style of its people and how the team members work together. Each team has its own style, approach, dynamic, and ways of communicating which are different from those of other teams. For example, some teams are serious, formal, and businesslike, whereas others are more informal, casual, and fun-loving. Some teams are composed of friends; others are not. Friendship, though desirable, is not a necessary requirement for a team to be successful. Commitment to a common purpose and to working together is required. Regardless of style or personality, a team must have professionalism. A team's professionalism ensures that its personality promotes productive progress toward its purpose.

## TEAM ATTRIBUTES

The successful team should have the following attributes:

1. A common goal or purpose. Team members are individually committed to that purpose.
2. Leadership. Though one member may be appointed or voted as the team leader, every team member should ideally contribute to the leadership of the team.
3. Each member makes unique contributions to the project. A climate exists in the team that recognizes and appropriately utilizes the individual talents/abilities of the team members.
4. Effective team communication. Regular, effective meetings; honest, open discussion. Ability to make decisions.
5. Creative spark. There's excitement and creative energy. Team members inspire, energize, and bring out the best in one another with its can do attitude. This creative spark fuels collaborative efforts and enables a team to rise above the sum of its individual members.
6. Harmonious relationships among team members. Team members are respectful, encouraging, and positive about one another and the work. If conflicts arise, there are peacemakers on the team. The team's work is productive and fun.
7. Effective planning and use of resources. This involves an appropriate breakdown of tasks and effective utilization of resources (people, time, money).

## TEAM MEMBER ATTRIBUTES

Individual team members of a successful team should have the following attributes:

1. Attendance. Attends all team meetings, arriving on time or early. Dependable, faithful, reliable. Communicates in advance if unable to attend a meeting.
2. Responsible. Accepts responsibility for tasks and completes them on time, needing no reminders or cajoling. Has a spirit of excellence, yet is not a perfectionist.
3. Abilities. Possesses abilities the team needs, and contributes these abilities fully to the team's purpose. Contributes oneself. Actively communicates at team meetings.
4. Creative and energetic. Acts as an energy source, not a sink. Brings energy to the team. Conveys a sense of excitement about being part of the team.

Has a can do attitude about the team's task. Has creative energy and helps spark the creative efforts of everyone else.

5. Personality. Contributes positively to the team environment and personality. Has positive attitudes and encourages others. Acts as a peacemaker if conflicts arise. Helps the team reach consensus and make good decisions. Creates a team environment that is productive and fun. Brings out the best in the other team members.

## GROWTH STAGES OF A TEAM

Simply forming teams to solve problems may seem like a good idea, given the evidence and cases from the previous section, but teams are not a panacea. Teams require nurturing. Successful team development is a process. Whether in a corporate, academic, or any other setting, all teams must pass through several developmental phases before they truly become productive. These stages are forming, storming, norming, performing, and adjourning. Unfortunately, successfully completing the first four stages and becoming a productive, performing team is not guaranteed. Many teams, in practice, never reach the performing stage; they remain mired in the forming, storming, or norming stages. Every team's challenge is to grow through these stages and achieve performance.

## Stage 1: Forming

The first stage of team development is forming. In this stage, typical dialogue may consist of the following: "Nice to meet you. Yeah, I'm not sure why we are here either. I'm afraid this might be a lot of work." In the formation stage, team members get acquainted with one another, with the leader (or they choose a leader if one is not pre-appointed), with the team's purpose, and with the overall level of commitment (workload) required. Team members learn one another's personalities, abilities, and talents and one another's weaknesses and idiosyncrasies. Individual team members are typically shy, reserved, self-conscious, and uncertain. The leader's role in this stage is to lead team ice-breaker activities, facilitate discussion, and encourage everyone to speak while quieting some who might dominate the conversation. Another role is to focus the team on the task at hand.

## Stage 2: Storming

The second stage of team development is storming. In this stage, typical dialogue (or private thoughts) may consist of the following: "Do I have to work with this team? What did I do to deserve this? There clearly are no super-heroes on this team, including that dizzy leader. How are we supposed to solve this messy problem?" During the storming stage, the enormity and complexity of the task begins to sink in, possibly sobering and discouraging the participants. "We are supposed to do what? By when?" Teams are rarely formed to solve easy problems, only difficult and complex ones. Typically, time schedules are short, and budgets are inadequate. Complicating the issue is that teammates have learned enough about their fellow team members to know they have no super-heroes, no saviors they can count on to do it all. (One person doing all the work is a team failure.)

Some team members may not initially hit it off well with the others. Cliques or factions may emerge within the team, pitted against other. Since the leader's weaknesses (all leaders have weaknesses) are by now apparent, some individuals or factions may vie for team leadership. Though possibly under siege, the leader's role is critical during the storming stage. The leader must focus the team on its collective strengths, not its weaknesses, and direct their energies toward the task. To be a successful team, team members do not have to like one another or to be friends. A professional knows how to work productively with individuals with widely differing backgrounds and personalities. Everyone must learn constructive dialogue and compromise.

## Stage 3: Norming

The third stage of team development is norming. In this stage, typical dialogue (or private thoughts) may consist of the following: "I think we can do it. True, we have no super-heroes, not by a long shot, but once we stopped fighting and started listening to one another, we discovered we had good ideas. Now if we can just pull these together. . . ." Norms are shared expectations or rules of conduct. All groups have some kinds of norms, though many times unstated.

Do you recall a time you joined a group or team and felt a subtle influence to act, dress, look, speak, or work in a particular way? The more a team works together, the more it tends to converge toward some common perspectives and behaviors. During the norming stage, team members begin to accept one another instead of complaining and competing. Rather than focusing on weaknesses and personality differences, they acknowledge and utilize one another's strengths. Individual team members find their place in the group and do their part. Instead of directing energies toward fighting itself, the team directs its collective energy toward the task. The key to this shift of focus is a collective decision to behave in a professional way, to agree upon and adhere to norms. Possible norms include working cooperatively as a team rather than individually, agreeing on the effort level expected of everyone, conducting effective discussions and meetings, making effective team decisions, and learning to criticize one another's ideas without attacking the person. One commanded norm is that all team members are expected to be at all meetings, or to communicate clearly in the event that they cannot attend.

During the norming stage, feelings of closeness, interdependence, unity, and cooperation develop among the team. The primary role of the leader during the norming phase is to facilitate the cohesion process. Some team members will lag behind the team core in embracing norms. The leader and others on the team must nudge individuals toward group accountability and a task focus (at this stage leadership is beginning to emerge from others on the team as well).

## Stage 4: Performing

The fourth stage of team development is performing. In this stage, typical dialogue (or private thoughts) may consist of the following: "This is a fun team. We have a long way to go, but we have a great plan. Everyone is pulling together and working hard. No super-heroes, but we're a super team." In this stage, teams accomplish a great deal. They have a shared vision. Responsibilities are distributed. Individual team members accept and execute their specific tasks according to the planned schedule. They are individually committed, and hold one another accountable. On the other hand, the roles blur. Team members pitch in to help one another doing whatever it takes to make the team successful. In a performing team, so many team members have taken such significant responsibility for the team's success that the spotlight is rarely on a single leader. Typically, whoever initially led the team becomes indistinguishable from the rest of the team.

## Stage 5: Adjourning

The fifth and final stage of team development is adjourning. Because teams are typically assembled for a specific purpose or project, the time comes when the team disbands and it has accomplished the goal. If the team was successful, it will have a feeling of accomplishment, even euphoria. On the other hand, an underperforming team will typically feel anger or disappointment upon adjournment.

These stages, originally set forth by Bruce W. Tuckman (1965), appear in most books, which discuss teams. Some leave off the final step, adjourning. Others switch the order and present the steps as forming, conforming (instead of norming), storming, and performing.

## TEAM LEADERSHIP STRUCTURES

### Traditional Model

Occasionally, engineering design teams are asked to create an organizational chart. Preparing this chart gives the team the opportunity to define roles and to give thought to the relationship between the team and the team leader. Most student teams, in part because they do not know any better, default to a traditional team structure. This structure is easy to imagine, but it may or may not represent the best team operation. It implies a strong leader who largely directs the actions of the group, possibly with little participation or discussion from team members. It suggests separation between the leader and the other team members.

### Participative Model

The second typical structure emphasizes a participative leadership/team model in which the leader is positioned closely to all members, with short, direct communication paths. The figure implies direct accountability of the leader to all members and dependence of the leader on their participation.

## Flat Model

A third structure differs from the second because it emphasizes the leader's role as a working team member. It suggests a flat structure where the leader is an equal, rather than a hierarchical structure with the leader above the team. For this structure, it is easy to visualize the leadership function shifting around the ring of members as situations require the individual members' expertise.

## Consultant Model

The fourth structure shows the relationship between a student team and its instructor. The student team may be a design group, a research team, or a collaborative study unit. The instructor, though not part of the team, will be nearby and will serve as an important team resource. He may advise the team on administrative issues, act as a technical consultant, or assist in intervention or disciplinary action for a non-performing team member. All teams—in industry, in universities, and in private organizations—operate under the leader's or administrator's authority.

There is no one right structure. None of these structures is inherently good or bad. Examining these team structure examples gives the team the opportunity to define roles and to give thought to the relationship between the team and the team leader. A team needs to choose a structure that effectively models how it wants to behave. The ultimate success of the team hinges upon the team and its leader(s) working together to accomplish the team's purpose. As a team works, they may find that their operational structure may shift periodically in response to different situations and challenges.

## HOW EFFECTIVE TEAMS WORK

Every team makes decisions in order to accomplish its work. The ability to make high-quality decisions in a timely manner is a mark of a great team. The process of how decisions are made affects the quality of these decisions. Unfortunately, most teams arrive at decisions without even knowing how they did it. In this section, we present many ways in which decisions are made and discuss their advantages and disadvantages. Though we recommend teams decide in advance what decision-making process they will use, the best teams use many decision-making means depending on the circumstances. The following classification of decision-making is summarized from, *Why Teams Do not Work,* by Harvey Robbins and Michael Finley (1995).

## MODES OF TEAM ACTION

*Consensus:* A decision by consensus is a decision in which all the team members find common ground. It does not necessarily mean a unanimous vote, but it does mean that everyone has an opportunity to express their views and to hear the views of others. The process of open sharing of ideas often leads to better, more creative solutions. Unfortunately, achieving consensus can take time and become unwieldy for larger teams.

*Majority:* Another way to make a decision is by majority vote. The option that receives the most votes wins. The advantage here is that it takes less time than reaching consensus. Its disadvantage is that it provides for less creative dialogue than consensus, and the minority, who lose the vote, may become alienated.

*Minority:* Sometimes a small subset of the team—a subcommittee, for example—makes the decision. The advantage here is that it may expedite the decision. The disadvantage is that team communication is less, and some team members may be prevented from contributing to the decision.

*Averaging:* Averaging is compromise in its worst form. It's the way Congress and some committees arrive at decisions. Averaging is often accomplished with haggling, bargaining, cajoling, and manipulating. No one is happy except the moderates. The advantage of averaging is that the extreme opinions tend to cancel one another out. The disadvantages are that there is little productive discussion (as in consensus), and that the least informed can cancel the votes of the more knowledgeable.

*Expert:* When facing a difficult decision, expertise is always best. If an expert is on the team, he or she may be asked to make a decision. If a team lacks an expert, which is often the case, the best teams recognize this and seek expert advice. The advantage of this is that the decision is made with accurate, expert knowledge. The disadvantage is that experts who, with the same information, will disagree on the best course of action. How can a team know the expert has given them the best advice? Additionally, team members may be divided on which expert to consult, as well as on their assessment of the expert's credentials.

*Authority Rule without Discussion:* This occurs when a strong leader makes decisions without discussing the details with or seeking team advice. This works well with small decisions, particularly of an administrative nature, with decisions that must be made quickly, and with decisions the team is not well qualified to contribute to. Of the many disadvantages, the greatest is that the team's trust in its leader will be undermined. The team members will perceive that the leader doesn't trust and is circumventing them. If one person is continually making all the decisions, or if the team is abrogating its responsibility to act together and, by default, forcing the decision onto an individual, then this is a group, not a team. It is a loose aggregation of individuals, a dysfunctional team.

*Authority Rule with Discussion:* This can be an effective way of decision making. One person, the leader or a delegated decision maker, makes the final decision but the leader first seeks team input. The team meets and discusses the issue. Many, but perhaps not all, air their viewpoints. Now, more fully informed, the appointed individual makes the decision. The advantage of this method is that the team members, being part of the process, feel valued. They are more likely to be committed to the result. This type of decision-making process requires a leader and a team with excellent communication skills and a leader who is willing to make decisions.

## GETTING GOING IN A TEAM SETTING

If you have an opportunity to become part of a team, how should you approach it? What should your attitudes be?

1. Determine to give your best to help the team grow and accomplish its purpose. Do not just tolerate the team experience.
2. Do not expect to have perfect teammates. You are not perfect; neither are they. Do not fret that your friends are not on your team. Make friends with those on your team.
3. Be careful about first team impressions. Some team members may seem to know all the right things to say, but you may never see them again. Others might at first look like losers but become the ones who attend all meetings, make steady contributions, and ensure the team's success. Be careful about selecting a leader at the first team meeting. Sometimes the person who seems to be the most qualified to lead will never make the commitment necessary to be the leader. Look for commitment.
4. Be a leader. If you are not the appointed leader, give your support to the leader and lead from your position on the team. Do not be scared off by past team experiences where the leader did all the work. Be one who encourages other leaders.
5. Help the team achieve its own identity and personality. A team is a kind of corporation, a living entity with special chemistry and personality. No team is like any other team. Every team is formed at a unique time and setting, with different people and a different purpose. Being part of a productive team that sparkles with energy, personality, and enthusiasm is one of teamwork's rewards. Being part of a great team creates great memories.
6. Be patient. Foster team growth, and give it time to grow. Teammates are at first aliens, unsure of one another and of the team's purpose. Watch for and help the team grow through the stages of Forming, Storming, Norming, and Performing.
7. Evaluate and grade yourself and your team's performance. Document its successes and failures. Dedicate yourself to be the one who understands the team and who acts as a catalyst, i.e., one who helps the team perform at a high level. Record team experiences in your portfolio, especially times when you took a leadership role. Prepare to communicate these experiences to prospective employers.

## THE CHARACTER OF A LEADER

Every successful human endeavor involving collective action requires leadership. Great teams need great leadership. Without leadership, humans tend to drift apart, act alone, and lose purpose. They may work on the same project, but their efforts, without synchronization and coordination, interfere with rather than build on one another. Without coordinated direction, people become discouraged, frustrations build, and conflicts ensue. Money is wasted, time schedules deteriorate, but the greatest loss is that of human potential. A failed team effort leaves a bitter taste not easily forgotten. Incomparably sweet, on the other hand, is the thrill of team success.

## Leader Attributes

Leaders ensure that the team remains focused on its purpose and that it develops and maintains a positive team personality. They challenge and lead the team to high performance and professionalism. With one hand they build the team, with the other they build the project. In support of these objectives, leaders must do the following:

1. Focus the team on its purpose.
2. Be a team builder. Leaders may actively work on some project tasks themselves, but their most important task is the team, not the project. They build, equip, and coordinate the team efforts so it can accomplish its purpose and succeed.
3. Plan well and utilize resources (people, time, money) effectively. Leaders effectively assess and use team member abilities.
4. Run effective meetings. They ensure that the team meets together regularly and that meetings are productive.
5. Communicate effectively. Leaders communicate the team vision and purpose. They praise good work and improve substandard performance.
6. Promote team harmony by fostering a positive environment. If team members focus on one another's strengths instead of weaknesses, conflicts are less likely. Conflict is not necessarily evil, however. The effective leader must not be afraid of conflict. He or she should view conflicts as an opportunity to improve team performance and personality and to refocus it on its purpose.
7. Foster high levels of performance, creativity, and professionalism. Have a high vision. Challenge team members to do the impossible, to think creatively, and to stimulate one another to high performance.

Every team needs all the leadership it can get. Most teams have a single voted or appointed leader. In the best teams, however, many members contribute leadership in ways that support and complement the appointed leader. Like the appointed leader, every member should focus on the team's purpose, build the team, plan, recognize the gifts of others, contribute to effective meetings, communicate well, promote a harmonious environment, and be creative. Every member, together with the leader, should build the team with one hand and labor on the project with the other hand. Noted author John Maxwell defines leadership as influence. Every team member, from his or her team position, can influence the team's performance. It is possible to have an outstanding leader yet have the team fail if the team chooses to avoid following the leader, to undermine the leader, and to promote team disunity. Conversely, a team with a relatively weak appointed leader can be greatly successfully if the team members pitch in and build the team. The entire team, the members and the appointed leader, must work together to build a great team. It takes teamwork to make a great team.

## Leadership Styles

According to John Schermerhorn, author of *Management* (1996), "Leadership style is a recurring pattern of behaviors exhibited by a leader." It is the tendency of a leader to act or to relate to people in a particular way. Two categories of leadership style are task-oriented leaders and people-oriented leaders. Task-oriented leaders are concerned about the team's purpose and task at hand. They like to plan the schedule, define the work, assign the task responsibilities, set clear work standards, urge task completion, and monitor results. People-oriented leaders are warm and supportive toward team members, develop team rapport, respect their followers' feelings, are sensitive to followers' needs, and show trust in followers.

You may wonder whether one leadership style is better than another. A successful team needs both styles of leadership in some proportion. Earlier we stated a team has a purpose and a personality. The team must be task-oriented, but it must cultivate a team personality and a positive environment in which team performance can flourish. So, the individual team members must provide leadership that complements the appointed leader's command. If the appointed leader tends to be more task-oriented, the team can balance this with a people-orientation. If the appointed leader is people-oriented, he will likely need individuals on the team to assist with the task management details. The best leaders welcome complementary team leadership.

## TEAM GRADING AND REPORTS

The primary aim of this chapter has been to give you tools and insights to increase the likelihood that you will be part of (and an active leader in) a great team. You likely will be involved in many teams during your school, in your professional life, and in volunteer capacities. Teams are everywhere, but few know how to make teams work. For teams, Voltaire's words are applicable: "Common sense is not so common." If you work hard to be sensible about teams, you will increase your value to an employer. Approach every team experience as a learning opportunity and grow as a team member and leader. An excellent way to gain the most benefit is to evaluate each team experience. Some suggested dimensions for evaluating teams are as follows:

1. Did the team accomplish its purpose? Did it get the job done?
2. Did it do the job well? Were the results of high quality? If not, why?
3. Did the team grow through all of the developmental stages (Forming, Storming, Norming, Performing)? Were there any detours?
4. Reflect on the team's personality. Did the team enjoy working together? Did team members inspire one another to greater creativity and energy?
5. Evaluate team members on a report card like the one shown in Figure 13.2.

6. Evaluate the team leader(s). Was he or she an effective leader? Why or why not? How could he or she be more effective?
7. Honestly evaluate your contribution to the team.

In addition to learning more about teams, another reason to document team performance is to inform your instructor (when you are in school) or your supervisor (when at work) about the team's work. All teams have a supervisor. It is important that you report data to the instructor or supervisor. Some argue that student team members should each get the same grade, and in general, this is the best practice. But students should know from the start that if they do not participate in a significant way in the team's work, then their instructor may adjust the grade downward. In order for the instructor to make such determinations, he or she needs quantitative information on team member contributions (at midterm and at the end of the semester). Sometimes instructors ask team members to grade one another, but the reason for the grades remains unclear. The act of team members evaluating one another on team success criteria gives concrete evidence for team member grades. These criteria are more valuable to the team if they are held up as a target or goal at the beginning of the team activity. The team must understand these criteria will be the basis for the evaluation. In addition to the team member report card, the following should be valuable to your instructor:

1. Attendance records for all team meetings (usually weekly). Who attended? Was anyone habitually late or absent?
2. Actual contributions of each member. Include a table like the one presented in Figure 13.2 in the final report noting the each individual's contribution for each of the team's major projects and assignments. Add these percentages in order to quantify a team member's total contribution. Variation is expected in the members' contributions because balancing effort precisely is impos-

| Team Member Report Card | | | | | |
|---|---|---|---|---|---|
| **Criteria** | **Team Member** | | | | |
| | Ahmad | George | Jaime | Joe | Sarah |
| 1. Attendance: Attends all team meetings, arriving on-time or early. Dependable, faithful, reliable. | | | | | |
| 2. Responsible: Gladly accepts work and gets it done. Spirit of excellence. | | | | | |
| 3. Has abilities the team needs. Makes the most of abilities. Gives fully, doesn't hold back. Communicates. | | | | | |
| 4. Creative. Energetic. Brings energy, excitement to team. "Can do" attitude. Sparks creativity in others. | | | | | |
| 5. Personality: Positive attitudes, encourages others. Seeks consensus. Fun. Brings out best in others. Peace-maker. Water, not gasoline, on fires. | | | | | |
| **Average Grade:** | | | | | |
| **Grading Scale:** 5 - Always;  4 - Mostly;  3 - Sometimes;  2 - Rarely;  1 - Never | | | | | |

**Figure 13.2 Sample team member report card.**

| Team Members | Team Member Contributions on Primary Semester Projects | | | | | |
|---|---|---|---|---|---|---|
| | CADKEY Drawings | Project Construction | Instruction Manual | Final Report | Power Point Presen-tation | Total Indiv. Contri-bution |
| Sarah | 0% | 15% | 45% | 60% | 15% | **135 %** |
| Joe | 0% | 5% | 0% | 5% | 0% | **10 %** |
| George | 10% | 15% | 55% | 15% | 10% | **105 %** |
| Ahmad | 75% | 10% | 0% | 10% | 60% | **155 %** |
| Jaime | 15% | 55% | 0% | 10% | 15% | **95 %** |
| | **100%** | **100%** | **100%** | **100%** | **100%** | **400 %** |

**Figure 13.3. Team member contribution report.**

sible. In the example in Fig. 13.3, four members' contributions varied from 95 to 155 percent. Also included in the table (a sign of a healthy team) is that each of the four active team members took a leadership role on at least one aspect of the project effort: Sally took primary lead on the final report, George led in the creation of the Instruction Manual, Ahmad led in the CAD-KEY Drawings and the PowerPoint presentation, and Jaime led in the project construction. Though their contribution percentages vary, little evidence exists that anyone's grades should be adjusted downward. They each deserve the team grade. Joe, on the other hand, contributed nothing, and deserves a grade penalty. Chances are, the 10 percent is a gift from his teammates.

# REFERENCES

Bennis, Warren, and Biederman, Patricia Ward. Organizing Genius: The Secrets of Creative Collaboration. Boston, MA: Addison-Wesley, 1997.

Katzenbach, Jon R., and Smith, Douglas K. The Wisdom of Teams: Creating the High-Performance Organization. New York: HarperBusiness, 1993.

Nuese, Charles J. Building the Right Things Right: A New Model for Product and Technology Development Quality Resources, 1995.

Robbins, Harvey A., and Finley, Michael. Why Teams Do not Work: What Went Wrong and How to Make it Right. Peterson's/Pacesetter Books, Princeton, NJ, 1995.

Schermerhorn Jr., John R. Management, 5th ed. Hoboken, NJ: John Wiley and Sons, 1996.

Scholtes, Peter R., Joiner, Brian L., and Streibel, Barbara J. The TEAM Handbook, 2nd ed. Madison, WI: Oriel Incorporated, 1996.

Schrage, Michael. Shared Minds: The New Technologies of Collaboration. New York: Random House, 1990.

Tuckman, Bruce W. Developmental Sequence in Small Groups. *Psychological Bulletin,* Vol. 63, pp. 384-389, 1965.

Ullman, David G. The Mechanical Design Process, 2nd ed. New York: McGraw-Hill, 1997.

Whetten, David A., and Cameron, Kim S. Developing Management Skills, 3rd ed. New York: HarperCollins, 1995.

## LEARNING ACTIVITIES

### Activity 13.1    Airplane in a Box

Time:      1 hour

MARS SURVEYOR 98

## CASE STUDY

When a company designs something for production, many times they are not the ones who manufacture it. Everyone, who plays a part in the development and completion of a particular product, must be on the same page to achieve maximum results. If one team player makes a mistake, or does not understand the procedures for a project, the end result may be total failure. For example, NASA lost an unmanned module landing on Mars. The failure was attributed to something basic, something that may affect the way college students study science and engineering in the future. First, the satellite was composed of materials which were not sufficiently resistant to heat. Second, the weight distribution was uneven in the satellite, causing it to wobble excessively. The major blunder, though, was that the trajectory calculations were made using American units (pounds of force) by the subcontractor, Martin Marietta, whereas NASA assumed that they were metric units (Newtons).

You can see from this example that if one small piece of a project is overlooked or assumed, it can have catastrophic results.

In this case study, you will build one of the following out of one sheet of copy paper:

1. A paper airplane
2. A box

Your instructor will place you in two different rooms, or one group in the hallway and one in the classroom, to build your objects. If you are assigned to the paper airplane, please do not test your creation as this is not its intent.

## RESOURCES

http://mars.jpl.nasa.gov/msp98/news/mco991110.html NASA's report on the response to the loss of the Mars Climate Orbiter and the initial findings of the mission failure investigation board.

## Activity 13.2 Precision Directions

Time: 2–3 class periods

## CASE STUDY

Oral and written communications are important between teams of people working on projects. Mistakes or misrepresentations can cause the end product to be something different than intended. Therefore, any written communication should be done with neat and clear handwriting. Examples of the effects of bad handwriting range from criminals escaping jail time because of bad penmanship by police officers to patients that receive the wrong prescriptions from pharmacists because of a doctor's poor handwriting. When communicating with others, precision and clarity are paramount.

## Part A

You are responsible for developing a paper airplane prototype. Brainstorm a design for your prototype and write instructions so that someone else can replicate your design. Use one piece of paper to make your prototype airplane and one piece of paper to write the instructions on. Make sure that both pieces of paper have your name on them. Return your instructions sheet to your teacher and keep your prototype plane put away until asked to present it.

## Part B

Your instructor will provide you with a sheet of paper and a set of instructions from one of your classmates. Follow the instructions exactly to produce a paper air-

**Former President Ronald Reagan launches a paper airplane made from white house stationery from the balcony of his Los Angeles hotel room.**

plane. After you complete the paper airplane from the given instructions, your teacher will ask you to find the person that originally designed the plane and wrote the instructions. Compare the model you created from the instructions to the original prototype.

Answer the following questions:

1. Does the model made from the instructions look like the original prototype? If not, what are the differences?
2. What questions did you have about the directions written to make the model?
3. What could the person who wrote the directions have done to make it easier to make the model?
4. Rewrite the instructions for the original prototype with the creator.

## RESOURCES

The image used in this case study is courtesy of The Ronald Reagan Library, 40 Presidential Drive, Simi Valley, CA 93065. The library's website: http://www. reagan.utexas.edu

## Activity 13.3    Desert Survival

Time:    1–2 class periods

## CASE STUDY

A planeload of Uruguayan rugby players crashed into the Andes Mountains in 1972. Some survived but not all. As the days passed, they slowly realized that the tragedy wasn't over, and their lives depended on their ability to adapt and survive with what little they had. This is a true story. Books, articles, and movies have covered it. The books and the movies tell the story, but when it is broken down into its simplest form, it is about teamwork and problem solving. They had many more different troubles to unravel than the average person living in civilization; they had to figure out how to survive. For seventy-two days, those who survived the crash had to design their day with the next in mind. They had to work together, and work through their differences. Many engineering teams will have differences. It is human nature to fight for your own opinions and stick to your guns.

However, for a project to survive, or a group of people to survive, certain compromises have to be made. It is how a team, engineering or otherwise, goes though this process that is the valuable lesson learned.

*The key elements in the art of working together are how to deal with change, how to deal with conflict, and how to reach our potential . . . the needs of the team are best met when we meet the needs of individual persons. (Max DePree)*

You will start working individually and then come together in a team designated by your instructor to complete this case study. The situation is as follows:

At approximately 10 AM in mid-August, you have just crash landed in the Sonora Desert in southwestern United States. The light twin-engine plane containing the bodies of the pilot and co-pilot has completely burned. Only the airframe remains. None of the rest of you has been injured.

The pilot was unable to notify anyone of your position before the crash. However, he had indicated before impact that you were 70 miles from a mining camp, which is the nearest known habitation, and that you were approximately 65 miles off the course that was filed in your VFR Flight Plan.

The immediate area is quite flat and, except for occasional barrel and saguaro cacti, appears to be rather barren. The last weather report indicated that the temperature would reach 100 degrees that day, which means that the temperature at ground level will be 130 degrees. You are dressed in lightweight clothing: short sleeved shirts, pants, socks, and street shoes. Everyone has a handkerchief. Collectively, your pockets contain $2.83 in change, $85.00 in bills, a pack of cigarettes, and a ballpoint pen.

Before the plane caught fire, your group was able to salvage the fifteen items listed above. Your team's task is to rank these items in the order of their importance to your survival, starting with a one as the most important to fifteen as the least important.

Your team may assume:

1. The number of survivors is the same number as the number on your class team.
2. You are the actual people in the situation.
3. The team must stay together.
4. All items are in good condition.

## Part A

Each member of the team is to rank each item. DO NOT discuss the situation or the survival items until each member has finished the individual ranking. You will have 15 minutes to complete the reading and the rankings.

## Part B

After everyone has finished the individual ranking, rank the 15 items as a team. Once discussion begins, do not change your individual ranking. Your team will have 15 minutes to work on the collective rankings.

Add the individual scores in step four and divide the total by the number of people in your group. This is your team score. Compare that number with the total number in step five. Was your group ranking close to the team score?

| ITEMS | STEP 1 Individual Rank | STEP 2 Team Rank | STEP 3 Expert Rank | STEP 4 Difference Between 1&3 | STEP 5 Difference Between 2&3 |
|---|---|---|---|---|---|
| Flashlight (4 battery size) | | | | | |
| Jackknife | | | | | |
| Sectional air map of the area | | | | | |
| Plastic raincoat (large) | | | | | |
| Magnetic compass | | | | | |
| Compress kit with gauze | | | | | |
| .45 caliber Loaded pistol (loaded.45 caliber) | | | | | |
| Parachute (red & white) | | | | | |
| Bottle of salt tablets (1,000 tablets) | | | | | |
| 1 One quart of water per person | | | | | |
| Book entitled, *Edible animals of the desert* | | | | | |
| Pair of sunglasses per person | | | | | |
| 2 Two quarts of 180 proof vodka | | | | | |
| 1 One top coat per person | | | | | |
| Cosmetic mirror | | | | | |
| TOTALS | | | | | |

## RESOURCES

http://www.skillsusa.org/tqcpage.html The Total Quality Curriculum (TQC) was designed to help meet the needs of American business and industry because of changes in the global economy. The American workforce can no longer remain competitive using old methods. The TQC enhances SkillsUSA's Quality at Work movement by preparing students for the world of work.

## Activity 13.4    The String and the Ring

Time:     1–2 class periods

## CASE STUDY

When people are a part of a team, they need to learn to trust each other to do the job. The unit works in harmony with each component contributing to the cause.

For example, if you were to join a team of orange farmers in Florida, and you were originally from Wisconsin and owned an apple orchard, would some things stand in the way of your success in the orange business? More than likely, someone who has teamwork experience in one field can make the transition to another. This example can be used in different fields and provides the basic foundation for a teamwork philosophy: "Whatever you have done, whomever you have known, and however you have done things in the past, draw from those experiences to advance the team and complete your objective. In short, get it done." (Gomez, 2003)

In this case study, you will be part of team of students that has to accomplish a simple task. Move an ordinary tennis ball from the top of a one-inch section of PVC pipe that is ten inches off the ground. A set of strings will be attached to a 1.5-inch steel ring that is placed around the base of the PVC tube. Your team of students will have to work together to pick the tennis ball off the initial tube and transfer it to another tube that is at least 20 feet away. Team members may hold on to the end of the string only. If the tennis ball ever falls from the center ring, your team must place it back on the initial pipe and start over. One person should be designated to take notes on the different team ideas, the mistakes made, and the ways the team works together to accomplish the task. A short presentation to the entire class will then be completed as a debriefing.

## RESOURCES

The heart of this case study has been used by many corporations, ropes courses, and youth groups as a team-building activity. It has different variations and most are applicable to the classroom.

## Activity 13.5    Golf Course Development

Time:      10–14 class periods

## CASE STUDY

Everyday our lives take us past areas that we may never see or set foot on. People relax and enjoy their time there, whether this be in a city park, a neighborhood green space, a wooded area of town, or even a golf course. The environment is something we all should all respect a little more each day. This case study's intent is to provide perspective into the area of sustainable development.

Sustainable development is defined in Caring for the Earth (IUCN, UNEP, WWF 1991) as "improving the quality of human life while living within the carrying capacity of supporting ecosystems."

In this case study, three-student teams will design one hole of an 18-hole golf course. All vegetation, course required elements, and safety corridors must be strictly followed. All teams must design within the specified case study scale. Each team must establish a working relationship with the teams behind and in front of

them during the course construction. For example, hole three's team must communicate with hole two's team so a concurrent design flows from two to three. The same for team four's hole. For the entire project's development, each team must not only have a working relationship within their group, but also with other teams developing solutions.

Formulate criteria for site (hole) evaluation:

- List all on-site and off-site natural and cultural factors that influence the citing of lodge, lake, sport facilities, western town, golf course development, and access roads. Make efforts to understand the prevailing winds, slope orientation and percentage, views, solar orientation, etc. Think about them in terms of their potentials and constraints for site development.

List the square feet or length-width requirements for each site (hole) development. Develop this information into bubbles of approximate sizes.

- Lay these bubbles onto approximate location and discuss their potential and constraints. Mark each bubble's desired finish elevation for future grading reference. Holistically, route the vehicular circulation, walking trails, and major drainage ways.

Required elements:

1. Final drawing of your teams specified hole to scale of 1"=100'
2. Title block and credit information
3. Written summary and labels on drawing
4. Color-coded paper model of the team's hole (scale of 1"=60')
   - Should be three dimensional
   - Should match up correctly with hole previous and hole after
   - Green for tee boxes and putting greens
   - Red, blue, and white for designated tees
     - Red for furthest forward
     - White for next set back
     - Blue for next set back
     - Black for professional distance
   - Black for hole location
   - Light blue for water hazards
   - Light brown for sand hazards
   - Brown/gray for rock hazards
5. Vegetation as needed for design (use 3D objects to scale)
6. Presentation of your model to peers

The teams should pay strict attention to the surrounding environment and if a specific site is determined, care should be taken to sustain as much of the existing nature as possible.

## RESOURCES

www.agolfarchitect.com Tony S. Ristola is a fourteen-year PGA member and former tournament and teaching professional. Before designing and supervising the construction of his own projects, he was involved in building courses for some of the most recognized names in the industry. Tony's participation and his website have been a critical part of the development of this case study.

http://home.okstate.edu/homepages.nsf/toc/HsuHome

Professor Paul Hsu was born in Taiwan, ROC, and was naturalized as a U.S. citizen in 1985. Professor Hsu has contributed elements from his golf course development section within his landscape architecture program. His participation with this case study brings together the artistic and technical aspects of course development. He received Bachelor of Science Degree in Geography from National Taiwan University, Master of Landscape Architecture Degree from Cornell University, and has a Ph.D. candidate in Spatial Science and Engineering at University of Maine.

Carpenter, Jot. Handbook of Landscape Architectural Construction, Jot Carpenter, "Golf Course Design/Construction". Landscape Architecture Foundation. 1976. pp. 459-469.Edited by Mitchell Beazley International Limited.

The New World Atlas of Golf. London: Mitchell Beazley Publishers. 1989.

Jarrett, Albert. Golf Course & Grounds—Irrigation and Drainage. Englewood Cliffs, NJ: Prentice Hall, 1984.

Doark, Tom. The Anatomy of A Golf Course. New York, NY: Lyons & Burford.

Golf Course Development and Real Estate. Washington, D.C. : ULI, 1994.

*Chapter 14*

# Problem Solving

## Introduction

Seven-year-old Dan could build all kinds of things: toy cars, ramps, and simple mechanical gadgets. More amazingly, he could do all this with three simple tools: scissors, hammer, and a screwdriver. One Christmas, his father asked if he'd like more tools. "No thanks, Dad," was his response. "If I can't make it with these, it's not worth making." Soon after, Dan realized that the cool stuff he could create or the problems he could solve were dependent on the tools he had. So, he began gathering more tools and learning how to use them. Today, he still solves problems and makes many cool items like real cars, furniture, playgrounds, and even houses. His tools fill a van, a workshop, and a garage. Having the right tools and knowing how to use them are the critical components of his success.

Likewise, most people, when they are children, learn a few reliable ways to solve problems. Most fifth graders in math class know common problem-solving methods, such as "draw a picture," "work backward," or the popular "guess and check" (check in the back of the textbook for the correct answer, that is). Unfortunately, many people choose to stop adding problem-solving tools to their mental toolbox. "If I don't know how to solve it, it's probably not worth solving anyway," they may think. Or, "I'll skip this problem for now and get back to my usual tasks." This severely limits them. Engineers, by definition, need to be good at solving problems and making things. Therefore, filling the toolbox with the right tools and knowing how to use them are the critical components to becoming a successful engineer.

The road to solutions is just as important, if not more important, as finding the solution itself. We need to teach the learning process that students require so they can navigate that road. Although all students will not become engineers, they do need problem-solving skills for life in the technologically complex 21st century.

The goal of this chapter is to increase the number of problem-solving strategies and techniques commonly used, and to enhance you're the ability to apply them creatively in problem-solving settings. Hopefully, this will be the beginning of a lifelong process of gaining mental tools and learning to use them.

## ANALYTIC AND CREATIVE PROBLEM SOLVING

*In creative problem solving, there is no single right answer.*

The toolbox analogy is appropriate for engineering students. Practicing engineers are employed to solve many problems. In the engineering disciplines and the profiles chapters in this textbook we discuss how engineers can fill different roles in daily practice. These roles can require different problem-solving abilities. Problem solving can be Analytic or Creative. Most students are more familiar with Analytic problem solving where only one correct answer exists. In Creative problem solving, there is no single right answer. Your analytic tools represent what's in your toolbox. Your creative skills represent how you handle your tools.

To better understand the difference between these two kinds of problem solving, let's examine the one function that is common to all engineers: design. In the design chapter, we detail a design process with ten steps. Those ten steps are the following:

1. Identify the problem
2. Define the working criteria/goal(s)
3. Research and gather data
4. Brainstorm for creative ideas
5. Analyze
6. Develop models and test
7. Make the decision
8. Communicate and specify
9. Implement and commercialize
10. Prepare post-implementation review and assessment

*Engineering curricula provide many opportunities to develop analytic problem-solving abilities.*

By definition, design is open-ended and has many different solutions. The automobile is a good example. When you are riding in a car, count the number of different designs you see on the road. All of these designs were the result of engineers solving a series of problems involved in producing an automobile. The design process as a whole is a creative problem-solving process.

The other method, analytic problem solving, is also part of the design process. Step five, to analyze, requires analytic problem solving. When analyzing a design concept, only one response answer "Will it fail?" A civil engineer designing a bridge might employ a host of design concepts though at some point he or she must accurately assess the loads the bridge can support.

Most engineering curricula provide many opportunities to develop analytic problem-solving abilities, especially in the first few years. This is an important skill. Improper engineering decisions can put the public at risk. For this reason, it is imperative that proper analytical skills be developed.

As an engineering student you will be learning math, science, computer, and hands-on skills that will allow you to tackle complex problems in your career.

These critical skills are part of the problem-solving skill set for analysis. However, other tools are equally necessary.

Engineers also find that applications of technical solutions can become outdated, or prove dangerous, if not applied with the right judgment. Our creative skills can help evaluate the big picture; these skills determine a solution's long-term success. The solution of one problem can cause others if one neglects foresight, not easily avoidable in all cases. For instance, the developers of dynamite and atomic fission have publicly expressed guilt, and they regret their lack of foresight. Early Greek engineers invented many devices and processes, which they refused to disclose because they knew the devices would be misused. The foresight to see how technically correct solutions can malfunction or be misapplied takes creative skill to develop. What about fertilizers which boost crops but poison water supplies? Modern engineers need to learn how to apply the right solution at the right time in the right way for the right reason with the right customer.

*Engineers need to learn how to apply the right solution at the right time in the right way for the right reason with the right customer.*

Most people rely on two or three methods to solve problems. If these methods don't yield a successful answer, they become stuck. Truly exceptional problem solvers learn to use multiple problem-solving techniques to find the optimum solution. The following is a list of possible tools or strategies that can help solve simple problems:

1. Look for a pattern
2. Construct a table
3. Consider possibilities systematically
4. Act it out
5. Make a model
6. Make a figure, graph or drawing
7. Work backwards
8. Select appropriate notation
9. Restate the problem in your own words
10. Identify necessary, desired and given information
11. Write an open-ended sentence
12. Identify a sub-goal
13. First solve a simpler problem
14. Change your point of view
15. Check for hidden assumptions
16. Use a resource
17. Generalize
18. Check the solution; validate it
19. Find another way to solve the problem
20. Find another solution
21. Study the solution process

22. Discuss limitations
23. Get a bigger hammer
24. Sleep on it
25. Brainstorm
26. Involve others

In analytic and creative problem solving, different methods exist for tackling problems. Have you ever experienced being stuck on a difficult math or science problem? Developing additional tools or methods will allow you to tackle more of these problems effectively, making you a better engineer.

The rest of this chapter is a presentation of ways of solving problems. Some of the techniques have been effective. Each person has a unique talent set and will be naturally drawn to certain problem-solving tools. Others will find a different set useful. The important thing as an engineering student is to experiment and find as many useful tools as you can. This will equip you to tackle the wide range of challenges which tomorrow's engineers will face.

## ANALYTIC PROBLEM SOLVING

Given the importance of proper analysis in engineering and the design process, it is important to develop a disciplined way of approaching engineering problems. Solving analytic problems has been the subject of much research, which has resulted in several models. One of the most important analytic problem-solving methods for students is the **Scientific Method.** The steps in the Scientific Method are as follows:

1. Define the problem
2. Gather the facts
3. Develop a hypothesis
4. Perform a test
5. Evaluate the results

In the Scientific Method, the steps can be repeated if the desired results are not achieved. The process ends when an acceptable understanding of the studied phenomenon is achieved.

In the analysis of engineering applications, a similar process can be developed to answer problems. The advantage of developing a set method for solving analytic problems is that it provides a discipline to help young engineers when they are presented with larger, more complex problems. Just like a musician practices scales to set the foundation for complex pieces to be played later, an engineer should develop a sound fundamental way to approach problems. Fortunately, early in your engineering studies you will be taking many science and math courses suited to this methodology.

The Analytic Method we will discuss has six steps:

1. Define the problem and make a problem statement
2. Diagram and describe
3. Apply theory and equations
4. Simplify the assumptions
5. Solve the necessary problems
6. Verify accuracy to required level

Following these steps will help you understand the problem you are solving and allow you to identify areas with inaccuracies.

## Step 1: Problem Statement

Restate the problem you are solving in your own words. In textbook problems, this helps you understand what you need to solve. In real life situations, this ensures you are solving the correct problem. Write your summary, then check that your impression of the problem is the one that matches the original problem. Putting the problem in your own words is also an excellent way to focus on the problem you need to solve. Often, engineering challenges are large and complex, and the critical task is to understand what part of the problem you need to solve.

## Step 2: Description

The next step is to describe the problem and list all that is known. In addition to restating the problem, list the information given and what needs to be found. This is shown in Example 14.2, which follows this section. Typically, in textbook problems, you have all the information you need for the problem. In real problems, you have more information than you need to do the calculations. In other cases, information may be missing. Formally writing what you need and what you require helps you sort this.

Drawing a diagram or sketch of the problem will help you understand the problem. Pictures help many people to clarify the problem and what they need. They are a great aid when explaining the problem to someone else. The old saying could be restated as "A picture is worth a thousand calculations."

## Step 3: Theory

State explicitly the theory or equations needed to solve the problem. It is important that you write this out completely. Most real problems, and those you must solve as an undergraduate student, will not require exact solutions to complete equations. Understanding the equation parts to be neglected is vital to your success.

Commonly, you will have a routine of solving a simplified version of equations. An example is the flow of air over a body, like a car or airplane wing. At low speeds, the air density is considered a constant. At higher speeds, this is not the case; if the density were considered a constant, errors would result. Starting with full equations and then simplifying reduces the likelihood that you will overlook important factors.

## Step 4: Simplifying Assumptions

As mentioned above, not all engineering and scientific applications can be solved precisely. Even if they are solvable, determining the solution might be too costly. For instance, an exact solution might require a high-speed computer to calculate for a year to get an answer, and this would not be an effective use of resources, e.g., weather prediction.

To solve a problem in a timely and cost-effective manner, you must simplify assumptions. Simplifying assumptions can make the problem solving easier and still provide an accurate result. Record the assumptions along with how they simplify the problem. This documents the assumptions and allows the final result to be interpreted in terms of these assumptions. While estimation and approximation are useful tools, engineers are concerned with the accuracy and reliability of their results. Approximations are often possible if assumptions are made to simplify the problem. Therefore, an important concept for engineers to understand is the Conservative Assumption.

In engineering problem solving, "conservative" has a non-political meaning. A conservative assumption is one that introduces errors on the safe side. We mentioned that we can use estimations to determine the bounds of a solution. An engineer should be able to look at those bounds and determine which end of the spectrum yields the safer solution. By selecting the safer condition, you are assuring your calculation will result in a responsible conclusion.

Consider the example of a materials engineer selecting a metal alloy for an engine. Stress level and temperature are two parameters the engineer must consider. In the early design stages, you may not know these parameters. What you could do is estimate the parameters, done with a simplified analysis or prior experience. Often, products evolve from earlier designs with data. If such data exist, ask how the new design will affect the parameters you are interested in. Will the stress increase? Will it double? Perhaps a conservative assumption would be to double the stress level of a previous design.

After taking the conservative case, a material can be selected. An engineer should ask, "If a more precise answer were known, could I use a cheaper or lighter material?" If the answer is No, then the simple analysis might be sufficient. If yes, the engineer could build a case to justify a more detailed analysis.

Another example of conservatism is the design of a swing-set (See Fig. 14.1). In most sets, the swings hang from a horizontal piece, and an inverted V-shaped support is at each end. If you were the design engineer, you might have to size these supports.

One of the first questions would be, "For what weight do we size the swing set?" One way to answer this is to research the weight distribution of children, followed by research into swing use by children at different ages, and then analyze

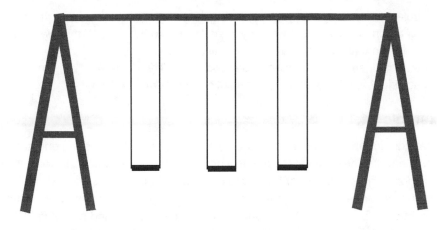

**Figure 14.1 Simple swing set.**

those data to determine typical weights of children who would use such swings. You might even need to observe playground activity to see if several children pile on a swing or load the set in different ways, e.g., by climbing on it. This analysis might take weeks, but you would have excellent data.

On the other hand, you could just assume a heavy load that would exceed anything the children would produce, say, 500 pounds per swing. That would assume the equivalent of two large adults on each swing. Make the calculation to determine the needed supports, and then ask if it makes sense to be more detailed.

To answer this, one has to answer, "What problem am I solving? Am I after the most precise answer I can get? Am I after a safe and reliable answer regardless of other concerns?"

In engineering, other concerns include such things as cost and availability. In the current example, the supports would most likely use common materials. Swing sets are typically made of wood or steel tubing. So, one way to answer the above questions is to check and see if making a more detailed analysis would be justified. Check the prices of the materials needed for your conservative assumption against potential materials based on a more detailed analysis. Does the price difference justify a more detailed analysis? Another way to ask this is to ask, "Does the price difference justify spending time to do the analysis?" As an engineer, your time costs money.

The answer may be, "It depends." It would depend on how much the difference is and how many you are going to produce. It wouldn't make sense to spend a week analyzing or researching the answer if it saves you or your company 500 dollars. If your company was producing a million sets and you could save $5 per set, the savings would justify the week's work.

Approximation can improve decision making indirectly as well. By using approximation to resolve a problem's minor aspects, you can focus on the more pressing aspects.

Engineers are faced with these types of decisions every day in design and analysis. They must determine the problem and how to solve it. Public safety is most important in the engineering profession. You might be faced with a short-term safety emergency where you need to prioritize solutions. Or you might be designing a consumer product where you have to take the long-term view. Swing sets are often used for generations, after all.

As an engineer, you will need to develop the ability to answer, "What problem am I solving?" and "How do I get the solution I need most efficiently?"

## Example 14.1

A manager for an aerospace firm had a method for breaking in young engineers. Shortly after the new engineer came to work in his group, the manager would look for an appropriate analysis for the young engineer. When one arose, he would ask the young engineer to perform the analysis. The engineer would embark on a path to construct a huge computer model requiring lengthy input and output files that would take a few weeks to complete. The manager would return to the engineer's desk the following morning, asking for the results. After hearing that all the new engineer accomplished was to map out the work for the next three weeks of work, the manager would erupt and explain loudly that the new engineer was supposed to be doing engineering analysis, not a "science project." He would show the young engineer how doing a simplified analysis that would take an afternoon to perform and would answer the question given the appropriate conservative assumptions. He would then leave the engineer to the new analysis method.

While his methods were questionable, his point is well taken. There are always time and cost concerns as well as accuracy. If conservative assumptions are made, safe and reliable design and analysis decision can often be made.

## Step 5: Problem Solution

Now the problem is set for you to perform the calculations. This might be done by hand or with a computer. It is important to learn how to perform simple hand calculations; however, computer applications make complex and repetitive calculations much easier. When using computer simulations, develop a means to document what you have done in deriving the solution. This will allow you to find errors more quickly, as well as to show others what you have done.

Step 6: Accuracy Verification

Engineers work on solutions that affect people's livelihood and safety. The engineer's solution must be accurate.

Engineers are responsible for verifying the accuracy of their own solutions. Therefore, the student must develop skills to meet any required standard.

A fascinating aspect of engineering is that the degree of accuracy is a variable under constant consideration. A problem or project may be solved properly within a tenth of an inch, but accuracy to a hundredth might be unneeded and difficult to control, rendering that kind of accuracy incorrect. The next step of the solution might require accuracy to the thousandth. Be sure of standards!

There are many different ways to verify a result. Here are some of the ways:

1. Estimate the answer.
2. Simplify the problem and solve the simpler problem. Are the answers consistent?

3. Compare with similar solutions. In many cases, other problems were solved similarly to the current one.
4. Compare to previous work.
5. Ask a more experienced engineer to review the results.
6. Compare to published literature on similar problems.
7. Ask yourself if the results make sense.
8. Compare to your own experience.
9. Repeat the calculation.
10. Run a computer simulation or model.
11. Redo the calculation backwards.

It may be difficult to tell if your answer is correct, but you should be able to tell if it is close or within a factor of ten. This will often flag any systematic error. Being able to back up your confidence in your answer will help you make better decisions on how your results will be used.

The following example will illustrate the analytical method.

## Example 14.2

A ball is projected from the top of a 35 m tower at an angle 25¡ from horizontal with an initial velocity v0 = 80 m/s. How much time will it take to reach the ground, and what will the horizontal distance be from the tower to the point of impact?

**Problem Statement:** Given:  Initial velocity v0 = 80 m/s

Initial trajectory = 25¡ from horizontal

Ball is launched from a height of 35 m

**Find:** 1) Time to impact

2) Distance of impact from the tower

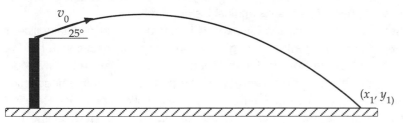

**Equations:** Found in a physics text:

$$(a)\ x_1 = x_0 + v_0 t \cos \theta + \frac{1}{2} a_x t^2$$

$$(b)\ y_1 = y_0 + v_0 t \cos \theta + \frac{1}{2} a_y t^2$$

**Assumptions:** Neglect air resistance

Acceleration $a_y = -9.81$ m / s$^2$ is constant in the vertical direction

Acceleration is zero in the horizontal direction, $a_x = 0$
Ground is level and impact occurs at y = –35 m

**Solution:** Begin with equation (b):

$$-35 = 0 + 80t \sin 25° - \frac{1}{2} 9.8t^2 \quad \text{or} \quad t^2 - 6.90t - 7.14 = 0$$

$$\therefore t = \frac{-b \pm \sqrt{b^2 - 4ac}}{2a} = \frac{6.9 \pm \sqrt{6.9^2 + 4 \times 1 \times 7.14}}{2} = 7.81 \text{ sec}$$

The solution must be 7.81 seconds since time must be positive. The distance is found using equation (a):

$$x_1 = 0 + 80 \times 7.81 \times \cos 25° = 566 \text{ m}$$

## ESTIMATION

Estimation is a problem-solving tool that engineers need to develop. It can provide answers to problems quickly and can verify complicated analyses. Young engineers are invariably amazed, when they begin their careers, at how the older, more experienced engineers can estimate so closely before the analysis is completed.

An engineer was responsible for doing a detailed temperature analysis on a gas turbine engine that was to go on a new commercial aircraft. He and his team presented the plan to the managers and people from the Chief Engineers Office. The plan called for two full-time engineers working for a month on the analysis.

One of the more senior engineers who had attended the presentation, went back to his office, made some calculations and sent the results. His note told the engineer to check their results against his numbers.

After the month-long analysis, which involved making a detailed computer model of the components, the engineer checked their results with his "napkin" calculations. He was close to the engineers' result! The detailed analysis did provide a higher degree of precision. What he had done was use suggestion #13 ("First solve a simpler problem) from the problem-solving methods. He used simple shapes with known solutions to approximate the behavior of the complex shapes of the jet engine's components.

Though estimation or approximation may not yield the precision required for an engineering analysis, it is a useful tool and can help check an analysis. Estimation can provide bounds for potential answers, especially critical in today's dependence on computer solutions. Your method may be correct, but one character mistyped will throw your results off. You must have confidence in your results by developing tools to verify accuracy.

Estimation can also be used to decide if a detailed analysis is needed. Estimating using best-case and worst-case scenarios can yield an upper and lower bound to the problem. If the entire range of potential solutions is acceptable, why do the detailed analysis? What if for our case, the whole range of temperatures met our design constraints? The month-long work would not have been justified. In this case, the added precision led to a design benefit. As engineers, you will be asked to decide when a detailed analysis is required.

Estimation can be a powerful tool to check accuracy and make analysis decisions. Senior engineers in the industry often comment that current graduates lack the ability to do an approximation. This has become more important with the dependence of computer tools and solutions. Computer analyses lend themselves to typos on inputs or arithmetic errors. Predicting the ballpark of the expected results can head off potentially disastrous results.

## CREATIVE PROBLEM SOLVING

Many complex engineering problems are open-ended and require creative problem solving. To maximize the creative problem-solving process, a systematic approach is recommended. Just as with analytic problem solving, developing a systematic approach to using creativity will pay dividends in better solutions.

*Dividing the process into steps allows you to break a large, complex problem into simpler problems where your various skills can be used.*

Revisiting our toolbox analogy, the creative method provides the solid judgment needed to use your analytic tools. Individual thinking skills are your basics here. The method we'll outline now will help you apply those skills and choose solution strategies effectively.

You can look at a creative problem-solving process in many ways. We will present a method which focuses on answering these five questions:

1.  What is wrong?
2.  What do we know?
3.  What is the real problem?
4.  What is the best solution?
5.  How do we implement the solution?

By dividing the process into steps, you are more likely to follow a complete and careful problem-solving procedure and will have more effective solutions. This also allows you to break a large, complex problem into simpler problems where your various skills can be used. Using such a strategy, the sky is the limit on the complexity of projects which an engineer can complete. It's astonishing and satisfying to see what can be done.

## DIVERGENCE AND CONVERGENCE

At each phase of the process, there is a divergent and a convergent part of the process, as shown in Figure 14.2. In the divergent process, you start at one point

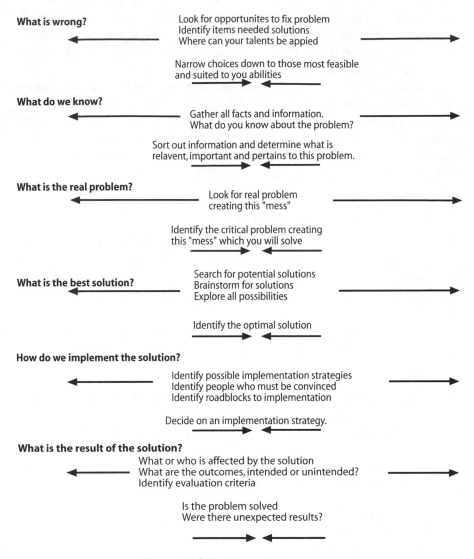

**Figure 14.2 Problem-solving process.**

and reach for as many ideas as possible. Quantity is important. Identifying possibilities is the goal of the divergent phase. In this portion of the process (indicated by the outward pointing arrows) use the brainstorming and idea generation techniques described in subsequent sections. Look for as many possibilities as you can. Often, the best solutions come from different ideas.

In the convergent phase (indicated by inward pointing arrows), use analytical and evaluative tools to narrow the possibilities to the one(s) most likely to yield results. Quality is most important, and the goal is to find the best possibility to move the process to the next phase. (One common method is to use a matrix to rate ideas based on defined criteria.) If one choice fails to produce satisfactory results, go back to the idea lists for another method to tackle the problem.

## Sample Problem: Raising Those Low Grades

In order to illustrate the problem-solving process, let's consider an example of a problem and work through the process. The problem we will use for illustration is a student who is getting low grades. This certainly is a problem that needs a solution. Let's see how the process works.

## What's Wrong?

In the first step of the problem-solving process, an issue is identified. This can be something stated for you by a supervisor or a professor, or something you determine on your own. This is the stage where entrepreneurs thrive while looking for an opportunity to meet a need. Similarly, engineers look to find solutions to meet a need. This may involve optimizing a process, improving customer satisfaction, or addressing reliability issues.

To illustrate the process, let's take our example of the initial difficulty that beginning engineering students can have with grades. Most students were proficient in their previous grades or schools and might not have had to study much. The demands of engineering programs can catch some students off guard. Improving grades is the problem we will tackle.

At this stage, we need to identify that a problem is worth solving. A good start would be to identify whether it is a problem. How could you do this? Total your scores thus far in your classes and compare them with the possible scores in your syllabus. Talk with your instructor(s) to find out where your level in the classes and if grades will be given on a curve. Talk to an advisor about realistic grade expectations. At the end of this process, you should know if you have a grade problem.

## What Do We Know?

The second step in problem solving is the gathering of facts. All facts and information related to the problem identified in the first step are gathered. In this information gathering stage, do not evaluate whether the data are central to the problem. As will be explained in the brainstorming and idea generation sections to follow, premature evaluation can be a barrier to generating sufficient information.

With our example of the student questioning his grades, the information we generate might include:

1. Current test grades in each course
2. Current homework and quiz grades in each course
3. Percent of the semester's grade already determined in each course
4. Class average in each course
5. Instructor's grading policies
6. Homework assignments behind
7. Current study times
8. Current study places
9. Effective time spent on each course
10. Time spent doing homework in each course

11. Performance of your friends
12. Previous school grades
13. Previous school study routine

Part of this process is to list the facts you know. To be thorough, you should request assistance from someone with a different point of view. This could be a friend, an advisor, a parent, or an instructor. He or she might come up with a critical factor you have overlooked.

## What Is the Real Problem?

This stage is often skipped but is critical to effective solutions. The difference between this stage and the first is that this step answers the question why. Identifying the initial problem answers the question of what is wrong. To fix the problem, a problem solver needs to understand why the problem exists. The danger is that only symptoms of the problem will be addressed, rather than root causes.

In our example, we are problem solving low class grades. That is the "what" question. To understand a problem, the "why" must be answered. Why are the grades low? Answering this question will identify the cause of the problem. The cause is what we must deal with to address a problem.

So, why are class grades low? Let's assume that poor test scores resulted in the low grades. This doesn't tell us why the test scores are low. This is a great opportunity for brainstorming potential causes. In this divergent phase of problem definition, don't worry about evaluating the potential causes. Wait until the list is generated.

Possible causes may include:

1. Poor test-taking skills
2. Incomplete or insufficient notes
3. Poor class attendance
4. Not understanding the required reading
5. Not spending enough time studying
6. Studying the wrong material
7. Attending the wrong class
8. Studying ineffectively (e.g., cramming the night before tests)
9. Failing to understand the material (this might reveal need for tutor or study group)
10. Using solution manuals or friends as a crutch to do homework
11. Not at physical peak at test time (i.e., up all night before the test)
12. Didn't work enough problems from the book

After you have created the list of potential causes, evaluate each as to its validity. In our example, there may be more than one cause contributing to the low grades. If so, make a rank-ordered list. Rank the causes in order of their impact on class performance.

Assume the original list is reduced to the following:

1.  Poor class attendance
2.  Studying ineffectively
3.  Not spending enough time studying
4.  Not understanding the material

Of these things, rank them in order of their impact. You may be able to do this yourself or you may need help. An effective problem solver will seek input from others when appropriate. In this case, a professor or academic advisor might be able to help you rank the causes and determine which ones will provide the greatest impact. For this example, say the rank order was the following:

1.  Not understanding the material
2.  Studying ineffectively
3.  Not spending enough time studying
4.  Poor class attendance

We determined that while class attendance was not perfect, it was not the key factor in the poor performance. The key item was not understanding the material. This goes with ineffective studying and insufficient time.

Identifying key causes of problems is vital, so let's look at another example.

## Example 14.3

"What was the confusion?" Defining the real problem is often the most critical stage of the problem-solving process and the one most often skipped. In engineering, this can be the difference between a successful design and a failure.

An example of this occurred with an engine maker. A new model engine was introduced and began selling quickly. A problem developed, however, and failures started to happen. Cracks initiating from a series of bolt holes were causing the engines to fail. A team of engineers was assembled and the part was redesigned. They determined that the cause was high stress in the area of the bolt hole. Embossments were added to strengthen the area. It was a success.

Until, that is, the redesigned engines accumulated enough time in the field to crack again. A new team of engineers looked at the problem and quickly determined that the cause was a three-dimensional coupling of stress concentrations created by the embossments, resulting in too high a stress near the hole where the crack started. The part was redesigned and introduced into the field. It was a success.

It was a success until the redesigned engines accumulated enough time in the field to crack yet again. A third team of engineers was assembled. This time, they stepped back and examined the real cause. Part of the team ran sophisticated stress models of the part, just as the other two teams had done. The new team members looked for other causes. What they found was that the machine creating the hole introduced micro-cracks into the bolt hole surfaces. The stress models predicted that further stresses would start a crack. However, if a crack were intentionally introduced, the allowable stresses were much lower and the part would not fail. The cause wasn't with cracking but with the wrong kind of cracking. The third team was a success because it had looked for the real problem. Identifying the real cause initially, which in this case involved the machining process, would have saved the company millions of dollars, and spared customers the grief of the engine failures.

## What Is the Best Solution?

Once the problem has been defined, potential solutions need to be generated. This can be done by yourself or with the help of friends. In an engineering application, it is wise to confer with experienced experts about the problem's solution. This may be most productive after you have begun a list of causes. The expert can comment on your list and offer his or her own as well. This is a great way to get your ideas critiqued. After you gain more experience, you will find that technical experts help narrow down the choices. Also, go to more than one source. This may provide more ideas as well as help with the next step.

In our example, the technical expert may be an instructor or advisor who is knowledgeable about studying. Let's assume that the following list of potential solutions has been generated.

1.  Get a tutor
2.  Visit the instructor during office hours
3.  Make outside appointments with instructors
4.  Visit help rooms
5.  Form study group
6.  Outline books
7.  Get old exams
8.  Get old sets of notes
9.  Outline lecture notes
10. Review notes with instructors
11. Do extra problems
12. Get additional references
13. Make a time schedule
14. Drop classes
15. Retake the classes in the summer
16. Go to review sessions

The list now must be evaluated and the best solution decided upon. This convergent phase of the problem-solving process is best done with input from others. Getting opinions of those with experience and/or expertise you are investigating is especially helpful. In an engineering application, this may be a lead engineer. In our case, it could be an academic advisor, a teaching assistant, or a professor. These experts can be consulted individually or asked to participate as a group.

When refining solutions, an effective tool is to ask yourself which of them will make the biggest impact and which will require the most effort. Each solution gets

## Effort

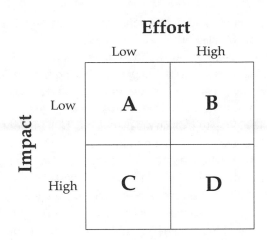

**Figure 14.3 A Problem-Solving Matrix.**

ranked high or low as to impact and effort. Ideal solutions are those that have a high impact yet require a low effort level. A visual tool which helps make this process easier is the problem-solving matrix shown in Figure 14.3. Solutions that fall in quadrant B should be avoided. These solutions require a great deal of effort for marginal payoff. Solutions that are the most effective are those in quadrant C. These are the ones that are easy to implement and will produce significant results. It may be necessary to include solutions in quadrant D, requiring high effort yet yielding high impact. These should be carefully evaluated to ensure that the investments are worth the reward.

## Implementing the Solution

Implementing the solution is a step that may seem trivial. This step, however, is a critical phase of the problem-solving process. In our scenario, the solution selected may have been to do extra problems.

To accomplish this solution, appropriate additional problems must be selected, done, and corrected. Implementation probably requires the assistance of the instructor to help select appropriate problems. To pick additional practice problems randomly could move the effort from quadrant C to D and even to B. To have the greatest chances of success, you would want to gain the assistance of your instructor. Other sources of assistance could come from a study group, which poses the challenges of getting them to do extra problems with you.

As with all the other steps, a divergent phase begins the step including such activities as brainstorming. The convergent phase of the step culminates with the selection of the implementation plan.

In engineering applications, implementation can be a critical process phase. Most of the solutions to problems require additional resources (i.e., money) or the cooperation of other groups (either not directly affected by the problem or not under your control). Especially early in your career, you will not have control over the people who need to implement solutions, so you will need to sell your ideas to the managers who do. An effective implementation plan will be critical to getting your solutions accomplished.

President Jimmy Carter was educated as an engineer and was one of the most respected presidents as a strong moral person. During his presidency, the country was in an energy crisis and action was needed to address the problems. The story goes that he summoned Tip O'Neil, the speaker of the House of Representatives, to the White House to explain his plan. Mr. O'Neil responded by giving the President a list of people who would need to be sold on the plan before it went public. President Carter who was new to Washington was going to skip the development of an implementation plan and enact the plan. He believed that with the country in such an energy crisis that it was obvious what was needed and cooperation was a given. Tip O'Neil was experienced in the Washington political system and understood that they needed to spend time gaining acceptance to their plan before implementing it.

Engineering situations can require a similar process. While most are not as political as Washington, D.C., most require gaining acceptance of others to implement a solution. In all of these cases, gaining the recognition of those involved is a critical part of the problem-solving process.

## Evaluating the Solution

Implementation does not necessarily end the problem-solving process. Just as the design process is cyclic with each design leading to possible designs, problem solving can also be cyclic. Once a solution has been found and implemented, an evaluation should be performed. As with the other steps in the problem-solving process, this step begins with a divergence phase as the problem solver asks what makes a successful solution and how to evaluate it. To evaluate a solution, a criterion for success must be established. Sometimes objective criteria for evaluation exist, but the criterion for success is often more subjective.

Once the criterion has been established, the evaluation process should be defined including who will evaluate the solution. Often it is desirable to get a neutral view of someone to be an evaluator and who was not involved in the formulation of the solution process.

If the solution is a success, the process may be complete. In engineering applications, solutions are often intermediate and lead to other opportunities. The software industry is a prime example. Software is written to utilize the current computer technologies to address issues. After it is completed, new technologies become available opening new opportunities or requiring new solutions.

Sometimes a success does not address the true problem as was the case of the turbine disk failures described earlier. In these cases, the evaluation process may need time to deem it a true success. In other cases, solutions have unintended outcomes that require a another solution even if the initial solution was a success.

Even if a solution is effective, taking time to reflect on the solution and its implications has value. This allows you, as a problem solver, to learn from the process and the solution. Critically evaluating solutions and opportunities is a valuable skill as an engineer and is discussed further in the critical thinking below.

# PERSONAL PROBLEM-SOLVING STYLES

The creative problem-solving model presented previously is one of the models used to solve the kind of open-ended problems engineers face daily. You may find yourself in an organization that uses another model. Many models for problem solving exist and each dissects process into slightly different steps. Isaken and Treffinger, for example, break the creative problem-solving process into six linear steps [Isaken and Treffinger, 1985].

These are the six steps:

1. Mess Finding
2. Data Finding
3. Problem Finding
4. Idea Finding
5. Solution Finding
6. Acceptance Finding

Dr. Min Basadur of McMaster University developed another model, a circular one, which he calls Simplex, based on his experience as a product development engineer with the Procter and Gamble Company. This model separates the problem-solving process into eight different steps [Basadur, 1994]

1. Problem finding
2. Fact finding
3. Problem defining
4. Idea finding
5. Evaluating and selecting
6. Action planning
7. Gaining acceptance
8. Taking action

All these problem-solving processes provide a systematic approach to problem solving. Each process has divergent phases where options need to be generated along with convergent phases when the best options need to be selected.

Basadur has created and patented a unique method to help individuals participating in the creative problem-solving process identify which parts of the process they are more comfortable with. This method, called the Basadur Simplex Creative Problem-Solving Profile, reflects your personal creative problem-solving style. Everyone has a different creative problem-solving style. Your particular style reflects your relative preferences for the different parts of the problem-solving process.

The Basadur Simplex Creative Problem-Solving Profile identifies four styles, and each style correlates with two of the eight problem-solving steps in the eight-step circular model of creative problem-solving above. Thus, the eight problem-solving steps are grouped into four quadrants or stages of the complete problem-solving process.

These four steps are shown as the quadrants in Figure14.4. The Basadur Simplex Creative Problem-Solving Process begins with quadrant one, the generation of problems and opportunities. It cycles through quadrant two, the conceptualization of the problem or opportunity and of potentially useful ideas and then to quadrant three, the optimization of solutions. It ends with quadrant four, the implementation of solutions. Each quadrant requires different kinds of thinking and problem-solving skills. Basadur describes the four different quadrants or stages in the following way.

## Generating

Generating involves getting the problem-solving process rolling. Generative thinking involves gathering information through direct experience, questioning, imagining possibilities, sensing problems and opportunities, and viewing situations from different perspectives. People and organizations strong in generating skills tend to focus more on creating options (diverging) than evaluating and selecting options (converging). They see relevance in almost everything and think of good and bad sides to almost any fact, idea, or issue. They dislike becoming too organized or delegating the complete problem but are willing to let others take care of the details. They are comfortable with ambiguity and are hard to pin down. They delight in juggling many projects simultaneously. Every solution they explore suggests several other problems to be solved. Thinking in this quadrant includes problem finding and fact finding.

## Conceptualizing

Conceptualizing keeps the innovation process going. Like generating, it involves divergence. But rather than gaining understanding by direct experience, it favors gaining understanding by abstract thinking. It results in putting ideas together, discovering insights that help define problems, and creating theoretical models to explain things. People and organizations strong in conceptualizing skills enjoy taking information scattered all over the map from the generator phase and making sense of it. Conceptualizers need to understand: To them, a theory must be logically sound and precise. They prefer to proceed only with a clear grasp of a situation and when the problem or main idea is well defined. They dislike having to prioritize, implement, or agonize over poorly understood alternatives. They like to play with ideas and are not overly concerned with moving to action. Thinking in this quadrant includes problem defining and idea finding.

## Optimizing

Optimizing moves the innovation process further. Like conceptualizing, it favors gaining understanding by abstract thinking. But rather than diverge, an individual with this thinking style prefers to converge. This results in converting abstract ideas and alternatives into practical solutions and plans. Individuals rely on mentally testing ideas rather than on trying things out. People who favor the optimizing style prefer to create optimal solutions to a few well-defined problems or issues. They prefer to focus on specific problems and sort through large amounts of information to pinpoint what's wrong. They are usually confident in their ability to make a sound logical evaluation and to select the best option or solution to a problem. They often lack patience with ambiguity and dislike dreaming about additional ideas, points of view, or relations among problems. They believe they know the problem. Thinking in this quadrant includes idea evaluation and selection and action planning.

## Implementing

Implementing completes the innovation process. Like optimizing, it favors converging. However, it favors learning by direct experience rather than by abstract thinking. This results in getting things done. Individuals rely on trying things out rather than mentally testing them. People and organizations strong in implementing prefer situations in which they must somehow make things work. They do not need complete understanding in order to proceed and adapt quickly to immediate changing circumstances. When a theory does not appear to fit the facts they will readily discard it. Others perceive them as enthusiastic about getting the job done but also as impatient or even pushy as they turn plans and ideas into action. They will try as many different approaches as necessary and follow up or "bird dog" as needed to ensure that their procedure will stick. Thinking in this quadrant includes gaining acceptance and implementing.

## Your Creative Problem-Solving Style

Basadur's research with thousands of engineers, managers, and others has shown that everyone has a different creative problem-solving style. Your particular style reflects your relative preferences for each of the four quadrants of the creative problem-solving process: generating/initiating, conceptualizing, optimizing,

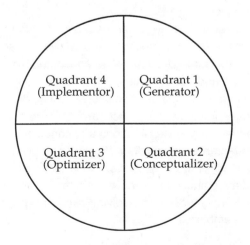

**Figure 14.4 The four quadrants of Basadur's Simplex Creative Problem-Solving Process (reprinted with permission from the Center for Research in Applied Creativity).**

and implementing. Your behavior and thinking processes cannot be pigeonholed in any single quadrant. Rather they're a combination or blend of quadrants: You prefer one quadrant in particular, but you also have secondary preferences for one or two adjacent quadrants. Your blend of styles is called the creative problem-solving style. Stated another way, your creative problem-solving style shows which particular steps of the process you prefer.

As an engineer, you will be working as part of a team or larger organization. Basadur's research has shown that entire organizations also have their own problem-solving process profiles. An organization's profile reflects such things as the people it hires, its culture, and its values. For example, if an organization focuses almost entirely on short-term results, it may be overloaded with implementers but have no conceptualizers or generators. The organization will show strengths in processes that deliver its current products and services efficiently. But it will show weaknesses in processes of long-term planning and product development that would help it stay ahead of change. Rushing to solve problems, this organization will continually find itself reworking failed solutions without pausing to conduct adequate fact finding and problem definition. By contrast, an organization with too many generators or conceptualizers and no implementers will continually find good problems to solve and great ideas for products and processes to develop. But it will never carry them to their conclusion. You can likely think of many examples of companies showing this imbalance in innovation process profiles.

Basadur suggests that in order to succeed in creative problem-solving, a team requires strengths in all four quadrants. Teams must appreciate the importance of all four quadrants and find ways to fit their members' styles together. Team members must learn to use their different styles in complementary ways. For example, generating ideas for products and methods must start somewhere with individuals scanning the environment, picking up data and cues from customers, and suggesting possible opportunities for changes and improvement. Thus, the generator raises information and possibilities, usually not fully developed but in the form of starting points for projects.

Then the conceptualizer pulls together the facts and idea fragments from the generator phase into well-defined problems and challenges and more clearly developed ideas worth further evaluation. Good conceptualizers give sound structure to fledgling ideas and opportunities. The optimizer then takes these well-defined ideas and finds a practical best solution and well-detailed, efficient plans for proceeding. Finally, implementers must carry forward the practical solutions and plans to make them fit real-life situations and conditions.

Skills in all four quadrants are equally valuable. As an engineering student, you will find yourself working on group projects. Often, these projects are with open-ended designs and are suitable for creative problem solving. Use these opportunities to practice utilizing your problem-solving skills and preferences as well as those of your team members.

## Example 14.4

What is your own preferred problem-solving style? You can discover your preferred problem-solving style by taking the Basadur Simplex Creative Problem-

Solving Profile. This profile and additional information about the Simplex model can be obtained from the Center for Research in Applied Creativity at: Center for Research in Applied Creativity, 184 Lovers Lane, Ancaster, Ontario, Canada, L9G 1G8 or http://www.basadur.com.

## Why Are We Different?

What causes the main differences in people's approaches to problem solving? Basadur's research suggests they usually stem from inevitable differences in how knowledge and understanding, i.e., learning are gained and used. No two individuals, teams, or organizations learn in the same way. Nor do they use what they learn in the same way. As shown in Figure 14.5, some individuals and organizations prefer to learn through direct, concrete experiencing (doing). They gain understanding by physical processing. Others prefer to learn through more detached abstract thinking (analyzing). They gain understanding by mental processing. All individuals and organizations gain knowledge and understanding in both ways but to differing degrees. Similarly, though some individual and organizations prefer to use their knowledge for ideation, others prefer to use their knowledge for evaluation. Again, all individuals and organizations use their knowledge in both ways but to differing degrees.

How individuals or organizations combine these different ways of gaining and using learning determines their problem-solving process profile. When you understand these differences, you can shift your own orientation to complement the problem-solving process preferences of others. Equally important, you can take various approaches to working with people. You can decide on the optimum strategy to help someone else learn something. And you can decide whom to turn to for help in ideation or evaluation.

Understanding these differences helps you interact with other people to help them make the best use of the creative problem-solving process. For example,

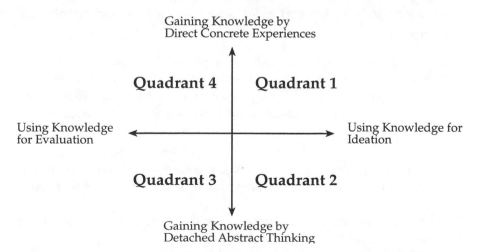

**Figure 14.5 The two dimensions of the Basadur Simplex Creative Problem-Solving Profile. (The Basadur Problem-Solving Model and Profile are copyrighted and are used with the permission of Dr. Min Basadur.)**

you can help strong optimizers discover new problems and facts or present new problems and facts to them. You can help strong implementers better define challenges or present well-defined challenges to them. You can help strong generators/initiators evaluate and select from among solutions and make plans, or present to them evaluated solutions and ready-made plans. You can help strong conceptualizer to convince others of the values of their ideas and push them to act on them, or help them push their ideas through to acceptance and implementation [Basadur, 1994]

## Brainstorming Strategies

*The goal of brainstorming is to stimulate your mind to trigger concepts or ideas that normal problem solving might miss.*

Alex Osborn, an advertising executive in the 1930s, devised the technique of stimulating ideas known as brainstorming. Since that time, countless business and engineering solutions have come about as a result of his method. Basically, brainstorming is a technique used to stimulate as many innovative solutions as possible. These then can be used in analytic and creative problem-solving practices.

The goal of brainstorming is to stimulate your mind to trigger concepts or ideas that normal problem solving might miss. This has a physiological basis. The mind stores information in a network. Memories are accessed in the brain by remembering the address of the memory. When asked about ideas for potential solutions, some would come just by thinking about the problem. A portion of your brain will be stimulated, and ideas will be brought to the surface. However, areas of your brain will not be searched because the pathways to get there are not stimulated. In brainstorming, we get the brain to search as many regions of the brain as possible.

*Let your mind wander and write down any ideas that come into your head.*

At times, generating ideas is difficult. Isakssen and Treffinger (1985) suggest using the method of idea lists. The concept is simple. Write down the ideas you have currently. Then, use these ideas to branch out into new ideas. To help stimulate your thinking in new directions, a set of key words was developed by Alex Osborn. Here are nine of them:

**Adapt:** What else is like this? What other ideas does this suggest? Are there any ideas from the past that could be copied or adapted? Whom could I emulate?

**Put to other uses:** Are there new ways to use the object as is? Are there other uses if modified?

**Modify:** Could you change the meaning, color, motion, sound, odor, taste, form, or shape? What other changes could you make? Can you give it a new twist?

**Magnify:** What can you add to it? Can you give it a greater frequency? Can you make it stronger, larger, higher, longer, or thicker? Can you add extra value? Could you add another ingredient? Could you multiply or exaggerate it?

**Minify:** What could you subtract or eliminate? Could you make it smaller, lighter, or slower? Could you split it up? Could you condense it? Could you reduce the frequency, miniaturize it, or streamline it?

**Substitute:** What if I used a different ingredient, material, person, process, power source, place, approach, or tone of voice? Who else instead? What else instead? What if you put it in another place or in another time?

**Rearrange:** Could you interchange the pieces? Could you use an alternate layout, sequence, or pattern? Could you change the pace or schedule? What if you transpose the cause and effect?

**Reverse:** What if you tried opposite roles? Could you turn it backward, upside down, or inside out? Could you reverse it or see it through a mirror? What if you transpose the positive and negative?

**Combine:** Could you use a blend, an assortment, an alloy, or an ensemble? Could you combine purposes, units, ideas, functions, or appeals?

Creativity expert Bob Eberle took Osborn's list and created an easy-to-remember word by adding "eliminate" to the list. Eberle's word is SCAMPER. These are the words used to form SCAMPER:

> **S**ubstitute?
> **C**ombine?
> **A**dapt?
> **M**odify? Minify? Magnify?
> **P**ut to other uses?
> **E**liminate?
> **R**everse? Rearrange?

Using the SCAMPER list may help to generate new ideas.

## Individual Brainstorming

Brainstorming can be done individually or in a group. Brainstorming individually has the advantage of privacy. There may be times when you deal with a professional or personal problem that you don't want to share with a group.

Some research even shows that individuals may generate more ideas by working by themselves. Edward de Bono attributes this to the concept that "individuals on their own can pursue many different directions. There is no need to talk and no need to listen. An individual on his or her own can pursue an idea that seems mad at first and can stay with the idea until it makes sense. This is almost impossible with a group."

When brainstorming individually, select a place that is free from distraction and interruption. Begin by writing down your initial idea. Now you will want to generate as many other ideas as possible. Let your mind wander and write down any ideas that come into your head. Don't evaluate your ideas; just keep writing them down. If you start to lose momentum, refer to the SCAMPER questions and see if these lead to any other ideas.

There are different formats for recording your ideas when brainstorming. Some find generating ordered lists works best for them. Others will write all over a sheet of paper or in patterns. Kinesthetic learners find that putting ideas on small pieces of paper, which they can move around, help the creative process. Auditory learners might say the ideas outloud as they are generated.

Good brainstorming means getting everything down on paper in thumbnail sketches or drawings that capture the thinking. Ideas not expressed in a communicable medium will exist in the designers' mind alone. Focus on generating ideas for 20 minutes at a time, no more. Probably 75% of your ideas are going to be useless, but that's normal. So, only one out of four ideas will be developed into a final solution.

During the final development toward the solution, pay close attention to the designs that you thought would never work; pieces of these designs may become parts of the final design.

When a prototype fails, a designer has to go back to the drawing board." The 75% of the useless ideas generated during the brainstorming time can be used. Examine all of the ideas originally sketched to develop a new design for the solution. This method may involve taking different parts and systems out of all of the thumbnail sketches and grouping them together to produce a design. Every subsequently developed design should be tested just as thoroughly as, if not more, the first prototype.

## Group Brainstorming

Group brainstorming's goal is the same as with individual brainstorming, which is generating as many potential solutions as possible without judging any of them. The power of the group comes when each member of the group is involved in a way that uses their creativity to good advantage.

Group brainstorming has a few advantages. The first is that additional people will look at a problem differently and bring fresh perspectives. When brainstorming is done correctly, these different views trigger ideas that you would not have come up with on your own.

Another advantage is that others get involved in the problem-solving process early. A key component in problem solving is implementing your solution (see following section). If people whose cooperation you need are in the brainstorming group, they will be much more agreeable to your final solutions since it was partly their idea.

To run an effective group brainstorming session, use basic guidelines. Those outlined in this section have been proven effective in engineering applications. They are helpful for use in student organizations and the committees that engineering students are encouraged to join. This process not only covers the generation of ideas but also provides a quick way to evaluate and converge on a solution. In some situations, you may want to stop after the ideas are generated and evaluate them later.

The guidelines for this process include:

1. Pick a facilitator
2. Define the problem
3. Select a small group
4. Explain the process
5. Record ideas
6. Involve everyone
7. No evaluating
8. Eliminate duplicates
9. Pick three

## Pick a Facilitator

The first step is to select a facilitator who will record the ideas and keep the group focused. The facilitator is also responsible for making sure the group obeys the ground rules of brainstorming.

## Define the Problem

It is important that all the participants understand the problem you are looking to solve before generating solutions. Once you start generating ideas, distractions can bring the definition process to a grinding halt. The group would hamper idea generation if it defined the problem they want to solve at the wrong time. The definition discussion should happen before the solution idea generating phase.

## Select a Small Group

The group size should be kept manageable. It is recommended that group size be limited to three people for brainstorming. If you need to brainstorm with a larger group, break it into smaller subgroups and reconvene after the groups have come up with separate ideas.

## Explain the Process

Providing the details of the process the group will follow is important. It gives the participants a feeling of comfort knowing what they are getting into. Once the process begins, going back and discussing ground rules is counterproductive. This can stymie idea generation just like a mistimed problem definition.

## Record Ideas

Visibly record ideas for the whole group. Preferably, arrange the group in a semi-circle around the person recording. A chalkboard, flip chart, whiteboard, or newsprint taped to a wall all work well. Record all ideas even if they seem silly. Often the best ideas come from a silly suggestion that triggers a good idea. By writing the suggestions visibly, the participants can use their sight as well as hearing to absorb ideas which may trigger additional ideas. Multiple senses help stimulate more ideas. Also, having all the ideas in front of the participants guarantees recollection and allows for new ideas, triggered by earlier ideas. Using the SCAMPER questions is a great way to keep ideas flowing.

## Involve Everyone

Start with one idea from the facilitator or another volunteer and write it down. It is easier to get going once the paper or writing surface is no longer completely blank. Go around the group, allowing each person to add one idea per round. If a person doesn't have an idea, let him or her pass, and move on. It is more important to keep moving quickly than to have the participants feel like they must provide an idea every time.

*The power of group brainstorming lies in taking advantage of the creative minds of every member of the group, not just the ones who could dominate the discussion.*

By taking turns, all the participants will have an equal opportunity to participate. If the suggestions are taken in a free-flowing way, some participants might monopolize the discussion. The power of group brainstorming lies in taking advantage of the creative minds of every group member, not just the ones who could dominate the discussion. Often, the best ideas come from the people who are quiet and would get pushed out of the discussion in a free-flowing setting. When or how the ideas come does not matter as long as they surface.

Continue the process until everyone has passed at least once. Even if only one person is still generating ideas, keep everyone involved and keep asking everyone in turn. The one person who is generating ideas might stimulate an idea in someone else at any point. Give everyone the opportunity to come up with a last idea. The last idea may be the one you use.

## No Evaluating

This may be hard for the facilitator or some participants. Telling someone that their suggestion is dumb, ridiculous, or from left field, or making any other negative judgment makes that person (and the rest of the group) less likely to speak. The genius of brainstorming is the free generation of ideas without deciding if they are good until the correct stage. There are countless times when the wackiest ideas are brought up right before the best ideas. Wacky suggestions can trigger ideas that end up being the final solution. Write each idea, as crazy as it may be. This makes participants feel at ease. They will feel more comfortable making unusual suggestions and will not fear being made fun of or censored.

Avoiding negative comments is one thing, but if you are facilitating, you need to watch for more subtle signals. You might be communicating indirectly to someone that his or her ideas are not good. If you are writing the ideas down, it is natural to say "good idea" or "great idea." What happens, however, if one person hears "great idea" and the others don't? What if just one person doesn't get the "great idea" comment?

The participants need to feel at ease during brainstorming. Providing negative feedback in the form of laughter, sarcastic remarks, or even the absence of praise will work to dampen the comfort level and, therefore, the creativity. It is safest not to make any value judgments about any ideas at this point. If you are facilitating, respond to each idea the same way and keep order. If you are a participant, not make value statements or jokes about other ideas.

## Eliminate Duplicates

After generating ideas, you may want to study the suggestions. In some cases, it is to your advantage to sort out a solution or the top several solutions. Doing this at the end of group brainstorming can be easy and also serves to involve the group members in the decision process.

The first step is to examine the list of ideas and eliminate duplicates. There is a fine line between eliminating duplicates and creating limiting categories. This step is designed to eliminate repeated ideas. Ask if they are identical. If they are similar, leave both. Creating categories only reduces the possible solutions and defeats the purpose of brainstorming.

The next step is to allow the group to ask clarifying questions about suggestions. This is still not the time for evaluation, only clarification. Don't let this become a discussion of merits. Clarification may help in eliminating and identifying duplicates.

## Pick Three

Once all the members understand the ideas, have them evaluate the suggestions. This can be done quickly and simply. First, ask all the members to pick their three top choices. (Each person can select more if you need more possibilities or solutions.) After each member has determined these in his or her own mind, let each member vote outloud by marking the three top choices. Do this with a dot or other mark but not with a number. The top ideas are selected based on the total number of marked votes they receive. The winners can be forwarded to the problem-solving phases to see if any survive as the final solution.

Note that with this method of voting, the group's valuable time is optimized. Time is not wasted determining whether an idea deserves a low mark or a high mark. Also, no time is spent on the ideas no one liked. If all ideas are discussed throughout the process, the ideas that are not chosen by anyone or which are least preferred often occupy discussion for quite a while. It's best to eliminate these time sinkholes quickly.

Another benefit of not ranking choices at first is that each member could think that his or her idea was the fourth on everyone's list, which might be frustrating. How might people feel if their ideas were ranked low by the other members? Such a discouraged person could be one who would have valuable input the next time, or this person could be one whom you need as an ally to implement the group's decision. Either way, there is no need to cast any aspersions on ideas not adopted.

## CRITICAL THINKING

Since engineering is so dominated by systematic problem solving, it is easy to get caught up in the methodologies and look at everything as a problem with a solution. Problem-solving methods teach a systematic manner for analyzing problems which is necessary to tackle engineering problems, but many of life's issues are more complex and have no single solution.

Many issues and challenges you face will have complexity and require a critical perspective. Engineering issues with human, environmental, or ethical impact are among the issues with complex ramifications and require a critical perspective. The issue of poverty is a good example. Many programs have been used, many involving engineering-based solutions, but poverty has remained as a pressing social issue. Solutions for poverty are complex and vary. Though engineers continue to work in this area, programs or solutions are not being brought to bear on an issue that can be solved.

It is easy to get caught up in the processes and methodologies and move quickly to solutions and implementation plans. Many times in your career, you will need to evaluate what you are doing or what you are working on. Learning to be able to evaluate why or how you are doing something by taking a step back to gain perspective is an important skill. At each step, one of the divergent questions you could ask is "why?"

Sometimes when you critically evaluate what you are doing, you discover deeper meanings or larger issues. An example is toy design for physically challenged children. In the (Engineering Projects in Community Service (EPICS) program (at http://epics.ecn.purdue.edu), students develop engineering-based solutions to community problems. As these students work with the children, families, and therapists, they gain an appreciation for the many complex issues facing the children. When they design toys, they must design them so the toys can be used within the children's physical limitations, which is fairly straightforward.

However, the students have learned they can help address many other issues, such as socialization issues. Some of the toys are now designed so that the physically challenged children can play right along with those who are not challenged. Students have also learned to design the toys so physically challenged children can control the play. Normally, the children who are not physically challenged control the games and allow other children to participate. The engineering students have learned about the socialization of the children and its impact on their self-esteem.

When working with issues, such as disabled and physically challenged children or adults, simple or straightforward answers do not often exist. We do not encounter simple problems that must be solved but rather complex issues to be recognized and addressed. By working through these issues and the multiple dimensions of the issues, the students have become aware of the social and emotional aspects of their work. They have become better designers, but more importantly, they have become better educated about relevant and current issues of the day. When discussions about new facilities for the disabled or about mainstreaming children occur, the students are able to participate.

As engineers, you are entrusted with a significant responsibility, as discussed in the chapter on ethics. The fruits of your work will have great potential for good and for harm. Though you must maintain the highest ethical standards, you must also focus on your work's context and direction. Take opportunities to reflect and evaluate what and why you are doing, which is as important as any problem-solving methodology.

At the beginning of the creative problem-solving process, we discussed the importance of getting the question right. Similarly, when evaluating an issue, the question is important. When evaluating an issue, you must become aware of the slant you bring based on your background.

Thomas Edison wanted people around him who would not limit their creative thinking by jumping to conclusions. He would take every prospective engineering employee to lunch at his favorite restaurant and recommend the soup. Of course, all ordered the soup unaware he was testing them. Edison wanted to see if they would salt their soup before tasting it. Salting without tasting, he reasoned, was an indication of someone who jumps to conclusions. He wanted people who were open to questioning and not limited by their preconceptions.

Our own preconceptions make us vulnerable to salting before tasting in many situations. Periodically, ask yourself if you are jumping to conclusions based on your background or personality or thinking style?

Critical thinkers ask the Why and What questions. Why are we approaching the problems this way? What are the implications of our work? What are the deeper issues? Asking these questions about yourself is a healthy part of being a good critical thinker. Why are you studying to become an engineer? What will you do as an engineer to help with your personal fulfillment? Becoming a critical thinker will make you a better problem solver and engineer and make you more engaged in and aware of society and its issues.

## LEARNING ACTIVITIES

### Activity 14.1   Destructive testing and mathematical modeling

Time:     2–3 class periods

## CASE STUDY

Destructive and nondestructive tests are used for determining the quality of materials. For example, you might test a cookie to see if it is good. In the process, the cookie is consumed. That would be a destructive test. Amniocentesis and ultrasound techniques are used for determining the sex of unborn children. These are nondestructive tests. X-rays were once used to determine the sex of a baby, but it is now believed that high-energy radiation can destroy some living cells. Although X-rays may not completely destroy living things, they are considered partially destructive.

Manufacturers produce a huge number of items daily. In industries where only a few items are produced at a time, it is possible to inspect each item. Even then, some defects may be difficult to detect. Science and technology have developed many different tests to help detect hidden defects. Nondestructive tests are frequently used on expensive items or on items that are difficult to replace.

Since destructive tests usually destroy the item being tested, the tests are only used on a few out of possibly thousands of produced items. A type of mathematics called statistical analysis determines the probability of finding a defective product. This type of analysis is the job of a quality control (QC) person. Another job done by a QC person might be to determine the life span of a product. This is frequently accomplished by putting parts under repeated abuse until the parts fail due to fatigue. Statistical analysis is again applied in order to make a good estimate of the life of the part. You will use some of these techniques in this experiment.

You and a partner will use destructive testing to determine the strength and durability of two types of paper clips. A good question to ask yourselves is if the paper clips in your experiment are of the same material as another group across the room. Could slight differences exist? Will these differences be enough to produce startling results? The procedure for this experiment is below.

1.  Lay a paper clip flat on the table or desk top and hold the smaller loop with one hand. Grasp the larger loop with your other hand and bend the clip open one quarter of a turn. Keep the loops flat. You should end with an "L" shape. Note if the paper clip was easy to open into the L shape.

**Step One**

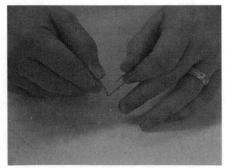

**Carefully bend the paper clip open.**

**Step Two**

**Lay the clip with the small loop on the table and the large loop in your hand.**

2. Lay the paper clip at the edge of your desk or table so that the small loop is on the table and the large loop is at the edge of the table and projecting up.

3. Grasp the large loop with your dominant hand and hold the small loop firmly right at the table edge with your other hand.

4. Bend the large loop all the way down so that it is pointing down. Count this as one bend. Then bend it back to the position where it is pointing up. Count this as bend number two.

5. Continue to bend the paper clip up and down. Keep count of each bend. Record the number of bends it took to break the paper clip.

6. You and your partner should destructively test ten paper clips. Record your results for each paper clip.

7. Repeat the destructive test with another type of paper clip. Again test ten paper clips until they fail. Record your results.

8. Record your results using a spreadsheet like Microsoft Excel so that you can share data with your classmates. Record the entire class' data on paper from the chart on the board created by your instructor.

9. Use a spreadsheet to enter all the class data. Make a series of graphs and curves according to your instructor's directions for each of the two types of paper clips.

## ASSESSMENT

1. Did you record all observations?

2. What type of testing did you do on the paper clips? List two other materials that would use this type of testing.

3. Why didn't all of the paper clips give the same results?

4. If you were going to sell and guarantee the paper clips, what do you think would be a reasonable guarantee on the number of bends before breaking?

5. If you changed your production techniques, how many tests do you think you would have to make before you could determine a new guarantee? Explain your answer.

## RESOURCES

ENERGY CONCEPTS, INC. (1999) Materials Science Technology: Solids. Pg. 4.9–4.12

http://www.pnl.gov/education/mst.htm This is the Department of Energy's Materials Science Curriculum web site. The site also gives a glimpse into the full document with information and selected activities in .PDF format.

http://www.energy-concepts-inc.com Energy Concepts is a national vendor that sells the reorganized Materials Science Curriculum in a more teacher friendly form. This curriculum covers many topics beyond the DOE curriculum. Energy Concepts, Inc. 404 Washington Blvd. Mundelein, IL 60060: Phone 847-837-8191; Fax 847-837-8171

## Activity 14.2   The Catch

Time:     4–6 class periods

## CASE STUDY

Materials for the students include copy paper, three-quarter-inch masking tape, standard rubber bands, scissors, rulers, and a few golf balls. The golf balls are used to simulate the test as they are about the same size and weight. Students can use these in class to get a better idea of how the structure will react but should not be allowed to drop eggs in class until test day.

The egg drop activity has been used in many elementary and middle school technology and science courses where students learn about forces and motion by dropping a packaged egg from a window. This learning activity will focus on using engineering design concepts to create optimal solutions for protective packaging of a raw egg.

Students will be required to utilize attributes of the design process (listed in the standards) in order to build their prototype. Documentation including sketches should be kept throughout the activity.

The objective of this activity is to design a free-standing prototype structure to capture and contain a raw egg from a six-foot free fall using nothing more than two sheets of copy paper, six inches of three-quarter-inch masking tape, and one rubber band.

### Brainstorming

|  | Points |
|---|---|
| • Three or more detailed sketches that reflect design requirements | 30 |
| • Three detailed sketches that reflect most design requirements | 25 |
| • Two or three sketches with little detail and attention to requirements | 15 |
| • One or two sketches, with no detail and effort | 5 |

### Structure

|  | Points |
|---|---|
| • Contains and protects eggs within design requirements | 50 |
| • Contains and protects eggs; meets some of requirements | 40 |

- Contains but does not protect egg; meets requirements    30
- Contains but does not protect egg; meet some requirements    20
- Structure does not achieve the desired solution; reflects
  lack of attention to design concepts    10

## Overall Creativity

                                                    Points

- Creative and innovative    15
- Limited creativity    10
- Use publicized or common solution    5

## Additional Exercises and Activities

1. A new school has exactly 1,000 lockers and 1,000 students. On the first day of school, the students meet outside the building and agree on the following plan: the first student will enter the school and open all the lockers. The second student will then enter the school and close every locker with an even number. The third student will then reverse every third locker. If the locker is closed, he or she will open it; if it is open, he or she will close it. The fourth student will then reverse every fourth locker, and so on until all 1,000 students in turn have entered the building and reversed the proper lockers. Which lockers will remain open?

2. You are stalled in a long line of snarled traffic that hasn't moved at all in twenty minutes. You're idly drumming your fingers on the steering wheel, when you accidentally started tapping the horn. Several sharp blasts escaped before you realize it. The driver of the pickup truck in front of you opens his door, gets out and starts to walk menacingly toward your car. He looks big, mean and unhappy. Your car is a convertible and the top is down. What do you do?

3. Your school's team has reached the national championship. Tickets are difficult to get but you and some friends have managed to get some. You all travel a great distance to the site of the game. The excitement builds on the day of the game until you discover that your tickets are still back in your room on your desk. What do you do? What problem you must solve?

4. You are a manufacturing engineer working for an aircraft company. The production lines you are responsible for are running at full capacity to keep up with the high demand. One day, you are called into a meeting where you learn that one of your suppliers of bolts has been forging test results. The tests were never done, so no data exist to tell if the bolts in question meet your standards. You do not know how long the forging of the test results has been going on. You are only told that it has been for a while. Your whole production line including the planes ready to be delivered may be affected. You are asked to develop a crisis management plan. What do you do?

Some background information:

- Bolts are tested as batches or lots. A specified number are taken out of every batch and tested. The testing typically involves loading the bolts until they fail.
- The bolts do not have serial numbers so it is impossible to identify the lot number from which they came.
- The supplier who forged the tests supplies 30% of your inventory of all sizes of bolts.
- Your entire inventory is stored in bins by size. Each bin has a mixture of manufacturers. (To reduce dependence on any one company, you buy bolts from three companies and mix the bolts in the bins sorted by size.)
- Bolts have their manufacturer's symbol stamped on the head of the bolt.
- It takes weeks to assemble or disassemble an aircraft.
- Stopping your assembly line completely to disassemble all aircraft could put your company in financial risk.
- Your customers are waiting on new planes and any delays could cause orders to be lost.
- The FBI has arrested those who forged the test and their business has been closed temporarily.

5. How much of an automobile's tire wears off in one tire rotation?

6. A farmer going on a trip with a squirrel, acorns, and a fox had to cross a river in a boat in which he could only take one of them with him each time he crossed. Since he had to leave two of them together on one side of the river or the other, how could he plan the crossings so that nothing gets eaten, and they all get across the river safely?

7. Measure the height of your class building using two different methods. Compare your answers and indicate which is more accurate.

8. Pick a grassy area near your class building. Estimate how many blades of grass are in that area. Discuss your methodology.

9. How high do the letters on an expressway sign need to be to be readable?

10. Estimate the speed of a horse.

11. Estimate the maximum speed of a dog.

12. Assume that an explosion has occurred in your building. Your room is intact but sustained damage and is in immediate danger of collapsing on you and your classmates. All exits but one (as specified by your instructor) are blocked. The one remaining is nearly blocked. Develop a plan to get your classmates out. Who goes first and last? Why?

13. A child's pool is eight feet in diameter and two feet high. It is filled by a garden hose up to a level of one foot. The children complain that it is too cold. Can you heat it up to an acceptable temperature using hot water from the house? How much hot water would you need to add?

14. How else could you heat the water for the children in problem?

15. Without referring to a table or book, estimate the melting temperature of aluminum. What are the bounds of potential melting temperatures? Why? What is the actual melting temperature?

16. You are working on a project that has had many difficulties that are not your fault. However, you are the project leader and your manager has become increasingly frustrated with you and even made a comment that you are the kind of irresponsible person who locks your keys in your car. Your manager is coming to town to see why the project is having problems, and you are going to pick her up. You leave early to take time to relax. With the extra time, you decide to take the scenic route and stop the car next to a bubbling stream. Leaving your car, you go over to the stream to sit and listen. The quiet of the surroundings is what you needed to calm down before leaving for the airport. Glancing at your watch, you realize you must leave to pick up your manager. Unfortunately, you discover that you have locked your keys in the car. A car passes on the road about every 15 minutes. The nearest house is probably a mile from your location and the nearest town is 15 miles away. The airport is 20 miles away and the flight is due to arrive in 30 minutes. What will you do?

17. You are sitting at your desk in the morning reading your e-mail when your boss bursts into your office. It seems a local farmer was not happy with your company and has dropped a truckload of potatoes at your plant's entrance. The potatoes are blocking the entrance and must be moved. You have been selected as the lucky engineer to fix this problem. What would you do to move the potatoes and how would you get rid of them? Brainstorm ideas and select the best solution. (Note: You work at a manufacturing plant of some kind that does not use potatoes in its process. The potatoes also appear to be perfectly fine.)

18. You work in a rural area known for chickens. An epidemic has swept through the area killing thousands of chickens. The EPA has mandated that the chickens may not be put into the landfill nor can they be buried. Develop a plan to get rid of the chickens.

19. A fire has destroyed thousands of acres of forest. Seedlings have been planted in the area but are being destroyed by the local wildlife. Develop a way to reforest the region by preventing the wildlife from eating the seedlings.

20. You and seven of your friends have ordered a round pizza. You need to cut it into eight pieces and are only allowed to make three straight cuts. How can you do this?

21. With which part of the Basadur problem-solving process are you most comfortable? Write a one-page paper on how your preference affects your problem solving.

22. NASA has decided to send astronauts to the other planets in the solar system. Psychologists, however, have determined that the astronauts will need something to keep them occupied during the months of travel or they will develop severe mental problems that would threaten the mission. Your job is to devise ways to keep the astronauts occupied for the long journey. Remember that you are limited to what will fit in a space capsule.

23. Landfill space is rapidly running out. Develop a plan to eliminate your city's dependence on the local landfill. The city population is 100,000.

24. Prepare a one-page report on a significant engineering solution developed in the past five years. Evaluate its effectiveness and report (intended and unintended) on the outcomes of this solution.

25. Brainstorm ideas for toys for physically disabled children. Narrow your choices to the top three and identify the benefits of each design. What are the important issues that you would need to consider in your designs?

26. Early one morning is starts to snow at a constant rate. Later, at 6:00 A.M., a snowplow sets out to clear a straight street. The plow can remove a fixed volume of snow per unit time. In other words, its speed it inversely proportional to the depth of the snow. If the plow covered twice as much distance in the first hour as the second hour, what time did it start snowing?

27. While three wise men are asleep under a tree a mischievous boy paints their foreheads red. Later they all wake up at the same time and all three start laughing. After several minutes suddenly one stops. Why did he stop?

28. Using just a five-gallon bucket and a three-gallon bucket, can you put four gallons of water in the five-gallon bucket? (Assume that you have an unlimited supply of water and no measurement markings are on the buckets.)

29. A bartender has a three-pint glass and a five-pint glass. A customer walks in and orders four pints of beer. Without a measuring cup but with an unlimited supply of beer how does he get a single pint in either glass?

30. You are traveling down a path and come to a fork in the road. A sign lays fallen at the fork indicating that one path leads to a village where everyone

tells the truth and the other to a village where everyone tells lies. The sign has been knocked down so you do not know which path leads to which village. Then someone from one of the villages (you don't know which one) comes down the path from which you came. You may ask him one question to determine which path goes to which village. What question do you ask?

31. Four mathematicians have the following conversation:
    Alice: I am insane.
    Bob: I am pure.
    Charlie: I am applied.
    Dorothy: I am sane.
    Alice: Charlie is pure.
    Bob: Dorothy is insane.
    Charlie: Bob is applied.
    Dorothy: Charlie is sane.
    You are also given that:
    • Pure mathematicians tell the truth about their beliefs.
    • Applied mathematicians lie about their beliefs.
    • Sane mathematician's beliefs are correct.
    • Insane mathematician's beliefs are incorrect.
    Describe the four mathematicians.

32. Of three men, one man always tells the truth, one always tells lies, and one answers yes or no randomly. Each man knows which man is which. You may ask three yes/no questions to determine who is who. If you ask the same question to more than one person you must count it as a question used for each person asked. What three questions should you ask?

33. Tom is from the census bureau and greets Mary at her door. They have the following conversation:
    Tom: I need to know how old your three kids are.
    Mary: The product of their ages is 36.
    Tom: I still don't know their ages.
    Mary: The sum of their ages is the same as my house number.
    Tom: I still don't know their ages.
    Mary: The younger two are twins, seven years younger than their sibling.
    Tom: Now I know their ages! Thanks!
    How old are Mary's kids and what is Mary's house number?

# REFERENCES

Basadur, M. Simplex: A Flight To Creativity. Hadley, MA: Creative Education Foundation, Inc., 1994.

Beakley, G. C., Leach, H. W., &Hedrick, J. K. Engineering: An Introduction to a Creative Profession, 3rd ed. New York: Macmillan Publishing Co., 1977.

Berkemer, Dr. Robert. Stout, WI: University of Wisconsin-Stout DFI Course, 1994.

Burghardt, M.D. Introduction to Engineering Design and Problem Solving. New York: McGraw-Hill Co., 1999.

Eide, A.R., Jenison, R. D., Mashaw, L. H. & Northup, L. L., Introduction to Engineering Problem Solving. New York: McGraw-Hill Co., 1998.

ler, H. S. and LeBlanc, S. E. Strategies for Creative Problem Solving. Englewood Cliffs, NJ: Prentice Hall, Inc., 1995.

Gibney, Kate. Awakening Creativity. *ASEE Prism*. March 1998, pp. 18-23. http:www.prism-magazine.org for more information.

Gomez, Alan G. The 21st Century Experience. Madison, WI: University of Wisconsin-Madison, 2000.

Isaksen, Scott G. and Treffinger, Donald J. Creative Problem Solving: The Basic Course. Buffalo, NY: Bearly Limited, 1985.

Lumsdaine, Edward and Lumsdaine, Monika. Creative Problem Solving-Thinking Skills for a Changing World, 2nd ed. New York: McGraw-Hill Co, 1990.

Osborn, A. F. Applied Imagination: Principles and Procedures of Creative Problem-Solving. New York: Charles Scriber's Sons, 1963

Panitx, Beth. Brain Storms, *ASEE Prism*. March 1998, pp. 24–29.

*Chapter 15*

# Design and Modeling

## What Is Engineering Design?

Engineers create things. Engineers build things. To perform these tasks, engineers must be involved in design or in a design process. What is engineering design and an engineering design process? Webster's dictionary defines design as "to create, fashion, execute, or construct according to plan." It defines a natural process as "a natural phenomenon marked by gradual changes that lead toward a particular result: a series of actions or operations conducing to an end; a continuous operation or treatment, especially in manufacture."

This chapter may seem closely related to Chapter 14, "Problem Solving," and many similar topics and themes from that chapter are important in design. One critical area which has an impact on design is the issue of external constraints that can influence the outcome of the process.

Engineering design is an important and ongoing activity. Students who graduate with an engineering degree from an accredited program have had a significant amount of design experience as part of their education. The Accreditation Board for Engineering and Technology (ABET) has traditionally defined engineering design as follows:

> *Engineering design is the process of devising a system, component, or process to meet desired needs. It is a decision-making process . . . in which mathematics, basic sciences, and engineering sciences are applied to convert resources optimally to meet a stated objective. Among the fundamental elements of the design process are the establishment of objectives and criteria, synthesis, analysis, construction, testing, and evaluation. . . . it is essential to include a variety of realistic constraints, such as economic factors, safety, reliability, aesthetics, ethics, and social impact.*

ABET describes the basic educational components for a major engineering design experience that must in an engineering program:

> *Each educational program must include a meaningful, major engineering design experience that builds upon the fundamental concepts*

*of mathematics, basic sciences, humanities, social sciences, engineering topics, and communication skills. The scope of the design experience within a program should match the requirements of practice within that discipline. . . . all design work should not be done in isolation by individual students; team efforts are encouraged where appropriate. Design cannot be taught in one course; it is an experience that must grow with the student's development. A meaningful, major design experience means that, at some point when the student's academic development is nearly complete, there should be a design experience that focuses the student's attention on professional practice and is drawn from past course work. Inevitably, this means a course, or a project, or a thesis that focuses upon design. "Meaningful" implies that the design experience is significant within the student's major and that it draws upon previous course work, but not necessarily upon every course taken by the student.*

Many entering engineering students often confuse the design process with drafting or art-related work. To clarify this, ABET adds the following:

*Course work devoted to developing computer drafting skills may not be used to satisfy the engineering design requirement.*

Most engineering programs concentrate their engineering design coursework later in the student's program, which allows the student to apply much of the prerequisite background in math, science, and related fields to engineering problems. In most engineering programs, design courses comprise between 20% and 25% of the total curriculum.

## THE DESIGN PROCESS

Many uniform approaches to engineering design are followed by practicing engineers. Some firms approach engineering design as a short, simple process with only a few steps, while others use a more complex, multi-step method with several stages. No matter what process is used, engineering design is always continuous. The completion of one design, or the solution of one problem, may open opportunities for subsequent designs or modifications.

A design process is used whether a product is being developed for an ongoing manufacturing process, where thousands or even millions of a certain item will be produced, or for a one-time design, such as with the construction of a bridge, dam, or highway exit ramp.

## SIGNIFICANT FACTORS

Design is a process used by engineers to generate products, processes, and systems based on the recognition of a need. The following seven factors are significant in the design process that students should use in all case studies.

*Functionality*—The product or solution must fulfill its intended purpose. Imagine a student's table used for studying. If a person were to stand on the table, would it support the weight? Was this table intended to support a person's weight?" The table's original design was to support something other than a person's weight. The team that developed the table had not intended it for the excessive weight and the table was being used for something for which it was not intended.

*Quality*—The product or solution must be designed to meet certain minimum standards. Take, for example, the soles of shoes: Students were asked how long they thought their shoes would last. Most said that they would use them for less than two years. Why, if most of our day is spent walking or running on harder surfaces, such as concrete, are our shoes bottomed with a softer surface like rubber? The answer is comfort.

When a worker recently complained to his employer of being constantly tired and with chronic back pain, he was sent to a doctor for evaluation. The doctor found the problem to be his shoes. Improper support for his feet led to the back pain and fatigue. Quality has a relationship to the conditions of the item's intended use. This spin-off of functionality is important because quality is defined according to the item's proper or improper. Are casual shoes intended to be used to play basketball in, and if so, would quality issues occur when used in the incorrect environment? Quality has to be evaluated within the context of the items use and is evaluated when the context is correct for the item's intended use.

*Safety*—The product's design must comply with codes and regulations to provide safe user operation. Americans are becoming more safety conscious than ever, in everything from our cars to our homes. Air bags have become the norm in passenger vehicles since 1997; however, some have posed deadly problems. The force by which they are activated in an accident has injured many people, some of them fatally. These early air bags, intended to save lives, sometimes activated not as the result of an accident. Sometimes the driver's incorrect child carseat installation, (facing the windshield) injured and killed many small children from the inflating force of the air bag. How are drivers going to navigate their vehicles when an airbag has unnecessarily deployed, blocking their vision. The first generation airbags have been redesigned many times. For example, a switch exists to turn off the passenger seat air bag for different reasons ranging from a person who is taller to those who are short. A driver has the option to disable the air bag.

*Ergonomics*—The product must be designed so that the user can operate it with maximum ease and efficiency. Ergonomics is also called human factors engineering. Designing chairs for genuine long-term or short-term comfort is a part of ergonomics. One famous fast food chain restaurant designs the chairs so that customers are comfortable only for the time it takes to eat the kind of food served. Then the customer will leave so employees can clean the table and chair for the next person.

Imagine this scenario: You and your family are going to be traveling out east with your three children and want to buy a new conversion van. You are 6'3" and, your wife is 4'11" and petite. You test drive the van but she doesn't want to. You purchase the van and begin your trip out east. You begin the drive, until fatigue sets in, and turn the wheel over to your wife to get some sleep. When you wake from your nap, you noticed that you are not near the planned city you were supposed to be; she was driving at 45 MPH on the expressway. When you ask why she has been going so slowly, she replies "It was as fast as Icould go. My feet could not press any further down on the pedals."

The manufacturer had never considered that an adult only 4'11" in height would drive a full-sized conversion van. Manufacturers have begun to consider ergonomic factors in their designs and their marketing plans toward smaller people.

*Appearance*—The appeal of a product is based on the selection of materials, processes, finish, color, and shape. If you, the potential consumer, do not like what a finished product's looks, you are less likely to purchase it. For example, a "new" baseball hat is colorful and clean, with no rips or tears. However, in some stores new baseball hats have torn, pre-formed, and dirty bills. Some people have different definitions of what is new and stylish. Companies invest millions of dollars in major marketing campaigns to get consumers to buy into their ideas. Ultimately, the consumer decides to purchase the product.

*Environmental Considerations*—The product must be designed so that it does not adversely affect the environment. People think of oil dripping from their cars and smog over a city. Though both of these adverse conditions are in this category, many others exist.Planning commissions and city engineering departments continue to construct large buildings among large buildings and neglect to do any kind of wind analysis on the new structures or their surrounding environment. Because of a building's configuration and height, a five-mile-per-hour breeze could increase into a 15-mile an hour or more wind.

*Economics*—The product must be produced at the least cost without sacrificing safety. Engineers are poised to keep costs down to increase company profit. Often they take cut production and take short cuts that lead to unsafe products. Often, the public is not concerned about companies' actions until something drastic happens. Usually, it takes a death or a disaster to focus our attention.

For example, consider the Sam Poong department store in Japan that collapsed, killing 1,500 people. Store management left the building early in the morning after cracks appeared in the walls and ceiling, but they did not tell employees or customers of the danger. Shopping throughout the building went on as usual. Then the fifth floor collapsed onto the fourth, setting off a chain reaction all the way to the ground level. The building collapsed because of poor building codes and enforcement paired. In addition, the firm doing the concrete work added entirely too much water to the concrete mix to lessen the costs, creating a weak concrete structure.

## STAGE DESIGN

This chapter will present one design process (the ten-stage process) in use. We will apply this process to an existing product to analyze how each of the ten stages may have been applied in the product's realization.

The ten stages that make up the process are as follows:

Stage 1:   Identify the problem/product innovation
Stage 2:   Define the working criteria and goals
Stage 3:   Research and gather data
Stage 4:   Brainstorm and generate creative ideas
Stage 5:   Analyze potential solutions
Stage 6:   Develop and test models
Stage 7:   Make the decision
Stage 8:   Communicate and specify
Stage 9:   Implement and commercialize
Stage 10:  Perform post-implementation review and assessment

The actual process begins with the appointment of a project manager or team leader. This individual will be responsible for oversight of the entire process to ensure that certain key elements of each stage have been satisfied before the project moves on to the next stage. This person is responsible for recruiting team members of various backgrounds and expertise for each stage. The team will include non-engineers as well as engineers. Some of the team members will be used throughout the process, and others will be needed only for certain parts of the process.

## Stage 1: Identify the Problem/Product Innovation

Engineers are problem solvers; and the problems they solve are often identified as the specific needs and problems of customers. For example, a particular handicap may require a new prosthesis, increased gas mileage standards may demand higher efficiency engines, or a modified manufacturing process may need a new computer program. Therefore, the first stage to problem solving in engineering design is establishing the problem and identifying information sources to understand the problem's scope and nature.

The project manager will call upon people's resources to assist with these initial process stages. You may have a multitude of sources outside as well as within the organization to solve the problem. Many firms have a research and development unit made up of scientists and engineers who possess the training and expertise to assist with problem evaluation. In addition, sales engineers, who maintain consistent contact with outside individuals, can provide valuable input on problem identification. If the problem is for expanding a product line or for modifying a system to improve or make fit customer needs better, management will likely play a role in the problem's definition. Each group would be represented on the team at this stage.

External resources may come from trade shows, conferences, technical presentations, patent listings, and publications. Ideas generated from existing or prospective clients may also have value. Information gathered from external research agencies, private laboratories, and government-funded foundations can be useful as well. In addition, an awareness of the competition's products and services can aid in problem identification. At this early stage, establish a preliminary, formal problem statement. For example, the need for a new automobile safety device for infants might evolve into this preliminary problem statement: "Develop a better child restraint system that will protect children involved in automobile collisions." (Review the "real problem" section in Chapter 17.)

Before going ahead with the design process, the team must be certain that they have explored the correct issues and background, so they can develop the most appropriate solutions.

At this point, if the project manager is satisfied that all necessary issues have been resolved, the project will likely be passed on to the next stage to develop the project criteria and goals.

## Stage 2: Define the Working Criteria and Goals

Once the team has identified the problem, it must validate it throughout the design process. This requires the establishment of certain working criteria, or standards, which can be used in each of the ten stages to measure possible solutions. The ultimate objective in this stage is to establish preliminary goals, which will act as the team's focal point as it works through the process. The development of some working criteria provides a means to compare possible solutions. At this stage of the design process, everything is preliminary, so it is still possible for the team to modify the criteria.

Examples of working criteria could include answers to the following questions:

1. How much will it cost?
2. Will it be difficult to produce?
3. What will be the size, weight, and strength?
4. What will it look like?
5. Will it be easy to use?
6. Will it be safe to use?
7. Do any legal concerns exist?
8. Will it be reliable and durable?
9. Can it be recycled?
10. Is this what customers want?
11. Will our customers want to purchase it?
12. Will customers want to purchase this version instead of a competitor's product?

Once the team has established the preliminary working criteria, it has to develop process goals, which is an objectives statement to be evaluated as the design process evolves. Using the example of standards implemented to increase automobile gas mileage and reduce emissions, the design goals might be to develop an automobile engine which produces 25% fewer emissions while increasing gas mileage by ten percent.

Establishing the project goals provides a means of evaluating, monitoring and changing the process focus as it evolves through the ten stages. For the project managers, the criteria and goals become a checkpoint for progress assessment and will help them determine if the project is ready to move to the next stage. If not ready, the process needs to return to the first stage for re-evaluation.

## Stage 3: Research and Gather Data

This important stage affects all remaining design process stages. Having good, reliable background information is necessary for the team to explore all problem aspects. Consistent with the preliminary working criteria and the established goals, the selected team members (for this process phase) must determine what information they will need and the information's best sources:

1. What information has been published about the problem?
2. Does an available solution exist ?
3. If the answer to the above is yes, who is providing it?
4. What are the solution's advantages?
5. What are the solution's disadvantages?
6. What is the cost?
7. Is cost a significant issue?
8. What is the ratio of time spent compared to overall costs?
9. Do legal issues exist?
10. Must environmental concerns be considered?

The team members can use many resources to assist their research. A starting point may be an Internet search, providing useful material sources to focus the need for additional research. Examine other reference information sources:

1. Libraries
2. Professional associations (technical and non-technical)
3. Trade journals and publications
4. Newspapers and magazines
5. Market assessment surveys
6. Government publications
7. Patent searches and listings (U.S. Patent Office: www.uspto.gov.)
8. Technical salespersons and their reference catalogs
9. Professional experts including engineers, professors and other scientists
10. The competition's product (Disassemble and examine their product.)

The team members must keep detailed files, notes, pictures, sketches, etc., to assist them through the remaining design process stages. As the team discovers supporting information, it will add the material as additional reference resources.

Depending on the collected information, the team may want to review the established preliminary working criteria and the overall goals. To assist the project manager throughout the process, some modifications may be needed. For example, one or more of the criteria may not apply to the problem. Likewise, issues may surface necessitating additional criteria or goal modification. Have these issues resolved before moving onto the next stage.

## Stage 4: Brainstorm and Generate Creative Ideas

The basic concept involved at this stage is to develop as many potential creative solutions as possible. The more ideas generated, the better the likelihood of identifying a feasible solution. The project manager will want to gather a group from technical and non-technical backgrounds for their perspective on the problem. This group may include engineers, scientists, technicians, shopworkers, production staff, finance personnel, managers, computer specialists, and perhaps clients.

A major method of generating multiple ideas is called Creative Problem Solving and uses a technique called brainstorming. (See Section 15.6 for a thorough discussion of creative problem solving and brainstorming strategies.) With this method, a large group with varying backgrounds and training are assembled to solve a problem. Every group idea that is spontaneously contributed is recorded. The basic premise is that no idea is deemed too wild or illogical. No preliminary judgments are made about any member's idea, and no negative comments are allowed. The goal is to develop a complete list of possible alternative solutions. The group leader should be encourage participants to suggest random thoughts and ideas.

Some students may have had the opportunity to engage in brainstorming exercises. Brainstorming can be fun and highly stimulating to the creative process, e.g., how many ways can you suggest to use a piece of string and a Styrofoam drinking cup? What could be created from a trash bag that contains some old magazines, tape, and a ruler? Think of ways your student organization could earn extra funds for a field trip, etc. Brainstorming sessions should not last any longer than 20 minutes. The rationale for this is that the mind ends up focusing on single ideas, and people have a hard time thinking of additional ideas. Brainstorming sessions should be conducted on an individual basis to promote as many group ideas as possible.

A brainstorming session could continue to a second occasion to allow members time to consider other options. When the group reconvenes, members may have new ideas or perspectives.

At this stage's conclusion, the group should have a long list of potential solutions. At this stage, all ideas are being considered. The team will evaluate each idea but will keep all options. Once the project manager is satisfied that all possible solutions have been suggested, the project is ready for the next stage.

## Stage 5: Analyze Potential Solutions

In the early part of the analysis stage (Phase I), the generated ideas must be narrowed to a few ideas to be subjected to more sophisticated analysis techniques. (Chapter 15.3 discussed the Analytic Method in detail, which directly relates to this stage.) This early narrowing could include the following:

- Examine the list and eliminate duplicates. As discussed earlier, do not to create limited categories, but eliminate repeated ideas. If two are similar, both should remain.

- Allow the group to ask clarifying questions. This could help identify duplicate ideas.
- Ask the group to evaluate the ideas. The group members can vote for their top three ideas, and those with the most votes will be analyzed in more detail.

A small number of ideas will remain. These can now be analyzed using more technical and perhaps time-consuming analysis techniques (Phase II).

Many individuals should be involved at this stage, but the engineer will be of primary importance. The analysis stage requires the engineer's time and background. Here, one's training in mathematics, science, and general engineering principles will be extensively applied to evaluate the potential solutions. Some of the techniques in this phase can take time, but a thorough and accurate analysis is important before the project moves to the next stage. For example, if the problem was the development of an automobile bumper that could withstand a 20-mile-per-hour crash into a fixed object barrier, several forms of analysis could be applied. Here are six possibilities:

*Common sense*—Do the results seem reasonable when evaluated in a simple form? Does the solution make sense compared to the goal?

*Economic analysis*—Are cost factors consistent with predicted outcomes? Analysis using basic engineering principles and laws: Do each of the proposed solutions satisfy the laws of thermodynamics? Newton's laws of motion? The basic principles of the resistance of a conductor, as in Ohm's law, etc.?

*Estimation*—How does the performance measure up to the predicted outcomes? If the early prediction was that some of the bumper solutions would perform better than others, how did they perform against the estimate? (See Chapter 15.7 for a more thorough discussion of estimation.)

*Compatibility analysis*—Each of the possible solutions and their related mathematical and scientific principles are compared to the working criteria to determine their compatibility. For example, how would each bumper solution meet the criteria of being cost effective? What would be the size, weight, and strength of each of the proposed solutions? How easy would each solution be to produce?

*Computer analysis techniques*—One frequently used method is finite element analysis. With this method, a device is programmed on a computer and then numerically analyzed in segments. These segments are compared mathematically to other concept segments. In the bumper example above, the impact effects could be analyzed as a head-on crash and then compared to a 45-degree angle collision or a side-impact crash. As each section is analyzed, the worst-case scenario can be evaluated.

*Conservative Assumptions*—This technique can be most useful in analysis because it can build safeguards into the analysis until more data are generated (see Section 15.8).

After each of the working criteria have been examined and compared to the list of possible solutions, a process eliminates those not performing well under analysis. At this point, only three to five options from the original solution list will remain. The project manager will review these remaining options and likely authorize the project be cleared for the next stage. If not, the process will be terminated or will return to an earlier stage to correct the problem.

## Stage 6: Develop and Test Models

Once each of the prospective solutions has been analyzed and the list of feasible options has been narrowed, the specific models must be developed and tested. A strong background in engineering coupled with experience and sound judgment are important. However, this stage will also involve team members who are computer specialists, shop workers, testing technicians, and data analysts.

Modeling is a method of illustrating a solution to a practical problem. Students need to understand that engineers and technicians can use different forms of modeling to assess and develop a product. The types of modeling are descriptive, functional, mathematical, computer, and scale.

*Descriptive models*—These may include diagrams, graphs, flow charts, and block diagrams. Verbal modeling describes the behavior of systems. Examples of descriptive modeling from a project might include a three-dimensional computer-aided design (CAD) drawing of a structure, a verbal description of materials, heat flow formulas, and scale system models.

*Functional models*—These may include computer simulations that support investigations of system behaviors or physical models of real systems with moving parts.

*Mathematical models*—Various conditions and properties can be mathematically related as functions and compared to one another. Often these models will be computerized to assist in visualizing the changing parameters in each of the models. Mathematical modeling in the form of equations shows the relationship between variables in the systems.

*Computer models*—These models allow the user to create on-screen images, which can be analyzed prior to the construction of physical models. The most common computer modeling is referred to as CAD where models are designed and displayed as three-dimensional wire-frame drawings or as shaded and colored pictures. The computer can control equipment that generates solid models using stereo-lithography, (where quick-hardening liquids are shaped into models) or other rapid prototyping forms. These on-screen models, or the prototype models they produce, can then be used in the testing process. Animations of solutions can also be an effective way to communicate an idea. A variety of software can incorporate items created in a CAD program and import them to be represented in their intended environment.

*Scale models*—Typically, these smaller models have been built to simulate the proposed design but may not include all the particular features or functions. Scale models may also be included in descriptive models. These models are often called prototypes or mock-ups and are useful in helping engineers visualize the actual product. Such models may be used to depict dams, highways, bridges, new parts and components, or perhaps the entire body of a prototype automobile.

*Diagrams or graphs*—These models provide a tool for visualizing, on a computer or paper, the basic functions or features of a particular part or product. These diagrams or graphs could be the electrical circuit components of an operating unit or a visualization of how the components may be assembled.

Once the models have been developed and created, each of them must be tested. Performing various tests on each model allows for comparison and evaluation against the established working criteria and goals. Tests are done continually throughout a project including on early models, prototypes, and product quality as the product is manufactured or built. However, the testing results at this stage establish the foundation for the decisions that will be made about the project's future.

Examples of these tests include the following dozen categories:

*Durability*—How long will the product run in testing before failure? If the product is a structure, what is its predicted life span?

*Ease of assembly*—How easily can it be constructed? How much labor will be required? What possible ergonomic concerns are there for the person operating the equipment or assembling the product?

*Reliability*—These tests characterize the reliability of the product over its life cycle and to simulate long-term customer use.

*Strength*—Under what forces or loads is a failure likely, and with what frequency is it likely to occur?

*Functionality*—The product or solution must fulfill its intended purpose. Think of a kitchen table. If you stood on a kitchen table and it collapsed, would the manufacturer be at fault for your injuries? Was this table intended to support a human's dynamic weight?" The table's purpose is to support meals and light work related to food preparation, serving, and consumption, not a human's dynamic weight.

*Environmental Considerations*—Can the parts be recycled? The product must be designed not to affect the environment adversely. For example if a university grew, it would have to conduct studies to ensure that the land could support the building. However, wind analyses are not often done. As a result, the building's configuration and height could increase a five-mile-per-hour breeze into a 15-mile-per-hour wind. Another example is the housing industry, in which the land that once was used for farming or just an open field is now developed with 250 homes and roads. This drastically diminishes the land's ability to absorb water from rainfall with an increased chance for flooding.

*Quality and consistency*—Do tests show that product quality is consistent in the various stages? Is the design such that it can be consistently manufactured and assembled? What conditions need to be controlled during manufacture or construction to ensure quality? The product or solution must be designed to meet minimum standards.

Quality has to relate to the item's intended use. This spin-off of functionality is important because quality is defined according to the item's proper or improper use. If casual shoes were intended to be used to play basketball, what would the quality issues be when they are used in the incorrect environment?

*Safety*—Is it safe for consumer use? The product must comply with codes and regulations to provide safe use and operation. Americans are becoming more safety conscious than ever, in everything from our cars to our homes, so safety is an important issue.

*Consistency of testing*—This technique examines each of the testing methodologies against the various results obtained to determine the consistency among the various tests. In the bumper design example, the testing methodology might be different for a head-on impact test than for a 45-degree crash test in order to minimize testing inconsistencies.

*Ergonomics*—The product must be designed so that the user can operate it with ease and maximum efficiency. Ergonomics is also called human factors engineering. For examle, designing chairs for genuine long-term or short-term comfort is a part of ergonomics.

*Appearance*—The appeal of a product is based on material selection, processes, finish, color, or shape. For example, two friends were looking for a new baseball hat that would not stick up in the front near the forehead and one that had a bill that could be conformed to their liking. They found one particular hat that met their appearance standards.Their idea of a new hat was one that was colorful and clean with no rips or tears.This is an issue of personal preference. The store had a marketing plan and knew that some or many customers preferred the worn look.

*Economics*—The product must be produced at the least cost without sacrificing safety. Engineers must keep costs down to increase company profits.

The project manager will evaluate this information. If satisfied these results consistently meet the working criteria and goals, the project will be cleared for the next stage.

## Stage 7: Make the Decision

At this stage, it is important for the team members to establish a means to compare and evaluate the results from the testing stage to determine which, if any, of the possible solutions to use. The working criteria throughout the process will determine the advantages and disadvantages of the remaining potential solutions. One of the ways to evaluate the advantages and disadvantages of the proposed solutions is to develop a decision table to help the team visualize each solution's merits. Typically, a decision table lists the working criteria in one column. A second column assigns a weighted available point total for each criterion. (The team will need to determine the order of priority of each of the criteria.) The third column provides performance scores for each solution. A sample Decision Table might look like this:

**TABLE 15.1     A Decision Table**

| Working Criteria | Points Available | #1 | #2 | #3 |
|---|---|---|---|---|
| Cost | 20 | 10 | 15 | 18 |
| Production Difficulty | 15 | 8 | 12 | 14 |
| Size, Weight, Strength | 5 | 5 | 4 | 4 |
| Appearance | 10 | 7 | 6 | 8 |
| Convenience | 5 | 3 | 4 | 4 |
| Safety | 10 | 8 | 7 | 8 |
| Legal issues | 5 | 4 | 4 | 4 |
| Reliability/Durability | 15 | 7 | 9 | 11 |
| Recyclability | 5 | 4 | 3 | 4 |
| Customer Appeal | 10 | 7 | 8 | 9 |
| Total | 100 | 63 | 72 | 84 |

Based on this sample decision table, it appears that while none of the proposed solutions have scored near the ideal model, solution number three did perform better than the others. Using this information, the project manager and team leaders would decide whether to use this project. They may decide to pursue this solution, begin a new process, or scrap the entire project. Assuming they decide to pursue solution number three, the team would prepare the appropriate information for the next stage.

## Stage 8: Communicate and Specify

Before a part, product, or structure can be manufactured, a complete communication, reporting, and specification for the item's aspects must occur among the team member engineers, skilled craft workers, computer designers, production personnel, and other key individuals. These include detailed written reports, summaries of technical presentations and memos, relevant e-mails, diagrams, drawings and sketches, computer printouts, charts, graphs, and any other relevant and documented material. This information will be critical for those involved in determining the project's final approvaland for those involved in the product's final implementation. They must have total knowledge of all parts, processes, materials, facilities, components, equipment, machinery, and systems involved in the product's manufacture or production.

Communication, an important tool throughout the design process, is especially important in this stage. If team members cannot sell their ideas to the organization, and be able to describe the product's or process' details and qualities, then good solutions might be ignored. At this stage, creating training materials, operating manuals, computer programs, or other relevant resources will be important for use by the sales team, the legal staff, prospective clients, and customers.

If the project manager is satisfied that all necessary materials have been adequately prepared and presented, the project will pass to the next stage.

## Stage 9: Implement and Commercialize

The penultimate stage represents the final opportunity for revision or termination. At this point, costs begin to escalate, so all serious issues must be resolved now.

In addition to the project manager and team leaders, other individuals are involved at this stage, and they represent various backgrounds and expertise. While engineers are part of this stage, many of these activities are performed by others. Some of those involved in this stage may include these four groups:

*Management and key supervisory personnel*—These individuals will make the ultimate decisions concerning the proposed project. They are concerned with the long-term goals and objectives of the organization, determine future policies and programs that support these goals, and make the economic and personnel decisions that affect the organization's overall health.

*Technical representatives*—These may include skilled craft workers, technicians, drafters, computer designers, machine operators, and others involved in manufacturing and production. This group will have primary responsibility for getting the product out the door.

*Business representatives*—
  a. Human resource personnel for hiring new individuals
  b. Financial people to handle final budget details and financial analyses
  c. Purchasing personnel who procure the needed materials and supplies
  d. Marketing and advertising staff members who help promote the product
  e. Sales people who sell and distribute of the product

*Attorneys and legal support staff*—The legal representatives who will handle legal issues including patent applications, insurance, and risk protection analysis.

If all parties agree that all criteria have been satisfied and the goal has been achieved, production and commercialization will begin. However, one stage remains where the project activities and processes will be evaluated and reviewed. (Some stages of the process may be monitored differently depending on whether the project relates to a one-time product, i.e., a bridge or dam, or an ongoing manufacturing process.)

## Stage 10: Perform Post-Implementation Review and Assessment

At this point, the project is probably in full production. The project manager, key supervisory personnel, and team members who had significant input with the project are gathered together for a final project review and assessment. This stage involves the project team's termination since the product is a regular product offered in the firm's product line. The product's performance is reviewed, including the latest data on production efficiency, quality control reports, sales, revenues, costs, expenditures, and profits. An assessment report is prepared which will detail the product's strengths and weaknesses, outline what has been learned, and suggest ways that teams can improve the process quality. This report will be used as reference for future project managers and teams.

## CASE STUDY: USING THE TEN-STAGE PROCESS

In this section, we will discusse, analyze and evaluate a simple engineering problem to provide you with an appreciation for the stages involved in getting a product to market and the team effort required for implementation.

## Background

The ABC Engineering Company is a small to mid-sized engineering subsidiary of the XYZ Research Corporation. XYZ is a world-renowned in research and development for computer vision, real-time image processing and design, and other advanced video applications.

For many years, ABC has been a supplier of products and services for the banking industry. ABC has dominated the field in such areas as security systems, bank cards, and personal identification number (PIN) identification systems. The parent company, XYZ, an expert in computer vision systems, realizes that the future of face-to-face banking transactions is rapidly changing. The security of electronic transactions is a major problem. The challenge for XYZ and ABC is to develop an electronic security system to validate banking transactions. They decide to apply this technology to the automated teller machine (ATM) industry.

## Stage 1: Identify the Problem/Product Innovation

ABC appoints a project manager to oversee the program to launch a new product and resolve a particular problem. ABC's team must identify the problem and attempt to generate idea sources to understand the problem's scope and nature. In these beginning stages, the team will include engineers, drafters, and computer designers as well as manufacturing, production, management, sales, and finance personnel. In this case, the group discusses the increasing role of ATMs in the banking industry. The current method of conducting transactions relies on an individual PIN and a security access card. An individual approaches an ATM linked to a financial institution, inserts an access card, and enters a PIN. The machine then allows the user to begin a transaction.

The team realizes that fraudulent ATM use has increased. ABC's experience, with the banking industry and the growth in its field of computer vision, may provide an opportunity to expand its operations into this area.

Following this premise, the group begins to comprehend the problem and locates sources to focus the perceived problem. They begin by working with sources from within ABC. These include meetings with their research and development staff and input from sales and marketing personnel. In addition, managers and supervisors may have ideas regarding the expansion of product offerings through modifications and upgrades. External information sources include prospective banking industry customers, the Internet, libraries, trade shows, conferences, and competitions.

This information helps them define the problem. The group may develop a preliminary problem statement to guide them. Such a problem statement might be to eliminate fraudulent access to ATM accounts. Once the project manager is satisfied that all necessary issues have been addressed at this stage, the project passes to the next stage.

## Stage 2: Define the Working Criteria and Goals

Using the preliminary problem statement as a guideline, the team develops a list of working design criteria, which will allow them to evaluate solutions. Based on the information used for problem identification, the group establishes the following preliminary working criteria:

- Cost: The machine must be affordable to banks and other ATM machines buyers.
- Reliability: The solution must have near 100% accuracy.
- Security: The cardholder's security must be fraud-proof.
- Technical feasibility: Current technology must produce the solution easily.
- Convenience to the customer: The solution should be easy to use.
- Acceptance: The solution must not be too peculiar, or the public may not use it.
- Appearance: The solution must have a pleasing appearance.
- Environmental: The solution and its parts should be recyclable.

After discussions and meetings, the team decides to re-examine its preliminary problem statement. Following discussion and review of the working criteria, the project goal becomes to design and implement an alternate method of identifying a cardholder at an ATM, while focusing on reducing fraudulent entries.

Establishing this project goal provides a means of evaluating, monitoring, and changing the process focus as it evolves. For the project manager, these criteria and goal become a checkpoint, or a means for assessing the project's progress and helps determine when the project is ready for the next stage.

## Stage 3: Research and Gather Data

Now that the goal and working criteria have been developed, all information relevant to the problem must be collected. Specifically, the team needs information regarding the current PIN method of identification and data on ATM fraud. They must investigate field developments, including current product information. They may look at this information to determine if they can redesign or modification their current product or can explore the feasibility of developing a new product. They may want to explore public opinion and market research.

Sources for this information may include the Internet, libraries, newspapers, magazines, journals, trade publications, and patent searches. The team should maintain accurate notes, records, and files for the collected information. Sketches and diagrams of similar products or competitor's models may assist them as they develop their solutions.

The project manager determines that all necessary conditions have been met and gives clearance to move to the next stage.

## Stage 4: Brainstorm and Generate Creative Ideas

Using the collected and studied information, the team decides to return to, prior to the brainstorming session, the list of preliminary working criteria and to assign each factor a weighted percentage of importance. The team decides to eliminate the criteria of appearance and environmental from their list. The resulting working criteria and their respective assigned weights are as follows:

Cost               = 10%
Reliability        = 25%
Security           = 10%
Feasibility        = 15%
Convenience of use = 20%
Public acceptance  = 20%

The team enters into a series of creative problem-solving/brainstorming sessions. The session group consists of engineers, sales people, managers, supervisors, research and development staff members, production workers, computer specialists, and a selected banking clients. They must develop options for identifying ATM customers. They are instructed not to make any judgments or negative comments on the proposed ideas. They are not to consider any ideas ridiculous. They develop ideas to identify ATM customers:

1. PIN access code use modification: Adjust current access methods to tighten security.
2. Individualized signature verification: An ATM attendant would match the card holder's signature with one written "on the spot" by the person at the machine.
3. Machine recognition signature verification: A scanner inside the ATM would match the cardholder's signature with one written on the spot by the person at the machine.
4. Voice recognition: The ATM would be equipped with a speaker phone so people could speak their name and an internal computer could match the voice of the cardholder to a recording .
5. Speech pattern recognition: The person would speak a sentence or phrase to be analyzed through voice characteristics and speech patterns.
6. Fingerprint matching: The ATM would have a fingerprint pad where the customer would press on the pad and a computer would compare the print to one on file.

7. Blood matching: The ATM would have some device to prick the customer's finger for a small blood sample. The machine would analyze the blood, with an internal lab testing mechanism, for DNA compatibility.

8. Hair sample matching: The individual would submit a hair sample, with root attached, into the machine. The machine would analyze the hair, with an internal lab testing mechanism, for DNA compatibility.

9. Iris matching: The ATM machine would be equipped with a camera, which would take a picture of the iris structure and perform a profile comparison to one on file.

10. Breath analyzer matching: The individual would breathe onto a sensitized glass plate and the results would be processed into an internal computer to match them with those of the cardholder.

11. Dental identification matching: The ATM machine would be equipped with a sanitized plate that the individual would bite onto and a computer sensor would read the patterns and analyze for a match.

At this point, the project manager is satisfied that they have uncovered enough ideas. Authorization is given for the project to move to the next stage for analysis of these eleven solutions.

## Stage 5: Analyze Potential Solutions

The team begins to analyze each of the eleven ideas and compare them to the working criteria. In Phase I, the team must narrow the solutions, which will be subjected to more sophisticated analysis techniques (Phase II). The team can use many different analysis techniques . The results of this analysis of alternative ideas are as follows:

1. PIN access code use modification: Adjust current access methods to tighten security.
   a. Similar to current solution
   b. Not reliable or secure
   c. Neutral cost

2. Individualized signature verification: An ATM attendant would match the cardholder's signature with one written "on the spot" by the person at the machine.
   a. High cost
   b. Not customer-friendly
   c. Difficult to implement

3. Machine recognition signature verification: A scanner inside the ATM would match the cardholder's signature with one written on the spot by the person at the machine.
   a. Technology is currently available
   b. Public would probably accept this method
   c. Higher cost to implement
   d. Long-term reliability is untested

4. Voice recognition: The ATM would be equipped with a speaker phone so people could speak their name and an internal computer could match the voice of the cardholder to a recording.
    a. Technology is currently available
    b. Higher cost
    c. Easy to use
    d. Long-term reliability is untested

5. Speech pattern recognition: The person would speak a sentence or phrase to be analyzed through voice characteristics and pattern of speech.
    a. Expensive
    b. Technology not as well developed

6. Fingerprint matching: The ATM would have a fingerprint pad where the customer would press on the pad and a computer would compare the print to one on file.
    a. Technology is currently available
    b. Moderate cost
    c. Easy to use
    d. Long-term reliability is good

7. Blood matching: The ATM would have some type of device to prick the customer's finger for a small blood sample. The machine would analyze the blood, with an internal lab testing mechanism, for DNA compatibility.
    a. High cost
    b. Good reliability
    c. Completely unacceptable to the public

8. Hair sample matching: The individual would submit a hair sample, with root attached, into the machine. The machine would analyze the hair, with an internal lab testing mechanism, for DNA compatibility.
    a. High cost
    b. Good reliability
    c. Public acceptance would be poor

9. Iris matching: The ATM machine would be equipped with a camera which would take a picture of the iris structure, and perform a profile comparison to one on file.
    a. Technology is currently available
    b. Moderate cost
    c. Easy to use
    d. Long-term reliability is excellent

10. Breath analyzer matching: The individual would breathe onto a sensitized glass plate and the results would be processed into an internal computer to match them with those of the cardholder.
    a. Technology not fully developed
    b. Expensive to implement
    c. Public acceptance would be poor

11. Dental identification matching: The ATM machine would be equipped with a sanitized plate that the individual would bite onto and a computer sensor would read the patterns and analyze for a match.

     a. Technology not developed

     b. High cost

     c. Public acceptance would be poor

After examing each of the working criteria and comparing them to the possible solutions, those that have not performed well under analysis will be eliminated. Three to five options from the original solutions have survived. The project manager reviews these options and determines they satisfy the working criteria and goal.

## Stage 6: Develop and Test Models

At this stage, the team will take those proposed solutions and develop models for testing and evaluation. Based on the analysis stage results, the team has selected three proposals for further study: voice recognition (#4), fingerprint matching (#6), and iris matching (#9).

Initially, several diagrams and sketches of each idea concept are developed and reviewed. The team considers many styles, options, and units. They decide to develop different models, including computer-generated models and prototypes constructed by craft workers in the model shop.

Once the various idea concepts are prototypes, the testing begins. Based on collected research and information, and the information needed to compare to the working criteria, the team concentrates on specific tests including technical feasibility, quality performance, reliability, and ease of use. The test results indicate that all three perform well. For the iris matching , the following data were generated:

1. Technical feasibility test: The current technology of specialty cameras and computer systems can identify a human by scanning the iris for over 400 identifying features and is able to match them to a large database.

2. Quality performance test: Eye scans from 512 individuals who had their iris features stored in the database system were taken. The program correctly identified all those who participated. Those whose records were not on file were rejected.

3. Reliability test: The error rate on the eye scan reliability test was one in 131,578 cases. This solution had the lowest error rate among any other form of physical characteristic testing, including fingerprint matching and voice recognition. In addition, the iris scan would provide long-term reliability since the iris does not significantly change with age.

4. Convenience of use test: The technology is available so that a hidden camera can spot a person approaching and zoom in on the right iris. As the person draws closer, security access control can make a positive identification.

5. Cost estimation: While not a test per se, enough data have been collected to estimate the system cost at $5,000 per unit. While the initial estimate is high for machines, which typically sell for $25,000 to $30,000, the cost will drop with mass production.

6. Once all the test results have been gathered, the project manager evaluates the information and data. These results meet the working criteria and goal, so the project goes to the next stage.

## Stage 7: Make the Decision

A decision table is developed and the test scores and information sources are recorded. In this particular case, the final table looked like this:

**TABLE 15.2   The Decision Table**

| Working Criteria | Points Available | #4 | #6 | #9 |
|---|---|---|---|---|
| Cost | 10 | 6 | 8 | 8 |
| Reliability | 25 | 20 | 22 | 24 |
| Security | 10 | 9 | 7 | 9 |
| Technical Feasibility | 15 | 15 | 14 | 15 |
| Convenience of Use | 20 | 15 | 16 | 18 |
| Public Acceptance | 20 | 16 | 17 | 18 |
| Total | 100 | 81 | 84 | 96 |

The results confirm that the iris matching identification system performed well. Using this information, the project manager decides to move the iris matching solution ahead and the team prepares the appropriate information.

## Stage 8: Communicate and Specify

Team members, including engineers, skilled craft workers, computer designers, production personnel, and other associated individuals work together to develop the appropriate materials. The team members prepare detailed written reports, summaries of technical presentations and memos, relevant e-mails, diagrams, drawings and sketches, computer printouts, charts, graphs, and all other relevant and documented material. They provide complete and detailed information on all parts, processes, materials, space, facilities, components, equipment, machinery, and systems involved in product manufacture. They create training materials, operating manuals, computer programs, and other resources, which the sales team, the legal staff, and clients and customers can use.

The project manager is satisfied that all necessary materials have been adequately prepared and presented, and he or she passes the project to the implementation stage.

## Stage 9: Implement and Commercialize

ABC's management has given approval for the iris matching identification project to move into production. Questions concerning overall costs, financial and labor commitments have been resolved. The finance department has developed a project budget. The purchasing department has begun the process of obtaining bids for the parts and supplies. The legal staff is resolving legal issues, including patent applications and copyright materials.

A selected group of engineers will oversee the production startup, which includes obtaining production space, facilities, equipment, personnel, and timetables. Once this is in place, a pilot production process begins.

The marketing, advertising, and public relations staffs begin to contact prospective clients and customers to promote the iris matching ATM system. The sales people, involved in the actual selling and distribution of the product, are trained to work with clients and are given appropriate promotional literature including training and operations materials.

All the parties agree the criteria have been satisfied and the goal achieved. The actual production and commercialization begins. One stage remains where the project activities and processes will be evaluated and reviewed.

## Stage 10: Perform Post-Implementation Review and Assessment

The iris matching ATM system is in full production and commercialization. The project manager, key supervisory personnel, and team members (who had any significant input to the project) gather for a final project review and assessment. This stage involves the project team termination since the product is a regular offering in the firm's line. The iris matching ATM system's performance is reviewed, using the latest data on production efficiency, quality control reports, sales, revenues, costs, expenditures, and profits. A prepared assessment report will detail the product's strengths and weaknesses and will include a report detailing what was learned during the process and ways that future teams can improve the process quality. The working criteria are examined and suggestions for future groups will be included in this report. This report will be used as another reference resource for future project managers and teams.

Note: The Iris Matching Identification System is currently being piloted in several British and Western U.S. locations. The information presented in this case study is a simulated application of the ten-stage design process.

## LEARNING ACTIVITIES

### Activity 15.1   Chair Design

Time:     2–3 weeks

## CASE STUDY

The challenge for this case study is to design and build an ergonomically correct chair based on the measurements of the students in the class. The chair must be able to support a 200-pound person and will be made from corrugated cardboard. The criteria for the design is the following:

1. The chair must have a seat and a back. Arms are optional.
2. Measurements for the chair will be based on statistical class data.
3. Required angles:
   a. Back tilt = 45–60 degrees
   b. Seat lift under legs = 15 degrees
4. There must be at least three legs if they are separate
5. The chair must support a person who weighs 200 pounds
6. The chair must be comfortable to sit in.*
7. The chair must be made from corrugated cardboard
8. Glue can be used for the outer skin of the chair and to laminate sheets of cardboard only. Three sheets maximum.
9. Masking tape or duct tape can be used for temporary support only. None can be part of the final solution.
10. The exterior of the chair and overall appearance should be aesthetically pleasing.*
11. A cost analysis of the used products (even if not in the final solution) is required.
12. A research report is required. All appropriate information should be formatted and included in the report.

Students will begin this case study by completing initial project research for historical references, design ideas, and ergonomic and material strength data.

It is recommended that classes use recycled corrugated cardboard. Check for this symbol, or call your local recycling center for more information on how to obtain recycled cardboard. There are also many retailers that sell recycled cardboard.

---

* For these subjective areas, the class should determine a scale to evaluate the features.

This web-based search should gather all data, pictures, and graphs to detail this case study's basis. Teams of three will be formed at random. Job assignments within the team are as follows:

- **Structural Engineer Responsibility**
  1. The initial design and testing of materials, joints, and connections
  2. Computations of ultimate strength and safety margin of structur.
  3. The structural drawings (two cross-sections)
  4. Prototype's structural stability, structural journal, and analysis
- **Ergonomics Engineer Responsibility**
  1. Anthropometric data collection
  2. Ensuring dimensions meet all parameters and constraints on drawings
  3. Orthographic and isometric CAD drawings along with dimensions on models (preliminary and final prototype)
  4. The prototype's ergonomic journal and analysis
- **Product Design Engineer Responsibility**
  1. Brainstorming sketch compilation (three from each member).
  2. Final chair design solution and fabrication
  3. Insuring that the chair construction meets design/comfort criteria and constraints

Teams should collect anthropometric data or measurements from every team member. This data should be compiled in a chart and diagram. Statistical analysis (mean, median, mode, and range) should be defined on each chart or drawing. The data are recorded by ergonomic engineers on a class chart. Ergonomic engineers meet as a group and perform statistical calculations. They decide which statistical measure is best suited for each of the eight chair dimensions on the anthropometric handouts. Ergonomic charts are posted in the classroom, which describes each measurement and decisions made.

Structural engineers create and test designs (three designs minimum). The design and testing will include load capacity data on single-ply, double-ply and triple ply cardboard, and all findings will be recorded in structural journals (sketches must be included). Structural work, anthropometric data analysis, and design sketches will be created during the same class periods. Study models will then be created from corrugated cardboard to incorporate design with structure. (All team members will participate in model making.)

Drawings of the top, front, and side views, by hand or on CAD (ergonomic engineers), as well as section drawings: and/or other structural views (structural engineers). Final drawing will be completed, rendered, and animated if necessary. Chairs will be tested (teams meet for discussion, evaluations, and testing). Portfolios will be completed and presentations made.

## RESOURCES

http://www.nysatl.nysed.gov/ The mission of the New York State Academy for Teaching and Learning (NYSATL) is to improve students' achievement by enhancing teachers' abilities to provide quality instruction based on standards. This is adapted from a case study written by Gail Atlas, Garden City Central School District, New York.

## Activity 15.2   Ping Pong Launch

Time:      2–3 weeks

## CASE STUDY

The primary purpose of this project is to continue to develop teamwork concepts, and to involve students in the various aspects of open ended engineering design. Performance, economics, reliability, aesthetics, and manufacturing processes will be factors leading to your final design from the various alternatives.

Your three member team will design, document, and construct a self-propelled, rubber band powered, vehicle that will launch a ping pong ball towards a 3 ft. diameter target. Brainstorming sketches and a research report are major components of this case study. In order to evaluate the efficiency of the group's design, each team will compete on the course shown below.

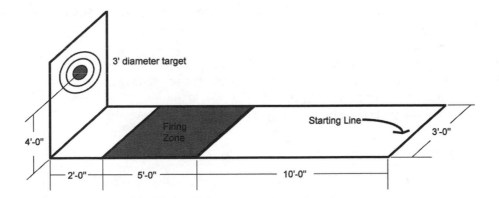

*Given:*

The floor will be located in the gym or the hallway
Three feet wide by 17 feet long lane (marked by tape)
Three-foot diameter target (see sketch)
Rubber bands, size #64
Ping Pong ball (True-Play Star brand Ping-Pong balls)

Each design team must submit a parts list with its design proposal/research report (see example below).

| Part # | Part Name | Dwg. # | Qty/Unit | Material | Size | Make or Buy | Cost |
|--------|-----------|--------|----------|----------|------|-------------|------|
| 1050 | Motor Shaft | 1000 | 1 | Mach. Screw | 8-32 × 1.0″ | buy | $0.12 |
| 2200 | Coupler | 1003 | + | Plastic | 1/16 ID Nylon Tube | Make | 1.42 |
| 3250 | Propeller | 1005 | + | Aluminum | 12 ga × 2 | Make | 0.18 |
| 3251 | Prop hub | 1000 | 1 | Aluminum | 14″ dia | Buy | 0.18 |
| 3252 | Hull | 1000 | 1 | Styrofoam | 9×4×3 | Buy | 0.24 |
| 3253 | DC Motor | 1000 | 1 | "See Drawing 1000" | | Buy | 4.99 |
| 3254 | Solar Cell | 1009 | 1 | "See Drawing 1009" | | Buy | 5.99 |

## *Design Restrictions:*

1. The device must leave the starting line under its own power. The devices may employ manual brakes (i.e., human fingers on the wheels or power source) to position them on the track. The operator may then let the device go without pushing.
2. Once the vehicle is started, no external communication, interaction, or influence of any kind is allowed (i.e., the system must be autonomous).
3. The device must fit into an 18-inch × 18-inch × 18-inch box.
4. Travel and launch capabilities must be accomplished with rubber bands only. The Ping-Pong ball must be launched within two minutes.
5. Any number and combination of rubber bands (not to exceed #64 in size) may be used.
6. No part of the launcher may be left behind at the start line.
7. The device must launch the Ping-Pong ball while in motion.
8. Only one shot per pass will be allowed. The pass is considered complete once the ball has been launched.
9. True-Play Star brand Ping-Pong balls will be the official balls.

## *Scoring:*

The objective is to fire the Ping-Pong ball within the designated launch area and hit the target. Scoring is based proximity to the bull's-eye the ball comes. Ten points will be awarded for each bull's-eye. Five points will be awarded for hitting the second ring, and two points will be awarded for hitting the outer ring. Five points will also be awarded for traveling (in bounds) the length of the ten feet approach lane, and five additional points will be awarded for simply launching from within the firing zone. In case of a tie, the highest average score for the three shots will be the winner. In case of an average score tie, the fastest average time from start to final stop will be the winner.

## RESOURCES

http://www.tcnj.edu/~asper/ This case study has been adapted from Fundamentals of Engineering Design, a course taught by Norman L. Asper, Ph.D. Dr.

Asper is a Department of Engineering professor at the College of New Jersey, P. O. Box 7718, Ewing, NJ, USA 08628.

## Activity 15.3    Scrambler Vehicle

Time:    2–4 weeks

## CASE STUDY

The contestants will design and build a purely mechanical device that will transport a large Grade A uncooked chicken egg a distance of eight to 12 meters along a straight track as quickly as possible and stop as close to the center point of a terminal barrier as possible without leaving a two-meter wide lane. The intent is to finish the run at the center of the straight lane. The distance will be announced at the event's beginning after the devices have been impounded. The device must be a single unit. The only source of energy allowed for moving and stopping the device must come from the falling of a mass not to exceed 2.00 kilograms. Scrambler systems will be impounded prior to the event. Safety goggles should be worn by the students during vehicle preparation and testing when warranted.

## CONSTRUCTION

The entire device in ready to launch configuration must fit inside a one-meter cube. The entire device must start behind the start line. The pointed tip of the egg must start even with the start line.

The falling mass must not exceed 2.00 kilograms. Any part of the device that has gravitational potential energy at the start of the run (such as swing arm and

**Mr. Alan G. Gomez, with the composite scrambler prototype vehicle. Its structure is 100% carbon fiber and Kevlar with a honeycomb core. This prototype placed 14th in the nation in 2002.**

any of its attachments) which decreases as the device moves down the course is to be considered part of the falling mass. The event supervisor will measure the mass. The mass must be able to be removed quickly and easily for measuring. The device should be impounded with the falling mass already detached.

An uncooked chicken egg (one per team, provided at the site) should be mounted on the front of the device such that the pointed end of the egg extends at least two centimeters in front of the most forward part of the transport. The egg must be five to 15 centimeters above the floor. It must be mounted so the egg will make first contact with the barrier in case of a collision. Tape will be provided to secure the egg to the device if needed, but no tape may be placed on the egg's front two centimeters. If any part of the device collides with the barrier, there will be a penalty.

The device must have a flat rigid unmodified backstop behind the egg made of at least a 5 cm × 5 cm piece of 3/4 ± 1/8-inch thick plywood or similar hard board. The egg mount must not be padded. The egg's rounded end must rest against the rigid backstop. No harness devices may cradle the egg. Once the falling mass has been set in motion, the vehicle may not be touched until it comes to a complete stop.

Any accessory equipment for the device (e.g., springs, rubber bands) initially must be in a completely relaxed state.

No chemical substances can be applied to the wheel surfaces.

All sighting or aligning devices must be permanently attached in a fixed position to the Scrambler. If a target is needed, it must be impounded with the device. The target cannot stay in the track but must be removed before the device starts its run.

## TESTING

The competition will be on a relatively smooth and level two-meter wide straight lane.

The Scrambler must finish as close as possible to the center of the terminal barrier without leaving the two-meter lane. The terminal barrier (for timing purposes) will be located eight to 12 meters from the starting line. This can be placed at the following intervals with varying difficulty to adjust the machine to: whole, half, and tenth of a meter. The entire device must be behind the starting line and within the track's two-meter width when the mass is released. The pointed tip of the egg can be placed anywhere along the starting line and must be even with it. Timing will begin when the falling mass is released. Timing will end when the device comes to a complete stop.

Students may not touch or guide the egg transport in any way once the falling mass has been released.

Stopping mechanisms must work automatically. The device may not be tethered or remotely controlled in any way to guide it or stop it. No electrical or electronic brakes may be used. Students will not be allowed to back the car up on or near the track in order to set the braking mechanism.

The contestants may hold and release the device, or they may hold and release the falling mass in order to start its motion. However, they CANNOT hold the device and the mass. The device cannot be pushed or pulled by hand. All of its propulsion must come from the falling mass. The uppermost part of the falling mass may not be higher than one meter from the floor.

## RESOURCES

http://www.soinc.org/ This case study was adapted from the rules of the National Science Olympiad competition. Science Olympiad, 5955 Little Pine Lane, Rochester, MI 48306. Tel: 248-651-4013; FAX: 248-651-7835.

## Activity 15.4    Solar Car

Time:      3–8 weeks

## CASE STUDY

Solar power comes from the energy of our Sun, a yellow dwarf star located 93 million miles from the Earth. It is a middle-aged, mid-size star compared to the billions of other stars in the universe.

The Sun's interior is a high-temperature region filled with dense gases. The Sun's core is estimated to be approximately 27 million degrees Fahrenheit. Heat and light from the Sun are produced through a process called nuclear fusion.

As early as 1877, basic solar power was used for heating buildings using the sun's rays to heat iron plates, which warmed up the air. Solar power was used in the industrial revolution when water was heated up for use in machines as steam. Solar power's widespread use started in the 1930s but was cut short when gas and fossil fuels became cheaper and took over the market. The cost of solar power was high and little interest was put into it until the 1960s and 1970s. Today, solar arrays and solar concentrators are the focus of major studies for use in space and on earth.

Sunlight is an excellent energy source and the future solar power use is exciting. The Sun's energy can be used to heat and cool buildings, generate electricity, operate communication and navigation systems, and power solar cars.

Solar-powered cars all get their fuel from the Sun. The cars use hundreds of photovoltaic cells to convert sunlight into electricity. Each cell produces about one-half volt of electricity.

When the American Solar challenge teams design their electrical systems, they have to allow for sunlight variations. The Sun's energy powers the car's motor and charges a battery for use when the Sun is hidden by a cloud. If a car is designed to put all of its energy toward driving and keeps nothing in reserve, it will

APS is evaluating the performance of the latest in Dish Stirling Solar power systems at its STAR Center in Tempe. This new technology, capable of producing 25 kW of electricity uses mirrors to focus sunlight onto a thermal receiver. The heat runs a Stirling heat engine, which drives an electric generator. It is one of four Dish Stirling power systems currently operating in the country. This highly efficient solar energy production system also can use alternative fuels instead of the sun's heat. So, power can be made anytime. Once perfected and manufactured on a large scale, this system has the potential to be one of

the cheapest solar energy technologies available. Photo taken by Bill Timmerman.

According to The American Heritage Dictionary of the English Language, the word "solar" originates from the Latin word: sol, meaning sun.

stop in cloudy weather. If too much energy is diverted to the battery, the engine runs too slowly to keep up in the race.

Engineers and scientists still have many questions and problems to tackle before solar power becomes an efficient and economical way to fuel vehicles. As the demand on fossil fuel resources increases, research will continue to search for alternative energy sources, including harnessing the Sun's energy to drive a vehicle. The most exciting part of using solar power as an energy source is that it is pollution free and inexhaustible.

In this case study, your team of three will design a solar racer and compete against other teams in your class. It is important that you first have knowledge in aerodynamics, chassis development, and photovoltaics. These pre-competition lessons should be done in class before your team starts the design process. Complete CAD drawings and electronic presentations should be complete prior to testing. Your team will be evaluated in the following areas:

1.  Time to go 20 meters
2.  Rolling resistance and aerodynamic drag
3.  Total vehicle weight

4. Acceleration
5. Top speed

## RESOURCES

http://www.nrel.gov/education/student/natjss.html U.S. Department of Energy's National Junior Solar Sprint. Managed for the U.S. Department of Energy by the National Renewable Energy Laboratory (NREL). This competition is intended for middle school students, but the program is so detailed and technical in nature that it can be used in the high school arena. Some of the vocabulary and ways to accomplish the task do need editing and rewording to fit the high school level so the reading level is accurate.

http://www.formulasun.org/asc/index.html The American Solar challenge is a student organization dedicated to educating students about solar electric vehicle design and construction by providing a hands-on environment in which students are encouraged to apply theories learned in the classroom.

*How Did We Find Out About Solar Power?* Isaac Asimov. New York: Walker and Company, 1981.

*Solar Power.* Ed Catherall. Morristown, NJ: Silver Burdett Company, 1982.

*Sun Power Facts About Solar Energy.* Steve Gadler and Wendy W. Adamson. Minneapolis, MN: Lerner Publications, 1980.

Sunrayce c/o General Motors, Attn: Bruce McCristal, GM Building, Detroit, MI 48202. (313) 556-2025

Electric Power Research Institute Transportation Program Office, P.O. Box 10412, Palo Alto, CA 94303.

U.S. Department of Energy Conservation and Renewable Energy, 1000 Independence Avenue, S.W., Washington, DC 20585

## Activity 15.5    Fling It!

Time:     2–3 weeks

## CASE STUDY

The history of catapults goes back to medieval times. The Romans, however, were the first group to make catapult design an art of war. Catapults could be made to sling projectiles over 100 meters but they could only be used a small number of times before the components (mostly oak wood) began to fatigue. A trebuchet is similar to a first-order lever with the counter weight providing the effort and the projectile supplying the load. All this is on a fulcrum.

Your three-person team has been charged to design and build a mighty siege machine—called a trebuchet—that will fling a water balloon across a far distance. You have been provided with some materials to build your trebuchet: You must use 2 × 4s, selected plywood scrap, wood screws, 1″ black pipe for the axle, flat steel where needed, and twine. Your missile will be a water balloon.

In celebration of Saint Barbara Day, patron saint of artillery, the Marines of the 11th Marine Regiment decided to take part in a rather strange yet challenging competition at Camp Pendleton. Marines from 3/11's survey section put the finishing touches on the bowling ball throwing catapult. With the throwing arm in its fully erect position, the trebuchet is 21 ft. tall. Photo by: Sgt. Ken Griffin

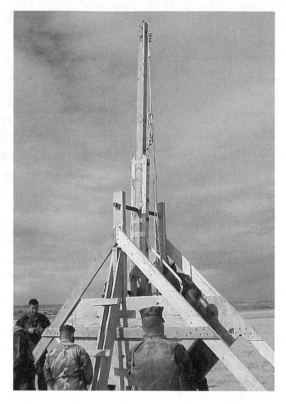

You will need to construct the following parts of a working trebuchet:

- Two triangular pieces for the sides of the frame; these will need to be supported or braced to stay upright. Parameters for base and height of fulcrum point will be 2′ wide by 4′ long, fulcrum point shall be no higher than three feet.
- A long, throwing arm pierced by the axle; the short end of the throwing arm should have a small, heavy counterweight that will allow the throwing arm to swing freely without touching the sides of the frame or ground.
- A sling that will hold the water balloon during the upswing and release it at the top of the arc.

You team is responsible to research safety when using trebuchets as well. Make notes of any issues from outside sources and be sure to include a section on safety in your final presentation. Each member of the teams should do individual research on catapults, trebuchets, and the systems associated with these machines.

m1 = the mass of the counterweight and beam (in Kg)
m2 = the mass of the projectile and beam (in Kg)
h = the distance that the counterweight falls (in meters)

Here is one trebuchet and its calculations:

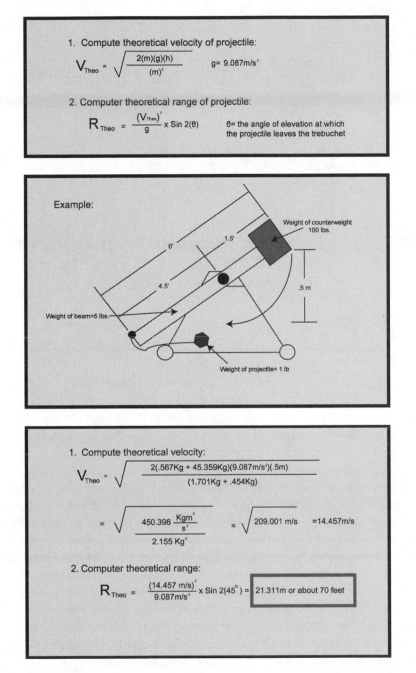

1. Compute theoretical velocity of projectile:

$$V_{Theo} = \sqrt{\frac{2(m)(g)(h)}{(m)^2}}$$   g= 9.087m/s$^2$

2. Computer theoretical range of projectile:

$$R_{Theo} = \frac{(V_{Theo})^2}{g} \times Sin\ 2(\theta)$$   θ= the angle of elevation at which the projectile leaves the trebuchet

Example:

Weight of counterweight 100 lbs.

6'

1.5'

4.5'

.5 m

Weight of beam=5 lbs.

Weight of projectile= 1 lb

1. Compute theoretical velocity:

$$V_{Theo} = \sqrt{\frac{2(.567Kg + 45.359Kg)(9.087m/s^2)(.5m)}{(1.701Kg + .454Kg)}}$$

$$= \sqrt{\frac{450.396\ \frac{Kgm^2}{s^2}}{2.155\ Kg^2}} = \sqrt{209.001\ m/s} = 14.457m/s$$

2. Computer theoretical range:

$$R_{Theo} = \frac{(14.457\ m/s)^2}{9.087m/s^2} \times Sin\ 2(45°) = \boxed{21.311m\ or\ about\ 70\ feet}$$

# RESOURCES

http://www.pbs.org/wgbh/nova/lostempires/trebuchet/ The companion website to the NOVA program "Medieval Siege." In the film, which is a part of the NOVA series, Secrets of Lost Empires, a team of timber framers and other specialists design, build, and fire a pair of trebuchets, devastating engines of war popular in the Middle Ages.

http://www.trebuchet.com/sim/ The ATreb Simulator Program. You can add the effects of air drag and friction, do stress analysis on your arm, sling, pin, axle, . . . and work out all the finer details of your trebuchet design before you even buy a single piece of lumber. It also includes a metric conversion calculator, release pin optimizer, and you can save hundreds of design parameters and simply load them from disk to work on different projects at the click of a button.

Special thanks to Mr. Vern Jordan and Mr. Johnson at Fort Atkinson High School in Fort Atkinson, Wisconsin, for the initial design and testing of this case study.

## Activity 15.6    Project Recycling and Energy Conservation of Ylem Compounds using Lightweight Equipment (Project RECYCLE)

Time:    2–3 weeks

## CASE STUDY

For many years, teams of mechanical engineering students have visited the distant planet of Gondwana to gain valuable work experience. On each occasion, they have been able to apply their knowledge and skill to devise ingenious solutions to help the Gondwanans develop their planet. Over the years, these students have been explaining to the Gondwanans the benefit of recycling as a means of conserving resources, even on the under-developed planet of Gondwana.

Consequently, this year the Gondwanans have asked the students to develop prototype machinery for recycling Ylem compounds which are found in limited quantity on the planet and have valuable properties in many of their industrial processes. The material must be moved in relatively small batches from a community collection point because larger quantities develop spontaneous reactions that destroy the material. It must be moved across an open level area (and must be kept away from this surface to avoid pollution) and placed in the elevated intake hopper of the reprocessing plant. The higher the hopper can be made, the greater will be the efficiency of the recycling process. Economical design of the equipment is important in order to conserve scarce construction materials and to limit ground pressure on the potentially unstable planet surface.

Can your team of three design and build a lightweight device that combines speed and lift in order to make the recycling process work as efficiently as possible?

Teams of three engineering students are required for the competition. Each team may enter one device only. Resources on Gondwana are limited, so the students must manufacture the device themselves using commonly available materials and components. The competition site will consist of two horizontal sheets of

2400 × 1200 × 19 mm Medium Density Fiberboard (MDF) joined together and arranged end to end. A starting line will be drawn across one sheet 400 mm from the free end. A finishing line will be drawn across the second sheet 3000 mm from and parallel to the starting line. All surfaces of the MDF sheets will be treated with one to two coats of Urethane sealer.

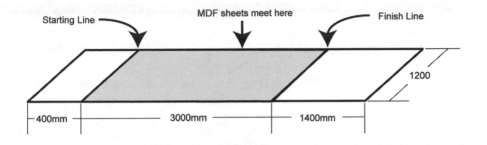

1. Prior to a run, the team must nominate the mass of dry White "Minute" brand medium grain rice, representing the Ylem to be recycled (to a maximum of five kilograms), and load this into a hand-held receptacle supplied by the team. This receptacle is not part of the device and cannot be carried on or with the device during the run.
2. Prior to its release, the device may have maximum plan dimensions of 300 × 300 mm, a maximum height of 500 mm, and its mass must not exceed five kilograms. The device must be placed entirely behind the starting line and setup ready for a run within two minutes of a request to do so. The Ylem must be loaded into the container in the device during this two-minute period.
3. At the signal to start, one team member must release the device by a single action that does not impart motion to the device.
4. The run finishes when the device is stationary on the competition site beyond the finishing line and the Ylem is elevated to its delivery height. The run must be completed within two minutes. The Ylem must remain elevated for one minute or, at the judges' sole discretion, for a lesser time. For a run to be valid, at least 100 grams of Ylem must be successfully transported.
5. After the device is released, it may not come into contact with anything other than the upper surface of the competition site, and its motion may not be influenced by any person.
6. Any spillage of Ylem will incur a twenty-second time penalty.
7. Each team will attempt two runs and the competition score will be the average of the scores from these two runs. The device may be modified between runs.
8. In the case of a tie for first place, a repeated head-to-head run between these teams will occur until only one team remains. These repeated runs will be conducted at five-minute intervals.
9. Devices that are deemed by the judges to be hazardous will not be permitted to run. In particular, devices using combustion or which damage the site are prohibited.

10. The judges' decisions on all competition matters are final.
11. Lighter than air gases (e.g., hydrogen, helium) may not be used in the device.

## Acknowledgment

Project RECYCLE was originally devised and run as the 2001 Warman Student Design-and-Build Competition for second-year students in Mechanical Engineering courses in Australian universities. The competition has been run annually since 1988 by the National Committee on Engineering Design (part of the Mechanical College of Engineers Australia) with sponsorship from Warman International, now known as Weir Warman Limited. Further details of the Warman Competition may be obtained from http://www.ncedaust.org

## Activity 15.7   Emergency/Homeless Shelter

Time:      2–3 weeks

## CASE STUDY

One of the most significant problems in metropolitan cities today is homeless people. During the cold winter months of the northern Midwest, outside temperatures can drop to 60 degrees Fahrenheit below zero. If humans were outside with no protection, they would perish. On a smaller scale, but just as critical, is when people are stranded in an area from which they cannot be recovered or rescued via airlift, etc. A viable temporary solution to this problem would be to drop them a structure in which they could survive in for a few days until a ground rescue team arrived.

Your team of three students has been selected to design a shelter for people to survive in cold weather. The people that are going to use the shelter must keep the shelter heated with nothing other than their own body heat. Two of the most important things your team must keep in mind are the following:

1. Insure that the human-generated heat is retained and most heat does not escape the shelter.
2. Insure enough air is circulating so the people do not suffocate.

This case study will involve many problem-solving methods. The team must work together and assign members specific jobs. The rules for the shelter development are listed below. These must be followed by every team. These rules are called Request For Proposals, or RFPs, which is how companies, the government, and other agencies outline their needs for projects.

"An RFP in its most formal sense is a specification of requirements that is sent out to suppliers who reply with proposals. Although common with large companies, the idea can be usefully applied at varying levels of sophistication to small

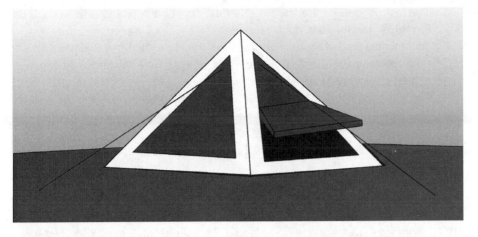

**This is a sample rendering of a shelter. This is not one of the required elements in the assessment; however, it is a good tool to aid in a formal presentation.**

and medium organizations as well. Used properly, it is a tool that supports and protects the buyer." (Lilley Information Systems)

## RFP

TO:               Engineering teams of three
SUBJECT:      Request For Proposals
PROJECT:      Shelter, portable, Homeless/Emergency supply, cold weather type
AGENCY:       Anytown, USA Disaster Agency
PROJECT DESCRIPTION: Design a low cost structure made out of 90% cardboard that will allow people to survive outside during cold weather. Because it is made of cardboard, it will biodegrade.

1. Shelter capacity: Protect five people for a period of up to three days
2. Construction materials: Standard 1/8" corrugated cardboard in multiple layers as needed, plus any additional materials for insulation, waterproofing, and protection. Other hardware as required for fastening, anchoring, packaging, etc.
3. Inside temperature: Keep at exactly 60 degrees Fahrenheit when the outside temperature is 20 degrees Fahrenheit below zero.
4. Durability: The shelter must survive a 200-foot drop without significant damage. Note: the shelter, when tested for this requirement will be done so only after the temperature testing is complete.
5. The shelter must be assembled in fewer than 15 minutes and without any tools.

6. The instructions for the shelter assembly must be in pictograms or cartoon format, so language is not a problem.
7. To minimize cost of an individual shelter, the ideal shelter would have an exterior surface area of 120 square feet or less. The thickness of the wall should not exceed five inches.
8. The shelter must allow proper ventilation to allow all five people to breathe.
9. The shelter (when empty) must be held down to the ground so it does not blow away with gusts of wind up to 25 miles per hour.
10. The structure must be able to withstand two feet of wet snow.

## RESOURCES

*HVAC Principles and Applications Manual.* Thomas E. Mull New York: McGraw-Hill, 1997.

An on-the-job handbook for beginning or experienced engineers, providing information on design, applications, and code compliance without delving into theory and complex mathematics. Includes such topics as basic scientific principles, climatic conditions, infiltration and ventilation, external heat gains and cooling loads, acoustics and vibrations, human comfort, fans and central air systems, an introduction to electrical systems, and controls for air distribution systems. Appendices contain tables of technical matter such as the thermal properties of building materials and the average winter temperature for major U.S. cities.

http://www.lilleyinfosys.co.uk/ Lilley Information Systems is an independent Information Systems consultancy. Established in 1980, it is based in London, UK. In addition to its systems work, it also carries out Public Relations (PR) work.

## Activity 15.8    Rube Goldberg Machines

Time:      2–3 weeks

## CASE STUDY

Students of all ages are participating in national engineering contests, transforming everyday material into their own wacky and innovative machines. The Machine Contest allows wild and ingenious inventions to be shown and provides a learning engineering experience. Students will design, build, and test a *Rube Goldberg* Machine. Best known for his INVENTION cartoons that use a string of outlandish tools, people, plants, and steps to accomplish everyday simple tasks in the most complicated way, Pulitzer Prize winning cartoonist Rube Goldberg's drawings point out that people are often overwhelmed by complicating their lives.

In this activity, students will design, build, and test a machine that will select, clean, and peel an apple in a minimum of ten steps. This gives students the

chance to use everyday materials and transform them into a wacky, innovative machine that accomplishes the task. Your assignment will also include a step-by-step description of the machine's systems and operation, development of working drawings, and a prototype.

## RESOURCES

http://rubegoldberg.com/ The official Rube Goldberg website.

http://www.uwm.edu/CEAS/rube/ The University of Wisconsin- Milwaukee's home page. UWM hosts a state competition every year for high school students. Many states and local areas run competitions check with your local engineering college for more information.

http://www.sciam.com/ Scientific American's web site. The graphic and description was featured on this website (with permission of Rube Goldberg, Inc.) in April 1998

A new book on Rube Goldberg and his work, plus instruction book on the Rube Goldberg Machine Contest, is now in progress. The only publication in print is published by Stewart, Tabori, and Chang (under license from Rube Goldberg, Inc). It's a postcard book with 30 color postcards illustrating Rube's INVENTIONS. The ISBN is 1-55670-524-7. You can order it through your local bookstore.

Marzio, P.C. Rube Goldberg, His Life and Work. New York: Harper & Row, 1973.

Kinnaird, C. Rube Goldberg vs. The Machine Age. Fern Park, FL: Hastings House, 1968.

## Activity 15.9     Crude Residue

Time:     1–2 weeks

## CASE STUDY

High Density Polyethylene (HDPE) that is contaminated with oil or other water in-soluble material poses a special problem for recyclers. Quart-sized motor oil bot-tles and bottles which previously contained other automotive materials or addi-tives are the most common source of this type of waste. Many curbside-recycling programs still cannot accept emptied motor oil bottles for recycling. Currently, few plastic recyclers can handle this type of material.

The primary problem companies face when recycling these materials is the washing process and the permitting associated with the accompanying waste-water discharge. Some compounding processes can use limited amounts of oil-contaminated HDPE directly in a mix with non-contaminated plastics, but for most applications the material must be washed first.

According to the Environmental Protection Agency (EPA), Americans pur-chase 2.5 billion gallons of lubricating oil and generate 1.3 billion gallons of used oil every year. Of the 1.3 billion gallons of used oil generated each year, approxi-mately 800 million gallons are recycled through a managed energy recovery sys-tem; approximately 500 million gallons are improperly discarded. By providing used oil collectors and recyclers with information concerning sources of used oil that are available for collection, used oil is being recycled and reprocessed as a fuel oil.

In this activity, students will design a low cost system or alternative container to safely house automobile oil. This design should not allow any residue of oil to remain in the container when the installer has completed use of the container. The design can be made of any materials that are compatible and it must follow the restrictions listed in the Request For Proposal (RFP).

## RESOURCES

http://www.epa.state.oh.us/ Ohio's Environmental Protection Agency's website.
The above information is on their website; however, your local or state EPA should have the same data available

Many environmental websites and publications are available. Major oil manufac-tures will also supply information if requested.

**A list of the many different recycling codes and material abbreviations associated with each.**

# Chapter 16

# An Interview With A Practicing Engineer

## Introduction

The following interview presents a discussion of one engineer's career. Students who aspire to be engineers should follow their dreams. In life there likely will be obstacles in your path, but how you deal with these obstacles and overcome them will aid you in your development as a professional.

The interviewee, Brian Lisles, has had a wide variety of experiences during his career. He is now managing a professional racing team and guiding young engineers to their dreams through his current company, Neumann/Haas racing.

You are encouraged to contact engineers in your community to discover the paths they took in order that you may gain additional perspective on how to become a successful engineer yourself.

## Name: Brian Lisles
## Title: General Manager

Brian Lisles has been general manager of Neumann/Haas Racing since 2001. Prior to being promoted to general manager in January 2001, he spent two seasons as senior engineer and ten seasons before that in the role of race engineer. He is responsible for overseeing the race team, the engineering department, and general team administration.

In his first season with Newman/ Haas Racing in 1989, Lisles worked as Mario Andretti's race engineer. In

**Brian Lisles stands with one of the race cars that his team runs.**

1990, he was race engineer for Michael Andretti. In his third season with the team, Lisles' hard work came to fruition when Michael Andretti became the 1991 PPG Cup champion after a record eight victories and poles. In 1992, he rejoined forces with Mario Andretti, who won the Phoenix event. Lisles worked with Mario through his final season of Champ Car competition in 1994, and In 1993, Mario set a world record for his 234 mph qualifying speed at Michigan International. In 1995, Lisles worked with Paul Tracy, winning two events. For the 1996 season, Lisles engineered Fittipaldi's two podium appearances (Detroit and Vancouver) and near first victories in Detroit and Elkhart Lake. Fittipaldi led 80 laps during his first season with Neumann/Haas Racing. In 1996, Lisles took an active part in the design of the Swift Champ Car, which debuted with a historic victory in 1997 with Michael Andretti behind the wheel. In 1997, Lisles engineered Fittipaldi's car until the second race when Fittipaldi was hit from behind. Lisles then worked for six races with substitute driver Roberto Moreno until Fittipaldi had recovered from his injuries. In 1998, Lisles continued to engineer Fittipaldi's race car and assisted in his two third-place finishes. In 1999, he passed the torch to Todd Bowland, who was in the midst of his first season as a race engineer. Under Lisles' and Bowland's tutelage, Fittipaldi scored his first win and pole.

Lisles earned his automotive engineering degree from Loughborough University of Technology, located north of London. While a Loughborough student, Lisles began to take an active interest in motor racing, first with a Lotus Seven and later by building racing cars of his own design. After graduating, Lisles joined Chrysler UK and became a design analysis engineer in the Advance Design Group. After five years with Chrysler, Lisles joined The Department of Terotechnology at the British Steel Corporation. Answering a newspaper ad for a development engineer at the Tyrrell Formula One Team brought Lisles back into racing. During his 11 years at Tyrrell, Lisles became the chief engineer before leaving and moving to United States to join Neumann/Haas Racing in 1989.

The following interview took place in June of 2003 at the headquarters of Neumann/Haas Racing in Illinois.

### What inspired you to start into engineering, racing, or what have you from a young age?

I'm not really sure, of course, because I think for most people you have some point when you're growing up you discover what it is you enjoy most and what satisfies you most. And I guess for me, by about probably 6 or 7 years old, Sunday afternoons on a rainy day in England was Erector set time. I used to make all sorts of odds and ends, building to the plans and making my own things was something that I enjoyed doing. I guess one thing led to another.

When we first got a car in the early '50s, my family at that time didn't have enough money to take it to a garage to get it fixed, so we did the maintenance ourselves. So it was Dad working and son handing him the tools. Dad went to night school to figure out how to do it. I was right there alongside him when he came home in the evening to tell me all about it. It just seemed to be the thing to do.

Early on in my school career when in my early teens, I aspired to be a physicist or whatever, but I don't think I was quite academically well enough endowed to become this and some of those mathematics that were needed were difficult. That plus the fact that I started to build model cars with engines in, just seemed to be a real good thing to be doing. So, I just looked around and found that it was time to go to a university. I told myself, "I'm going to go do automotive engineering." There was a university in the United Kingdom that offered that program, and I applied and was lucky enough to get accepted. It was a four-year program and a year out in industry. This was great. My family had no history of engineers at all. My family looked back and told me that my great-grandfather was the head of some railway at some point—that's where they say I got it all from, but the truth of the matter is, all my immediate family were not engineers.

***It sounded like Dad was the problem solver. He figured out how to do things.***

Yes. It was just necessity. If he wanted to make the house nicer or keep the car going, he just had to do it himself. There was not enough money to pay someone else to do it, so we just did it. So, doing the four-year degree course was great because I had no family history of how to use machine tools and all that kind of stuff. Well, actually, there was a precursor to that. I went up to technical college to get the necessary qualifications to go to the university. Because it was a tech college, you didn't have to do all of those additional things like languages and so forth. Whichever way you'd gone anyway, I passed all those exams. But I just needed to get to my university entrance exams. I got a year ahead in school—I'd been bright enough early on to have a year in hand, so I wasted it  . . .  and went to Tech College and did just what I needed—the math, the physics—to get me to do engineering in university. I had all the spare time. I only took 12 hours a week lecture time. It was great. I did my course work during the day and then I still had tons of time left.

So, I tagged on and did some technician classes and went in and did machine tool theory and practice and learned how to operate them and know how to make exact inch cubes to within two thousandths (of an inch) and spent a little bit of time at a blacksmith school where the college was. I did a bit of foundry work. I learned not to pick up anything that was lying on the floor. Even if it wasn't black, it could still burn you. That's when I went off to the university and started the four-year automotive engineering program. It was a four-year course, three years of academic work and a year of practical stuff. And that was great. I went out and worked in the industry, worked in tool rooms, worked in billing shops, lay shops, learned how to use all those machines. Certainly if you're going to go into engineering where you're responsible for design work you need to know how the stuff you make actually gets made. That was a huge asset that is still with me. I certainly wouldn't go down and machine anything these days—I haven't done it for a long time, but when you know the practice, when you have the practice, it's like riding a bike, you might be a bit wobbly when you first get on it, but it comes back. You know what can be done and what can't be done.

*If you had to go back into your, let's say teenage years, up until 22 or 21, would the Tech College be one of the defining factors with many of the things that you've done?*

It gave me a lot of confidence to be able to talk with anybody on the practical side. I don't claim to have skills guys have here or out in a prototype shop, but I can at least talk to them. I know what it is they're talking about and understand what their problems are and understand if they're making problems out of something where there really isn't a problem and should be using some other way to solve the problem. Once you do it, you understand what can and can't be done.

*So, you can speak on the same terms and on the same level? And that's what's more important sometimes than doing the work.*

Yes. You can't be a master of everything. I also worked my way through college at a great place. We did automotive engineering, which kept it interesting. I always felt the thing my university program failed to teach well and discuss was the actual process of designing something. We did do design work. I designed and made a gear box in my third year and an engine in my fourth year and all the calculations for it, and drew it up on paper. But that essentially was really how they taught design. You just got to do a project. So where I was pretty lucky was I got a job with the Chrysler Corporation in the UK, which was really big. They bought a big British automotive company called the Roots Group, which was a collection of old names, Sunbeam, and several others, amalgamated during the '50s. I was lucky because one of the companies they purchased, even from the '20s, had a strong technical department.

Everything was not only drawn but was correctly represented. The gears were properly checked out and so forth. It had a strong analytical approach. And that department was integrated into the Roots Group and into the Chrysler UK Group. And I was lucky enough to end up being placed there. I did the usual year tour around the company and got sent off to postgraduate studies. When I came back, I was put in what was called the Design Analysis Department. And that was great because basically what happened was we were in a large open plant office and we were given free reign to go and talk to whomever we liked. All the drawings, produced for all the cars that came through the Design Analysis Group, were kept, so you actually put your name in a little box on the drawing of every car that was drawn to say that it had come through you and, from a technical point of view, it was okay. So, it was a wonderful opportunity to learn because I worked for people who understood how to analyze things and I was working with people who were drawing things and creating shapes and solving problems.

Within the Design Analysis Group, I worked with the development teams to solve their problems as well. So early on, I got drafted into the engine part of that group. We had a huge valve gear problem, they'd already written a multi-degree Freedom model, and I worked alongside those guys for a year. This was about 1969–70. We had a 24-degree Freedom model of the valve gear itself, but basically the development guys couldn't figure out why we were breaking valve springs and all sorts of other pieces. I was part of the group that wrote that model and we used it and actually solved the problems with it. That was pretty cutting edge stuff for that time. We also had a full finite element model of the body shop, a 3,000-element model as I remember.

### And that was in 1970?

That was about 1968–1970. Running on an IBM. We had a 360 and we had a couple of 370s. About 1970, time-share became available, and we had a direct time-share line to a remote host that was a commercial thing. And we had access to write our own programs in BASIC and run those. So, we were a forward looking, small group, just three of us, I think, doing the analysis. Three guys were doing the engine, one or two guys doing the gear box, and two or three guys doing the body. We analyzed everything. As I stayed there and got more confident with what I was doing after a few years, I did more work with the advanced group, which was pretty exciting. We were still drawing manually, of course, in those days. They were doing all of what we used to call the felt-tip pen schemes, where you were doing concept cars, not concept cars as in styling, but concept as in engineering. I used to go over with those guys, work with them, and we'd lay out all that we had and make sure we had enough room to put the correct springs in there to support the car and so on, as well as do a quick stress analysis. The parts were going to be the right size to do the job. So, we did almost every part of the original design process which was exciting and all by hand.

### It sounds like, between 1960 and 1970, you worked with a group that did things like firms do today only not with current technology.

Right. I was lucky because it was a small enough group, and a friendly enough group. It was small enough that there weren't really any big territories involved. Everybody just pitched in to do the job. Our boss was under quite a lot of stress. He wasn't successful in the marketplace. He wasn't successful financially, so we did the best we could with what we had in hand. I was lucky. I just fell into a place and worked for people who really were topnotch people.

### Before the term Concurrent Engineering was around, that team was already doing it?

Yes. I think the reality is that probably a lot of the design teams always worked that way. It's pretty sensible, right? Everybody pitches in and gets the job done. The trick is when you get to be organizations. When I was in it; the legislation was starting to impinge on the design and product process. All the routine stuff. I remember early on having to go through and do all the analyses of the federal and European standards. We did all the calculations for the door makers and produced papers showing they would withstand a 30Gs in any plane. We had to do an inertial check, the tests to insure that the door locks wouldn't open with 30Gs applied in any direction. There are other tests that have to be done like the headlight rule and a bumper impact rule. All those were starting to come in and impinge on the design process and added to the kind of paperwork and bureaucracy needed to get a car approved. Nevertheless, the core design process for a car is still the same as it was in 1920 really, for automotive vehicles, just that you have much more powerful tools available.

*These days if you were to put a design on the market, you would have a lot more political hoops to jump through, correct?*

Right, a lot more legislative hoops you have to be aware of. You need additional people to check and make sure you don't get out of whack. Nevertheless, the fundamentals have remained the same.

*Running a team as you do now, what kind of things do you do that are similar and different to the experiences that you had with that small company? There were no turf wars and it was small enough to talk to each other. How do you bring that to a current workplace? Or can you?*

What we do here is different to that because, although we have actually a large engineering group for the total number of people, the engineering group is 15% of our head count and are qualified engineers dedicated to their engineering tasks. Although we do design and manufacture things, it's a fairly small part of the overall car process. The things we do, design and manufacture, are done for one of two reasons, either to improve the performance of the car or to improve the life of a component. The other engineering thing we do sometimes is clearly related to cost effectiveness. Sometimes we have to make something ourselves instead of buying it somewhere. We need to produce drawings for that.

*Knowing a little bit about how the circuit runs and how things go, do you think that a person with the money can do more of that? Let's say, for example, when you probably spent a lot of time going head to head with Roger Penske's team. As far as I know, those people spent a lot of money and a lot of time on that process that you were speaking of and did a lot of things themselves versus buying it. Is that what you see more and more—customization of your systems to meet your needs?*

Yes, but as you said there is a budget imperative, too. We have always been financed well relative to the competition, certainly not to the Penske level because when push comes to shove, the Penskes will spend whatever it takes without fear of any consequences—we can't do that. We, as a team, do at least as much or more than our competitors.

*So, you're more fiscally responsible to the business end as well, and you do what you can within the business plan versus throwing money at things, which is the way the real world does it.*

Yes. Now, you have to be careful. We have enough funds that we could almost do more than we could accommodate into the team. You have to pick and choose what it is you do; otherwise, you just waste a lot of time and effort, and that always dilutes where you're going. Because, if budget size alone brought you success, then we could always submit our budgets at the beginning of the year and they could mark up the championship accordingly. But it's more complicated than that, and that's where the people issue comes in because the people make the decisions on how to spend the money, where to spend the money, whether to do a certain project or not, even if they can't afford to do it. You might not be able to fit the project in, and it's just going to be a distraction, so we don't do it. The classic clear-cut example of that, which is clouded a little bit by the driver they have, but probably the clear-cut example is the success in motor racing of Ferrari Racing. Still, Shumaker is clearly head and shoulders above the rest of the drivers. Part of his edge is what he does off the track as well as on the track. You have to look at the group of people who are running the company. You look at each of the histories they bring to the business. From talking to people I know, they say a big change has occurred at Ferrari, and they do less now in terms of projects. But what they do, they do properly and make sure it has an end result.

*So . . . work smart and not hard?*
Right. But that's obviously what we would all aspire to do.

*Do you do that yourselves as a team?*
Oh, yes. You have to focus on what it is that's going to win your races, or put your business in front, not necessarily what the latest super-gizmo is. You need to be aware of what the super-gizmos are and if they're going to make a difference and if that difference is going to show up in your race reports. Then you need to look at it and decide if you're going to go get one now or wait a week or next year. Of course, it may be an idea that's generated internally. With an internally generated idea, you tend to have more equity in it because it's your idea, whether your idea is created personally or collectively. If it's something new, there's a temptation to do it just because it's yours and created within: You want to get on and do it. At some point you have to stand back and say, no this isn't going to take us anywhere, or, yes, this is really going to take us somewhere and we need to suck it up and get on with it.

*The overall objective is to win the race, which can be metaphorical or an actual objective of a business like yours.*
Right. We don't have an incentive scheme like big companies do. They don't let you keep the rights to a patent. They might give you .001% of whatever the net result is or you get $500 or $1000 for every idea that gets incorporated into the product or whatever. We don't do that. With everyone here, that's what your job is. Your job is to make the car go better in some way or another. Everybody here from the guy who drives the van to pick the stuff up through to the driver. Everybody can actually negatively impact the process. If the van driver brings something late, you may miss the race. Or the late delivery causes everybody else to be in a panic to get everything done and then something gets missed through to the truck drivers laying out the pit equipment and not setting the pressure regulators right or the guns correctly. This goes on from the mechanics not doing things correctly to the engineers not doing their homework properly and getting the right parts on the car for that event through to the driver performing at the highest level. Everybody and anybody can trip the process up at any time. The challenge is to keep everybody with the mindset that you have responsibility that anything you see that's wrong, you either need to fix it yourself or you need to tell somebody who can fix it even if it's not your job responsibility. Simplest example: if you walk around the workshop and see something laying on the floor, you pick it up.

***Even if you didn't leave it there?***

Everybody contributes. You see something that's not right, you need to come and address it right there and then, or you need to come and see somebody who can fix it or can say I know it's not right but will accept it for the minute because fixing it is a huge process.

***It seems to be a closely knit, wise company and team that everyone and every piece of the puzzle is important and no one person towers above (of course, the boss always towers above), but everybody works together.***

That's really what you want. People have different skills, and everybody has a market value, and ultimately we pay our employees fair market value for their skills. And after that, everybody's contribution is considered to be worthwhile. If it isn't worthwhile, then you shouldn't have them.

***What kind of advice can you give to students who want to become engineers?***

One of the problems that I see—there are a couple of problems that I see or understand much better now I'm older. The whole truth is that people who are good at engineering don't always have the best interpersonal skills, social skills. There's a tendency to let the numbers speak for themselves, to be able to crunch the numbers or produce the drawings. We're lucky in engineering in that we have clear rules about what will and won't work. If you're lucky enough to have the gift as an engineer, you can pretty much see it in your mind's eye whether or not it's going to work before you even start. The fact of the matter is in many ways, many people can do that. Maybe there aren't so many who are genuinely creative in terms of producing the new solution to a problem. These days with the computers, tools, and the amount of things that you can use, you often get a novel solution even if you're not creative but if you use the tools correctly because they will push you in the right direction. You can run up simulations to get there by pure hard work, and these days it doesn't take long. It used to be that you just couldn't do that much work so you usually relied on the guy who had the mind's eye to do that himself and say "I think probably this is the way we need to do it because I've been thinking about it and I've seen it the right way."

***The thinker versus the doer. What kind of advice would you give to a student who might read this book and who may be good at the numbers but not be the thinker or have the personal or soft skills?***

I learned a lot about dealing with people on the workshop floor in the following ways. I went to work at Chrysler for five years. As an engineer with engineers, I didn't deal with production people. Occasionally, I went to a manufacturing plant if they had a particular problem that impinged on engineering, but basically I was a design office guy. The only people I dealt with who produced things were guys in the prototype shops who obviously were very skilled. Things weren't going well across the UK and essentially the prototype shops died out in the end, Chrysler UK got bought out and it's a production facility now. So, I did a big career change. I stayed as an engineer, but I went and worked for British Steel, which is at the other end of engineering, a heavy engineering turnabout. Everything's in thousands not millions. I worked for the Chief Engineer of British Steel. British Steel was a company, which was formed after World War II.

The steel industry suffered heavily during WWII to satisfy the war effort, and they came out on the other side basically with no money left, so they were then all nationalized. The government took over all the steel companies and called them British Steel. It was a big mess because there were all these different steel companies all over the United Kingdom, some of which were in good shape, some of which were in bad shape and never talked to each other or had been integrated into a big process. When I went there in the early '70s, they'd been 20 years into sorting this out. One of the things they realized was they kept building steel mills with the same mistakes that somebody else had done across the other side of the country, so they formed an engineering group called the Department of Technology, which was a buzzword in the early '70s, which was basically total time line cost management. Better to spend an extra hundred million dollars on a plant now because it's going to last 20 years and you'll save a billion dollars over the twenty-year process. Be sensible about the way you spend money. I did that for a couple of years, which is interesting, dealing with the economical engineering side. But within British Steel Corporation, if you wanted to make a career there, being an engineer was not the way to do it because it was a production company; they produced steel. Engineering was an unfortunate necessity to keep the plant running. If they didn't have to have engineering, they wouldn't have had it. They were forced to have it maintain the plant. They produced 15–20 million tons of steel product a year, and that's what paid the bills. So, I applied for and got moved into management.

They sent me off to Management College, which was actually hard work since I hadn't written an essay since I was about 18 and I was a little shocked at the system taking these guys who got MBAs, etc. The process also tried to turn you into part manager and wanted you to sit in a central location. They moved you when they had a need for an assistant plant manager, shift manager, or a plant manager for the whole plant. You have the whole plant where the stuff all comes in and gets laid out at the center plant where you take the iron ore and turn it into stuff called sinter to make it fit into the glass burners. You have the coke making plant, which takes coal and makes coke. Each one of these plants is 100-million-dollar to a billion-dollar facility but they're all on the steel works part of the continuous process and finally you end up with a finished product—some kind of a steel finished product. So, a plant manager would be in charge of a plant that is worth on average, say, a hundred million to 150 million pounds. If the plant manager got sent up on a course or was promoted and the company had no obvious candidate, you would be sent in for 4 weeks, 6 weeks, up to about 3 months of stand in. You just turned up on Monday morning and, whatever was going on, you had to keep the place going until you left.

**Explain a bit more about your trial by fire because many engineering students may experience this same phenomenon with new jobs.**
The first thing that happened on Monday morning would be the union guy coming in and I would go out because there's always some issues, little things. I

learned a lot about people doing that—a lot in a hurry. I learned that basically almost everybody goes home to a family, wife, and kids, so they're all the same whether they're floor sweeper or managing director of the whole. They all have the same basic fears and issues. 99.9% of the people you come across are all nice guys. But that also includes life's pressure which makes them do really weird things.

**What about family, house, or not really enough money?**
Some of these issues are pressures because people do not cope well with their problem. That was quite an education for an engineer who always worked in nice, clean facilities. That was a pretty good education. In fact, British Steel wasn't doing well, and I was in that once when I was being sent off somewhere in Shetfield. It was a dreadful summer. It must have been almost September. I was in this damned awful hotel. I opened up my copy of the paper and was reading it and flipped through the full sale and wanted ads and found a company that wanted a development engineer. I was so bored in my current job. I kind of just wrote a letter, and two weeks later, I was working for them.

In the meantime, while at college, I started with a guy who had a loaded race car. I raced it a little bit. We got a dedicated race car we ran for a year, then we modified it the following year, and then the year after that, which probably was about my first or second year after I got to college. We actually designed and built a whole car over the winter, our little club racing car. We did that for about 7 years. We'd do 20 to 24 races a year. From Easter through to mid- to the end of September and then in early September we'd start thinking what we were going to do next year and started drawing the plan up on weekday evenings after work. October and November time we'd still be drawing, but we'd start making bits. Come next Easter, we would build the car and run it again.

**So, it's your own drive that got you rolling on that. You're a self-motivated person.**
When I got that job with British Steel, I kind of dived into that. I guess I was suffering from motor racing withdrawal symptoms when I finally applied to my first racing team. When I applied and went there, it was kind of like racing for yourself really. All the same pressures, just you had more money and more people.

**What do you say to people or engineers who say they are in the long black tunnel, or the rut in their career? There are some engineers who are happy with what they are doing. Obviously, you are one of those people. And some people and engineers are in a career and they aren't happy. Some say that only a select few, 10% of the engineers, are out there doing the thing that they really love.**
I think part of it is a little bit of a choice that you make. I've seen some extremely talented people in jobs they may not have wanted. It is an element of luck and chance in anybody's life. None of us can control the future or predict it well. But I've seen people who I thought were hugely talented at what they did, some of them extremely bright and no openings existed for them within that company.

Maybe the economy was doing badly and there weren't very many other job opportunities. So, they'd get stuck though they could do a lot more, but there's just no avenue for them to go to. I've seen people whose first response is, "Well I just do my job." But their main focus typically becomes their family. They put a lot of effort into their family and do a great job with their family.

Sometimes, they just make a major career change. I've seen people who were tremendously talented engineering mathematicians who could solve almost any problem given a couple of weeks, paper, and a few pencils. To solve it mathematically, you sit down and analyze it, which doesn't get you far in the automotive world. I guess there are more people around who could do, but where are they are not well appreciated for what they can do for the company. Now they worry or aren't happy. First of all, they lower their expectations and focus on something else outside of work. If they still have a drive to be fulfilled with work, they change their career. I've seen a couple of people who got into teaching. These people went into the areas where there tends to be a shortage of skilled people. I've seen that a couple of times, and they're a lot happier. Typically, they'll move out of the industrial area to a place that's nice to live and have a nice lifestyle.

I don't know whether you're lucky or I'm lucky, but it is true that you have to realize that we all get promoted to our own level of incompetence. We end up where we are because that's about what we're able to do. If you don't have the drive, or if you're just doing something and the job opportunities don't seem to be coming your way, you need to ask, "What do I have to do to be better?" and go do it. If you're lucky and work for a good boss who will help you and help point out what is causing you not to do well, you will prosper. That's one of the things which a good boss should be doing. A boss should be counseling his or her people and saying, "If you want to do better, here are the things that are causing you not to do any better than you are." It's up to you whether you take notice and take action. If you don't take any action, you can't expect anything to get any better. In the end, you will get tossed around by the economic climate of our time, and I think all of us are exceedingly lucky to live in Western civilization, and we should never lose sight of that.

#### Do you think your bit of luck was that letter you wrote in the hotel room?

If I'd stayed at British Steel and decided it was not what I wanted to do, then I would have (written the letter). It's almost a subconscious process. I wasn't really looking for a job when I moved to British Steel, but I was reading the paper, and for some reason, the job advertisements became more interesting even though I wasn't consciously looking for a job. I saw this job at British Steel, and, boy this sounded pretty cool, so I applied and they gave it to me. I guess I applied for it because I wanted it, and that makes a difference.

#### What do you look for in an interview candidate?

When you interview people, you can tell if they want the job or not. If they want the job it's a big plus because they're going to bring that enthusiasm into the room. They need to have basic skills. I couldn't interview for a position as a brain surgeon even if I wanted to and even if I were enthusiastic during the interview. I'm sure I wouldn't get the job, but if you have the skills, carry enthusiasm, show you want to do it, and are sensible in the demands that you're asking in return for your skills, then you should have a good shot in getting the job. One thing that you need to tell your students preparing for an interview, is that people who interview you are ordinary people, and they have prejudices, likes, and dislikes like everybody else in the world. You have to realize that you may think you're the most wonderful person in the world, but the interviewers think the same of themselves, too. You must realize that you have to be sensible about the way you approach things. Going in ragged shorts, a tank top, and unshaven with flip flops on, on average, is not going to do as well as the same guy going in looking smart. You don't have to have a suit, collar, and tie on, but you have to look good. You have to look like you care about yourself. You need to be able to project that image that you care about yourself and care about how you interact with other people. Giving grunts in reply to questions doesn't really get you very far.

### What about away from the office, yet still in a professional setting?

When you're young, you're idealistic. I've been idealistic, too, but I had to learn. Maybe it's wrong to say that life is a game, but you have to understand the rules that make you successful or the guidelines make you successful. It's like I tell all my guys particularly when we go abroad. If you get on an airplane and fly to Europe, especially these days, and you look unkempt, be careful. If you look unkempt, you're much more likely to get pulled over by those guys in immigration than if you have a pair of khakis on, a nice pair of shoes, and a clean shirt before you get on the plane. It's a fact. You watch it happen all the time. Because they know on average those people do cause them more trouble once they come into their country than the guy who has a briefcase and a suit.

The thing I implore employers to consider is that they must look for somebody who has enthusiasm. If you have people who are unenthusiastic about their whole attitude toward work, that makes an employer think they don't want to be there. If you don't want to be there, that probably means you're not going to make it and do a good job in whatever you're supposed to do. So that's going to have a negative impact on how you're viewed, a negative impact on any chance of promotion, any chance of getting an increase—and, heaven forbid, a good chance of being downsized. Nobody wants a pain in the neck working for them.

### So what kind of things, aside from engineering do you like to do? You race cars on the side, too. Everybody says to get a job you love because then it won't be like a job. What kind of things do you do outside of work? Do you attend conferences? Do you interact with other engineers? Do you enjoy the setting you're in and interacting with people?

Over the years, I've found I prefer—as most do when they get more responsibility—to leave work completely behind at the end of the day. It's a little different here because when you're with a race team and on the road, you end up going to dinner with people you're working with, so you have unavoidable social interactions above and beyond the normal workday. Because you're all away and the people with you are the only people you know, if you're going to go out and have dinner, it's natural to go with them. Other than that, and it depends on the company you work for, but when I leave work, I like to walk away from it all and focus on a different world. That starts with your family. If you ask me what my hobbies are, my hobby is my family. What do I do for relaxation? Depending on how full of a year I have had, I enjoy traveling and going to see things. I enjoy art. I can't do it. My wife can. My wife paints very well. We enjoy that together. We enjoy travelling somewhere and then seeing the sights. Going somewhere in Europe or traveling in the States and going to some museums, looking around. I enjoy going to what you might call engineering museums, the Smithsonian and those kinds of things. I enjoy seeing how it is we got where we are. I enjoy reading about that. I enjoy reading philosophy, of which few books have been written on engineering philosophy. There are one or two which are interesting. Some of them are easy to read. I ought to give you the names of a couple I've come across. I think people, when they first start in engineering, get stumped with all these little things like Young's Modulus, ratio, static's and dynamics, etc.

New students don't understand how equations, etc., relate to the real world. Do you realize that fashion designers when they're working with different crafts are looking at Poisson's ratio? Poisson's ratio is how something drapes. You pull it this way, what happens to it that way. That's Poisson's ratio. How that cloth drapes when you use it in a fashion design. Does it cling to the shape it's on or does it hang and is immovable? Unfortunately, engineering is being made a dry topic, but it relates to everything we do. I like engineering not because it's engineering but because it tells me about the world, how we do things, how we can solve problems, and why things are the way they are. Every year, people develop more and better theories of this, that, and the other. Even if I could remember everything I was taught, I assume I couldn't pass my graduate engineering exams because things have moved on and people are being taught additional and different things.

One of the problems is there is so much knowledge around now that you have to work your way through the bare bones of it and see the truth sometimes. That's where practical engineering comes in. That's where part of the "hands on" is. To turn what it is you're learning into reality. That's why you do the laboratory experiments. It's all well and good doing stress and strain and fracture mechanics, but there's something about physically doing it yourself. The human experience. You learn a lot better. When you first do machine tool theory before going out on the shop floor, they tell you all the things you must not do. I guarantee you everybody goes and either does them or attempts to do them, and they learn the hard way. Most of us are lucky. Chances are that you don't get hurt; then again somebody does gets hurt badly. So, humans aren't good at learning off the written word sometimes.

### What's your favorite food?
Favorite food? I don't really have one. In this job, you have to like everything. I've eaten food in Japan, Hong Kong, Arabia, and France.

### What don't you like best?

I don't like stuff that's cold and gooey. Probably the food that I had the most diffi-culty ever eating was in Arabia. The delicacy, if you have lamb, is to be given the lamb's eyes. They give them to you just sitting on a plate. That was pretty hard.

### Are you known for a particular phrase?

Probably, but I don't what it is. You'd probably need to ask my wife. I'm sure she'd give you a whole list of irritating phrases that I use.

### What is your favorite color?

I guess it depends on the circumstances. There's a color suitable for most things. Part of that is understanding using the correct color in the right place, whether it be on a painting or in a room. I'm not good at that. I'm not good at choosing col-ors. Like all of us, when you walk into a room that is painted correctly or has the right colors coordinated, you think, hey this is a nice room but may not understand why. I recognize there is a process to doing that. It's not like I always have a red car or a green car. I do enjoy reading, but I don't have much time to do it these days. There's always stuff to read for work and then there's no other time leftover for reading. That's probably one of the problems, reading, which for me is a great pleasure to read a good book, and a good book could be anything that's well writ-ten. It could be a simple, well-constructed, well-written story that's easy to read, and you can almost speed read it and get a kick out of it. It could be something constructed well that gives you insight into what you didn't know or understand or thought you understood and turns out you really didn't. Books which raise one's curiosity and then satisfy it on some level are always remembered. So, I guess I'm one of those people who have what you might call a healthy interest.

### What was the university experience like when you attended?

When I went to the university, only 5–10% of people went to the university. It's changed a lot now. In the UK, it's more like it is in the United States now where most people aspire to go to some kind of college once they leave school. Because of the way that was and the people who went to the university, we were being trained to be a specific engineer if we took engineering. It was pretty clear. That's what I did. I did the classic course. I did that. I ended up essentially being an en-gineering mathematician early in my career all by doing engineering by numbers essentially.

Of the 5 or 10% who went to a university, a tiny percent went as engineers and the rest were doing science or history. A large number of people who were going into engineering would go and get some kind of apprenticeship. Apprenticeship wouldn't necessarily be a craft apprenticeship. The ones who had the aptitude would end up in the drawing office for a year or two, would get sent to night school, and would get a pretty high technical education, certainly 2 or 3 years of the uni-versity education. They went through all the basics of the numbers. But they were typically being trained, for instance, to be a design drafter. They were taught how to create the perfect piece of paper to get something made with no chance of error. When you go to the university, you don't get taught like that. That's not part of the curriculum. You do a little bit of drawing, but they don't teach you every system. They might touch on it, and you know kind of what it is, but you don't get taught that if you draw a hole in a design. You can't expect it to be better than 5–10 thousandths in tolerance. But if you ream it, you'll get to within 2 thousandths and if you need better than that, you're going to have to devise some other way. You will probably touch on certain things, but you never get to the nitty gritty of when you use and how you use, etc. These students go through all that in detail. When they come out of it, maybe they do a couple of years actually producing drawings. By the time they're in their mid-twenties, they make a drawing for the part, and they know that it will fit the other piece without question and that there's no way to make that part other than the way that they've shown it. Even in the UK, that whole mid-range of people are absolutely critical to the part of making anything in engineering.

Now everybody goes to a university. They do the highly academic course without much practical stuff and they come out. Where do you recruit your design draftsmen from? Well, that's the only source you have. But they don't know the job at all, and it's hard for a small company to teach somebody on the fly. It's all right if you're Ford or GM, you can send them off to GM Design School for 3 months, and they can design on the fly. The companies have all these standards and they'll go through them all page by page and when the students come out, at least they'll know what they should do. Whether they do it or not is it another matter. The other thing I've noticed is when you talk to draftsmen about drawing manually is that you'd better make sure that every line that you put on there is right because if you make a mistake, it's difficult to fix it. It costs you personal pain. You'd better plan ahead and know what it is you're doing. When you're drawing manually, and you know that there are certain features that you have to explain on the drawing to get the piece made the way you want it, you plan ahead and make sure you do the minimum amount of drawing to explain that because it takes a lot of time. Whereas in computer-aided design, you just project another view, another view, and another view. You end up with a drawing that's on a piece of paper four times the area it needs to be with tons of views. Going through and checking that in order to manufacture something is difficult. Small companies like ours have a serious disconnect in that area of engineering.

### Were you taught estimation in your university experience? How has estimation helped you in your career?

I always feel like somebody understands their job if you give them a problem that they can solve on one sheet of paper. One example was when I was working on the Volkswagen Golf that had just come out. It had the twisted beam rear suspension, so I was asked to look at that and work up a report on it. The first thing I did was research how the suspension affects the handling. We still didn't have that much computer power then, so I just did it the rough way, asking what the actual geometry was and how would it affect the car. We just did a few hand calculations on paper for that analysis. Then we got to the stage where we were using our finite element analysis (FEA) computer program extensively. I actually made the model and put it in the program. I remember that took me probably 7 working days. I remember my last day at work there with the report done, I thought, I better just make sure this model is realistic. So I sat down and did some hand calculations. I thought to myself this is roughly the section, this is roughly that section, and so forth. When I was finished with the hand calculations, it was within 5% of what the computer program said it would be. Well, given that I know my FEA program isn't accurate anyway, wouldn't it have been easier to have done the hand calculations in the first place? But, equally, I was pleased that the hand calculations and the FEA results were about the same. It told me that I understood what I was doing, and that's important. If you understand what you're doing, you can things done quickly.

Engineers always talk about being able to do what they love. The guy who is the head of Lockheed's Skunk Works is lucky. As described by former director Ben R. Rich, a thermodynamicist who took over the Skunk Works in 1975 and then retired in 1990, the Skunk Works is a "small, intensely cohesive group made up of about fifty veteran engineers and designers and around one hundred expert machinists and shop workers." Their purpose is to build a small number of technologically advanced airplanes for secret missions.

In 1933, Clarence Johnson, better known as Kelly, joined Lockheed Martin as a 23-year-old engineer to help design the twin engine Electra transport. Before starting the Skunk Works during World War II in Burbank, California, Kelly aided in the design of the Lockheed Constellation and the renowned P-38 Lightning. How did he do the U2 and the SR-71? He was the genuine Chief Engineer. Kelly understood his job. He was obviously a bright guy, he understood materials science, mechanics, aerodynamics, and he could bring all those together in his mind just with a few sketches. He knew his job well enough that he could process most of it. Then, of course, you have to do all the details. It's all very well having a picture in your mind, but you have to make the darn thing and make sure all the individual parts fit together and don't break. He understood what he was doing. Everybody in the aerospace world holds him in great awe and respect because he was a clever guy, had a tremendous impact. He understood his job and applied his understanding to whatever it was he had to solve at the time. He designed a few of the great aircraft, the U2 which flies 200–300 miles an hour and loiters around as a spy plane and the SR-71 which flies at mach 3 or greater. He's a guy who understood his job.

People who are good at design usually know what it is going to look like before they put a line on a piece of paper. Sometimes you have a problem to solve and you kind of solve it by drawing it. I'll put this here and that here, and it comes together. When I was involved with materials, we designed a car, manufactured it, and went to race it. But certainly on the design side, which I know is probably what you're particularly interested in, when we started the new car I pretty much was aware in my own mind's eye how it was all going to fit together before we even started. I knew where the gear shift rung was going to be. I knew how we were going to fit the clutch cable through. I had a good idea how we were going to the exhaust system to make it fit under the kind of body work the aero guys wanted. I knew how we were going to lay the radiators out and so on.

**So good engineers should be good "mental modelers."**
Yeah. Some people can think in three dimensions easily, can walk around what it is they see in their mind's eye. Some people can hardly do something in 2D let alone manipulate that in their mind's eye.

**You lead a pretty interesting life and have had plenty of experiences.**
Oh yeah.

Brian Lisles was born on October 9, 1945, in London. He and his wife Judy reside in Lake Zurich, Ill. The couple has two sons and two granddaughters.

# *Chapter 17*

# Computer Tools

## Introduction

Assume that you are a food process engineer, and a dairy company has hired you for a project:

> *"Design a process to manufacture high-quality powdered milk for a profit."*

This is no small undertaking! Table 17.1 shows a list of the steps that you must complete.

With this huge project ahead of you, you realize that you need help. So you turn to the computer. Computer tools can facilitate nearly every task you need to perform. These tools include operating systems, programming languages, and computer applications. For example, word processing applications enable computer users to write papers. To be able to use the computer to solve an engineering problem, you need strong computer skills and a working familiarity with the types of computer tools and applications that are available. Table 17.1 indicates the types of computer tools you would use to complete this project.

**Table 17.1 Steps Required to Complete the Powdered-milk Process Design**

| *Step* | *Computer Tool* |
| --- | --- |
| Research the dairy industry | Internet |
| Design the process | Mathematical Software |
| | Design Software |
| | Programming Languages |
| Select equipment | Internet |
| Perform economic analyses | Spreadsheet |
| | Mathematical Software |
| Analyze experimental data | Spreadsheet |
| | Mathematical Software |
| Document the design process | Word Processor |
| Present findings to management | Presentation Software |

The following sections of this chapter will highlight some of the computer tools that an engineer might use to solve problems. Many of the tools will be discussed in the context of solving the dairy design problem.

## THE INTERNET

### Introduction

The first step required to solve the dairy design problem is to fill gaps in your knowledge base. That is, you need to do research so that you have a better understanding of the design problem. Some things you should learn are:

- How is powdered milk manufactured?
- What are the consumer safety and health issues involved?
- Who manufactures equipment for dairy processing?
- What other dairy companies manufacture powdered milk?
- Who does dairy processing research?

Research, under the best conditions, can take a significant amount of time. So your aim is to avoid excessive trips to the library and the use of bound indexes. It is preferable to access as much information as possible via your computer. The Internet provides such capabilities.

### History of the Internet

The Internet is a worldwide network of computers that allows electronic communication with each other by sending data packets back and forth. The Internet has been under development since the 1960s. At that time, the Department of Defense was interested in testing out a concept that would connect government computers. It was hoped that such a network would improve military and defense capabilities (Glass and Ables, 1999). Before the end of 1969, machines at the University of Southern California, Stanford Research Institute, the University of California at Santa Barbara, and the University of Utah were connected.

The linking of those four original sites generated much interest in the Internet. However, extensive customization was required to get those original sites linked, because each site used its own operating system. It was apparent that the addition of more sites would require a set of standard protocols that would allow data sharing between different computers. And so the 1970s saw the birth of such protocols.

*Protocols* are used to move data from one location to another across the network, and verify that the data transfer was successful. Data are sent along the Internet in *packets*. Each packet may take a different route to get to its final destination. The *Internet Protocol* (IP) guarantees that if a packet arrives at its destination, the packet is identical to the one originally sent. A common companion protocol to the IP is the *Transmission Control Protocol* (TCP). A sending TCP and a receiving TCP ensure that all the packets that make up a data transfer arrive at the receiving location.

Internet applications were also developed in the 1970s and are still used today. The Telnet program connects one computer to another through the network. *Telnet* allows a user on a local computer to remotely log into and use a second computer at another location. The *ftp* program allows users to transfer files on the network. Both applications were originally non-graphical; they require that text commands be entered at a prompt.

*Internet IP addresses* were established in the 1970s. All organizations that become part of the Internet are assigned a unique 32-bit address by the Network Information Center. IP addresses are a series of four eight-bit numbers. For instance, IP addresses at Purdue University take the form 128.46.XXX.YYY, where XXX and YYY are numbers between zero and 255.

In the 1980s, corporations began connecting to the Internet. Figure 17.1 shows the growing number of sites connected to the Internet. With so many sites joining, a new naming system was needed. In came the *Domain Naming Service*. This hierarchical naming system allowed easier naming of individual hosts (servers) at a particular site. Each organization connected to the Internet belongs to one of the top-level domains (Table 17.2). Purdue University falls under the U.S. educational category; therefore, it has the domain name purdue.edu. Once an organization has both an IP address and a domain name, it can assign addresses to individual hosts. This author's host name is pasture.ecn.purdue.edu. "Pasture" is a

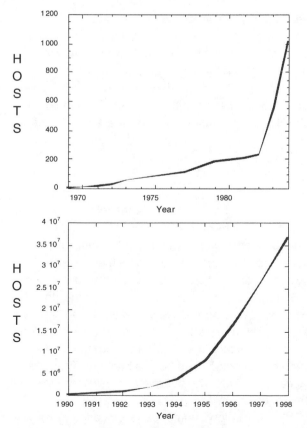

**Figure 17.1 Number of Internet hosts by year.**

**Table 17.2 Selection of Top-Level Domain Names**

| *Name* | *Category* |
|---|---|
| edu | U.S. Educational |
| gov | U.S. Government |
| com | U.S. Commercial |
| mil | U.S. Military |
| org | Non-profit Organization |
| net | Network |
| XX (e.g., fr, uk) | Two-Letter Country Code (e.g., France, United Kingdom) |

large computer (server) on the Purdue University Engineering Computer Network (ecn), a network of engineering department servers.

As *Internet Service Providers* (ISP), such as America On-Line and CompuServe, came into being in the 1990s, the Internet became more mainstream. The development of the browser named Mosaic made accessing information on the Internet easier for the general public. A *browser* is software that allows you to access and navigate the Internet. Mosaic displays information (data) using both text and graphics. Each page of information is written using the *HyperText Markup Language* (HTML). The most useful aspect of HTML is the hyperlink. A *hyperlink* connects two pages of information. The beauty of the hyperlink is that it allows an Internet user to move through information in a non-linear fashion. That is, the user can "surf" the *World Wide Web* (WWW). The Web, not to be confused with the Internet, refers to the interconnection of HTML pages. Today, the most common browsers are Netscape Navigator and Microsoft Internet Explorer. Updates to these browsers have allowed for the construction of more complicated web pages, such as multiple-frame pages. Sound and video can now be played through the use of helper applications. Greater user interaction with the Web has been made possible through HTML coded references to Java applets (defined later).

## Searching the Web

Let's return to the powdered-milk problem. You'd like to use the Internet to answer some of your research questions. First focus on searching for equipment vendors. Assume that you already know about one company that supplies food processing equipment, called APV. Check to see what kinds of equipment they have available for dairy processing. Using your Web browser, you access APV's web site by providing the URL (*Universal Resource Locator*) for APV's home page. In other words, you tell the Web browser what Web address to use to access APV's information. The URL is "http://www.apv.com." This URL consists of the two primary components of a Web address: the protocol (means for transferring information) and the Internet address. The protocol being used here is http (Hypertext Transport Protocol) which allows access to HTML. The Internet address consists of the host name (apv) and the domain (com).

APV's URL take you to APV's introductory (home) page. Once there, you can browse their Web site for information that pertains to your specific research problem. At one point, you find yourself on a page that describes a one-stage spray dryer designed specifically for temperature-sensitive materials such as milk. A spray dryer is a piece of equipment that converts liquid into a powder, or as in your application, milk to powdered milk. The URL for this page is http://www.apv.com/anhydro/spray/onestage.htm. This Web address is longer than that for APV's home page. The address contains a path name (/anhydro/spray/) and a file name (onestage.htm). The path name tells you the location within APV's directory structure of the file that you are accessing with your Web browser, just as you store files in directories on your hard drive.

APV is just one equipment vendor. You'd like to investigate others, but you do not know any other Web addresses. To further your research, you use a Web search engine. A *search engine* matches user-specified key words with entries in the search engine's database. The engine then provides a list of links to the search results. There are a number of search engines such as Yahoo! (www.yahoo.com) and Excite (www.excite.com). A keyword search performed using one search engine will yield different results than the same keyword search on another search. The reason for the different results lies in the type of information that a particular search engine tends to have in its database.

You decide to use the search engine called Lycos (www.lycos.com) to do a keyword search for "spray dryer." When your search results come up, you immediately notice that your query was not specific enough. You get back links to sites on laundry dryers! There are so many sites on the Internet that well-constructed queries are a must. Each search engine provides links to searching tips. In your case, you find that a keyword search for "spray dryer and food" improves your search results.

## Home Pages

*A personal reason for creating your own home page is to make available your resume and samples of your work to all Web users.*

*Home pages* are Web pages developed by users that contain information about topics of interest to the user, and often include links to the user's favorite Web sites. The language of all Web pages is HTML. An *HTML document* is a text file containing information the user wants posted to the Web, and instructions on how that information should look when the Web page is accessed by a browser. HTML code can be written from scratch using a text editor, or a composer can be used to generate the code. Many word processors and desktop publishers now have the capability of saving files in HTML format so that the document is displayed by the Web browser exactly as it was in the word processing program. The HTML code consists of instructional tags as seen in Figure 17.2. This code generates the Web page shown in Figure 17.3.

Your employer may ask you to design a Web page discussing the process you developed through the completion of the powdered-milk process design project. A personal reason for creating your own home page is to make available your resume and samples of your work to all Web users. This sort of visibility may result in employment opportunities such as internships and post graduation positions.

```
<HTML>
<HEAD>
<TITLE>FOODS BLOCK LIBRARY—BASIC ABSTRACT</TITLE>
</HEAD>

<BODY BGCOLOR="#FFFFFF">
<H1>A Computer-Aided Food Process Design Tool for
Industry, Research, & Education</H1>
<H2>Heidi A. Diefes, Ph.D.</H2>
<H2><I>Presented at the Food Science Graduate Seminar<BR>
Cornell University, Fall 1998</I></H2>
<HR ALIGN=LEFT NOSHADE SIZE=5 WIDTH=50%>
<P>
There is a great need in the food industry to link food science research with food
engineering and process technologies in such a way as to facilitate an increase in
product development efficiency and product quality. Chemical engineers achieve
this through industry standard computer-aided flowsheeting and design packages.
In contrast, existing programs for food processing applications are limited in their
ability to handle the wide variety of processes common to the food industry. This
research entails the development of a generalized flowsheeting and design pro-
gram for steady-state food processes which utilizes the design strategies em-
ployed by food engineers.
<P>
</BODY>
</HTML>
```

**Figure 17.2 HTML code for Web page shown in Figure 17.3.**

More information on creating home pages can be found on the Web or in any book on HTML or the Internet. You may find "HTML: The Complete Reference" by T. A. Powell (1998) to be helpful when writing HTML code.

## Libraries & Databases on the Internet

Not all information is currently made available through the Internet, especially highly technical information. Although the Internet may not make such information available in every instance, it can be used to locate the information. Most universities and many public libraries maintain on-line catalogs and links to electronic databases. For instance, Purdue University maintains Thor: The Online Resource (http://thorplus.lib.purdue.edu/index.html) which provides access to the school library catalog, indexes, and electronic databases. One of the many database links to which Thor provides access is ArticleFirst (Copyright © 1992–1998 OCLC Online Computer Library Center, Inc.), a database for journals in science, technology, medicine, social science, business, the humanities, and popular culture.

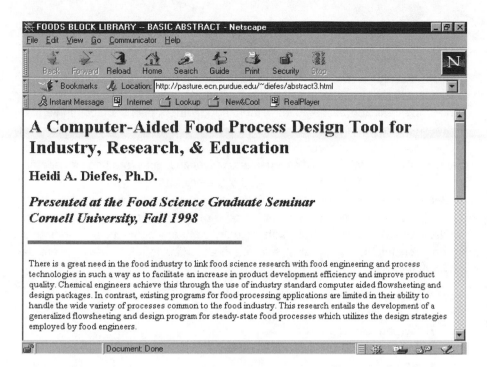

Figure 17.3 Web page generated using HTML code in Figure 17.2.

## Contacting People through the Internet via E-mail

*MIME will enable users to send formatted text messages, programs, audio files, encapsulated messages, multipart messages, videos, and image data via e-mail.*

The Internet can also be used to send *electronic mail* (e-mail) to individuals, groups, or mailing lists. E-mail is currently exchanged via the Internet using the *Simple Mail Transport Protocol* (SMTP). However, a new standard is currently under development called *Multipurpose Internet Mail Extension* (MIME). MIME will enable users to send formatted text messages, programs, audio files, encapsulated messages, multipart messages, videos, and image data via e-mail as easily as plain text (James, 1999). MIME is a mail system that identifies file contents by the file extension. For example, MIME can identify HTML files from the .html or .htm extension. Sending and receiving MIME files requires that both the sender and receiver have MIME-compliant mail tools.

To send e-mail, two things are required: an e-mail package and e-mail addresses. There are many e-mail packages from which to choose. PC and MacIntosh users might select PEGASUS or Eudora; UNIX users might select PINE or ELM. *Graphical user interfaces* (GUIs) and Web *browsers* provide their own e-mail tools. Regardless of the brand, e-mail tools have certain common functions including an ability to send mail, receive mail, reply to mail, forward mail, save mail to a file, and delete mail.

To communicate via the Internet, an *e-mail address* is required. All e-mail users are assigned a unique e-mail address that takes the form username@domain.name. However, it can be difficult to locate a person's e-mail address. One method of locating an e-mail address is to use one of the many people-finding Web sites. For example, the Netscape People Finder can be used to find both home and e-mail addresses. To learn more about other options for locating e-mail addresses, see *Introduction to the Internet* by Scott James (1999). A Web site that allows you to search many people-finding sites at once is http://www.theultimates/white. In addition, most colleges and universities have methods for searching for staff and students on the eb page of the school.

Once you have selected an e-mail address and e-mail software, and after you have signed up with an ISP, you can send e-mail. As part of your dairy design problem, you may keep in touch with the dairy company contacts and other engineering consultants via e-mail. A typical e-mail message might read like the example that follows.

**To:**   dairyconsult@purdue.edu
**cc:**   dairycontacts@purdue.edu
**Attachments:**   spdry.bmp
**Subject:**   Spray dryer design model
**Message:**
I have been working with the spray dryer vendor on a model of the system we intend to use. Attached is a current picture of the model. Please advise.
H. A. Diefes

The above e-mail message contains both the typical required and optional components. An e-mail is sent:

**To:** username@domain.name or alias. A copy (**cc:**) may also be sent to a username@domain.name or alias.
**Attachments:** are files to be appended to the main e-mail message.
**Subject:** is a brief description of the message contents.
**Message:** is the actual text of the message.

An *alias* is a nickname for an individual or group of individuals, typically selected by the "owner." An e-mail to an alias gets directed to the actual address. Address books, another common feature of e-mail packages, allow the user to assign and store addresses and aliases. For example, in your address book you may group all the e-mail addresses for your dairy company contacts under the alias "dairycontacts." If you put "dairycontacts" in the **To** or **cc** line of your mail tool, the e-mail message will be sent to everyone on the "dairy contacts" list.

## Words of Warning about the Internet

With millions of people having access to the Internet, one must be very discerning about the information one accesses. Common sense and intuition are needed to determine the validity of material found on the Web.

In addition, care must be taken not to plagiarize materials found on the Web. As with hard copy materials, always cite your sources.

Further, computer viruses can be spread by executable files (.exe) that you have copied. The virus is unleashed when you run the program. Text files, e-mail, and other non-executable files cannot spread such viruses. Guard against viruses attached to executable files by downloading executable files only from sites that you trust.

## WORD PROCESSING PROGRAMS

Throughout the powdered-milk process design project, you will need to do various types of writing. The largest document will be the technical report for the design. You also may have to complete a non-technical report for management. In addition, you may write memorandums and official letters, and you will need to document experimental results. The computer tool that will enable you to write all these papers is a *word processor.*

The most basic function of a word processor is to create and edit text. When word processors first became available in the early 1980s, that was about the extent of their ability. The user could enter, delete, copy, and move text. Text style was limited to bold, italic, and underline, and text formats were few in number.

"Word processing," as understood today, is really a misnomer, since the capabilities of word processors have become indistinguishable from what was once considered desktop publishing. Word processors such as Microsoft Word, Corel WordPerfect, and Adobe FrameMaker have features that are particularly useful to engineers. Equation, drawing, and table editors are features which make the mechanics of technical writing relatively easy. Many programs include spelling and grammar checkers and an on-line thesaurus to help eliminate common writing errors.

> *Equation, drawing, and table editors are features which make the mechanics of technical writing relatively easy.*

Figure 17.4 shows the graphical user interface (GUI) for Microsoft Word. All word processors display approximately the same components. The title bar indicates the application name and the name of the file that is currently open. The menu bar provides access to approximately 10 categories of commands such as file, edit, and help commands. Some word processors have a toolbar that gives direct access to commands that are frequently used such as "save," "bold," and "italic." A typical word processor GUI also has scroll bars and a status bar. The scroll bars allow the user to move quickly through the document. The status bar provides information on current location and mode of operation. The workspace is where the document is actually typed. The workspace in Figure 17.4 shows a snippet of the technical information related to the spray dryer of our example.

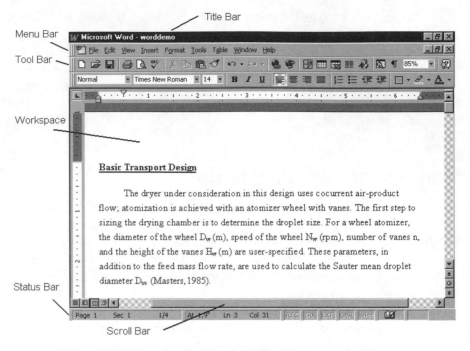

**Figure 17.4 Example of a Microsoft Word document in progress.**

This text you are reading was written using word processors. Throughout these pages you can find examples of tables, equations, and figures created using a word processor. A wide variety of typefaces, paragraph styles and formats are also in evidence. You also will see images and pictures that have been imported into this document. A user may import materials that are created using other computer applications. For instance, the photos of practicing engineers in the Profiles Chapter were image files sent to the author for inclusion in his chapter.

Documenting is an ongoing process. you will need to return time and again to your word processor to record all subsequent steps of the dryer design process. Periodically, you will use the word processor to write memos and official letters and to prepare experimental reports, technical documents, and non-technical papers.

## SPREADSHEETS

### Introduction

*The first spreadsheet came in 1978 from a student who was frustrated by the repetitive nature of solving financial-planning problems.*

Much data collection is performed during the research and design phases of an engineering project. Data come from a variety of sources including experiments, design calculations, product specifications, and product statistics. While designing the powdered-milk process, you have gathered cost information from your equipment vendors and collected experimental data on powder solubility. you may like to use the cost information in an economic analysis of your process; (recall that your process is supposed to make a profit). you may also attempt to analyze your laboratory data. Due to the tabular nature of these data sets, this sort of information is best recorded with a spreadsheet.

*Spreadsheet* packages have their origins in finance. The idea for the first spreadsheet application, called VisiCalc, came in 1978 from a Harvard Business School student who was frustrated by the repetitive nature of solving financial-planning problems (Etter, 1995). VisiCalc was originally written for Apple II computers. IBM came out with its own version of VisiCalc in 1981. In 1983, a second spreadsheet package was released by Lotus Development Corporation, called Lotus 1-2-3. While Lotus 1-2-3 remains popular today, there are a number of other spreadsheet packages from which to choose, including Microsoft Excel, Corel Quattro Pro for PC, and Xess for UNIX. One thing to note about the evolution of the spreadsheet is the increase in functionality. VisiCalc centered on accounting math: adding, subtracting, multiplying, dividing, and percentages. Spreadsheets today have more than 200 different functions.

## Common Features

A spreadsheet or worksheet is a rectangular grid which may be composed of thousands of columns and rows. Typically, columns are labeled alphabetically (A, B, C, . . . , AA, AB, . . . ) while rows are labeled numerically (1, 2, 3, . . . ). At the intersection of a given column and row is a cell. Each cell has a column-row address. For example, the cell at the intersection of the first column and first row has the address A1. Each cell can behave like a word processor or calculator in that it can contain formattable text, numbers, formulas, and macros (programs). Cells can perform simple tasks independently of each other, or they can reference each other and work as a group to perform more difficult tasks.

Figure 17.5 shows the GUI for Microsoft Excel. Notice that many of the interface features are similar to those of a word processor, such as the title bar, menu bar, toolbar, scroll bars, and status bar. Many menu selections are similar as well, particularly the "File" and "Edit" menus. The standard features of all spreadsheets are the edit line and workspace. The edit line includes the name box and formula bar. The name box contains the active cell's address or user-specified name. Within the worksheet in Figure 17.5 is the beginning of the equipment costing information and economic analysis of our example.

Today's spreadsheets have many features in common including formatting features, editing features, built-in functions, data manipulation capabilities, and graphing features. Let's look at each one of these areas individually.

As with word processing, the user has control over the format of the entire worksheet. Text written in a cell can be formatted. The user has control over the text's font, style, color, and justification. The user can also manipulate the cell width and height as well as border style and background color.

Editing features are also similar to a word processor. Individual cell contents can be cleared or edited. The contents of one or more cells can be cleared, copied, or moved. Excel even has a spell checker.

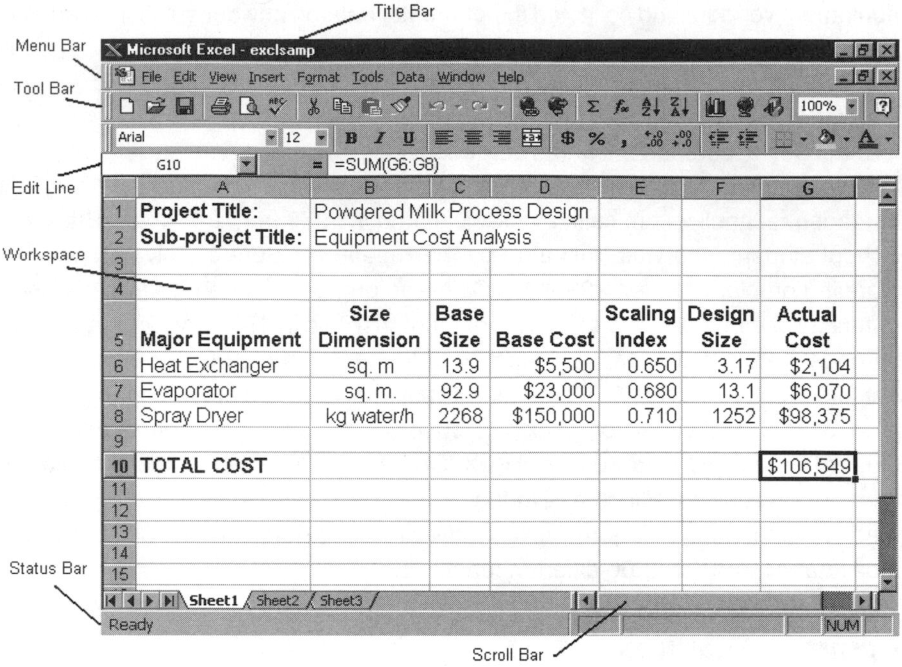

Title Bar

Menu Bar

Tool Bar

Edit Line

Workspace

Status Bar

**Figure 17.5 Example of a Microsoft Excel spreadsheet.**

Spreadsheets have categories of built-in functions. At a minimum, spread-sheets have arithmetic functions, trig functions, and logic functions. Arithmetic functions include a function to sum the values in a set (range) of cells (SUM) and a function to take an average of the values in a range of cells (AVERAGE). Trig functions include SIN, COS, and TAN, as well as inverse and hyperbolic trig functions. Logic (Boolean) functions are ones that return either a true (1) or false (0) based on prescribed conditions. These functions include IF, AND, OR, and NOT. You might use a logic function to help you mark all equipment with a cost greater than $100,000 for further design consideration. The formula to check the heat exchanger cost would be IF(G6>100000," re-design," okay"). This formula would be placed in cell H6. If the value in cell G6 is greater than 100,000, the word *true* is printed in cell H6. Otherwise, the word *false* is placed in cell H6.

Spreadsheet developers continually add built-in, special purpose functions. For instance, financial functions are a part of most spreadsheets. An example of a financial function is one that computes the future value of an investment. In addition, most spreadsheets have a selection of built-in statistical functions.

When a user wants to perform repeatedly a task for which there is not a built-in function, the user can create a new function called a macro. A *macro* is essentially a series of statements that perform a desired task. Each macro is assigned a unique name that can be called whenever the user has need of it.

*A database is a collection of data that is stored in a structured manner that makes retrieval and manipulation of information easy.*

Spreadsheets also provide a sort function to arrange tabular data. For instance, vendor names can be sorted alphabetically, while prices can be sorted in order of increasing or decreasing numeric value.

Graphing is another common spreadsheet feature. Typically, the user can select from a number of graph or chart styles. Suppose that you are interested in studying ways to reduce the energy it takes to operate the spray dryer. Figure 17.6 is a Microsoft Excel-generated *x–y* plot of such an analysis. Here you are investigating the impact that recycling hot air back into the dryer has on steam demand.

## Advanced Spreadsheet Features

Database construction, in a limited fashion, is possible in some spreadsheet applications such as Excel. A database is a collection of data that is stored in a structured manner that makes retrieval and manipulation of information easy. A database consists of records, each having a specified number of data fields. Typically, the user constructs a data entry form to facilitate entry of records. Once a database has been constructed, the user can sort and extract information. you could use the database feature to catalog all expenses during implementation of your powdered-milk process. This would require a record for each piece of equipment purchased. Data fields would include size, quantity, list price, discounts, and actual cost of each item.

Real-time data collection is another advanced feature of spreadsheets. This feature allows data and commands to be the received from, and sent to, other applications or computers. During the design phase of the powdered-milk process, you might use this feature to log data during pilot plant experiments. Later, you might use this feature to create a spreadsheet that organizes and displays data coming from on-line sensors in your powdered-milk process. This would enable line workers to detect processing problems quickly.

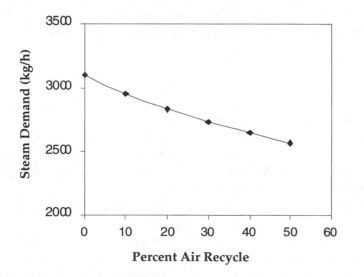

**Figure 17.6 An *x-y* plot generated using Microsoft Excel's Chart Wizard.**

## MATHEMATICS SOFTWARE

Mathematical modeling is a large component of the design process. For the powdered-milk process, you may need to develop a mathematical model of parts of the process. Problems of this nature are best solved using mathematical or computational software.

## MATLAB

*MATLAB enables the user to manage variables, import and export data, perform calculations, generate plots, and develop and manage files.*

MATLAB is a scientific and technical computing environment. It was developed in the 1970s by the University of New Mexico and Stanford University to be used in courses on matrix theory, linear algebra, and numerical analysis (MathWorks, 1995). While matrix math is the basis for MATLAB ("MATrix LABoratory"), the functionality of MATLAB extends much farther. The MATLAB environment enables the user to "manage variables, import and export data, perform calculations, generate plots, and develop and manage files for use with MATLAB" (Palm, 1999). Simply put, MATLAB can be used like a calculator; it can be used like a programming language; it can be used as a data analysis tool.

MATLAB is best explained through an example. Consider again the powdered-milk design. As with any food process, you must be concerned about consumer safety. The best way to ensure that the dairy company's customers do not get food poisoning from this product is to pasteurize the milk before drying. The first step in pasteurization is to heat milk to a specified temperature. Milk is typically heated in a plate heat exchanger where a heating medium (e.g., steam) is passed on one side of a metal plate, and milk is passed on the other side.

The required total surface area of the plates depends on how fast the heat will be transferred from the heating medium to the milk, and how much resistance there is to that heat transfer. Mathematically, this is represented by

$$A = Q \times R$$

where $A$ is the surface area of the plates (m$^2$), $Q$ is the rate of heat transfer (W), and $R$ is the resistance to heat transfer (m$^2$/W).

From your previous design calculations, you know that the rate of heat transfer is $1.5 \times 10^6$ W. The resistance to heat transfer range can be any value from $4 \times 10^{-6}$ to $7 \times 10^{-6}$ m$^2$/W. You may wish to study how the required surface area changes over the range of the resistance.

A typical MATLAB session would resemble that in Figure 17.7. The plot generated by these commands is shown in Figure 17.8.

Note that MATLAB has a text-based interface. On each line, one task is completed before going on to the next task. For instance, on the first line, the value of $Q$ is set. On the second line, the variable $R$ is set equal to a range of values from $4 \times 10^{-6}$ to $7 \times 10^{-6}$ in steps of $1 \times 10^{-7}$. On the third line, the arithmetic operation

```
>     Q = 1.5e6;
>     R = 4e-6:1e-7:6e-6;
>     A = Q.*R;
>     plot(R,A,'s-');
>     xlabel('Area m^2');
>     ylabel('Resistance m^2/W');
```

**Figure 17.7 MATLAB code for determining heat transfer surface area.**

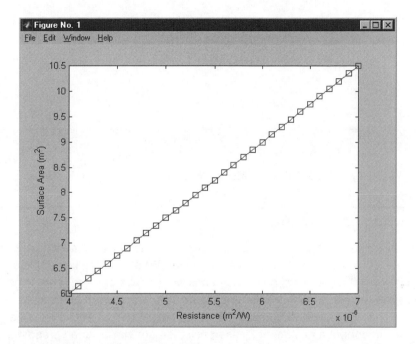

**Figure 17.8 MATLAB plot as seen on a UNIX platform.**

$Q \times R$ is performed. On command lines 4 through 6, MATLAB's built-in plotting commands are used to generate a plot of the results.

The commands shown in Figure 17.7 could be saved as a file with a ".m" extension (MATLAB file) and run again and again, or edited at a future date. MATLAB has a built-in editor/debugger interface that resembles a very simple word processor (Figure 17.9). The debugger helps the user by detecting syntax errors that would prevent the proper execution of the code. The editor shown in Figure 17.9 is displaying the code for the heat transfer area calculation.

MATLAB maintains, and periodically adds, toolboxes that are groups of specialized built-in functions. For example, one toolbox contains functions that enable image and signal processing. There is also a Symbolic Math Toolbox that enables symbolic computations based on Maple (see next paragraph). SIMULINK is another toolbox, a graphical environment for simulation and dynamic modeling. The user can assemble a personal toolbox using MATLAB script and function files that the user creates.

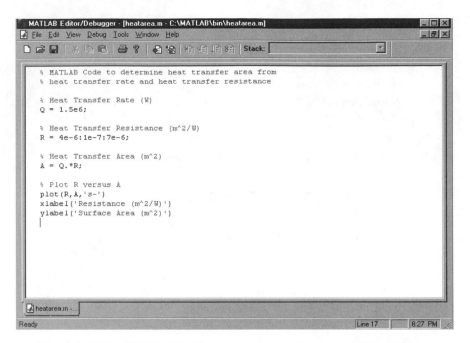

**Figure 17.9 The MATLAB editor as seen operating on a UNIX platform.**

## MathCad and Maple

MathCad and Maple are graphical user interface-based mathematics packages. Each is 100% WYSIWY*G*—computer jargon for "What You See Is What You Get." That is, what the user sees on the screen is what will be printed. There is no language to learn since mathematical equations are symbolically typed into the workspace. The workspace is unrestrictive—text, mathematics, and graphs can be placed throughout a document. These packages are similar to a spreadsheet in that updates are made immediately after an edit occurs. MathCad and Maple feature built-in math functions, built-in operators, symbolic capabilities, and graphing capabilities. Perhaps the greatest advantage of MathCad is its ability to handle units. In MATLAB, the user is charged with keeping track of the units associated with variables. In MathCad, units are carried as part of the assigned value.

You could work the same heat transfer problem in MathCad that you did using MATLAB. Figure 17.10 shows the GUI for MathCad and the text used to solve the problem.

## Example 17.1

The key to being able to use MATLAB to solve engineering problems effectively is to know how to get help with MATLAB's many commands, features, and toolboxes. There are three easy ways to access help.

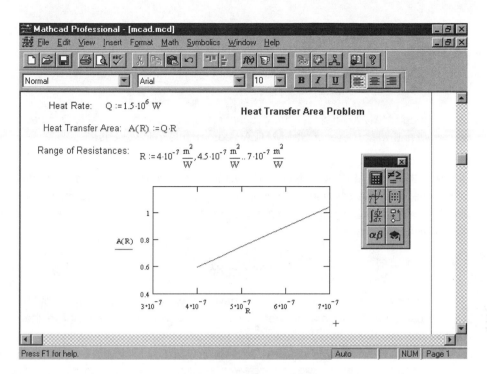

**Figure 17.10 Example of a MathCad window and problem solving session.**

**Method 1**—Access the hypertext documentation for MATLAB. At the MATLAB prompt type:

>> helpdesk

**Method 2**—Use the help command. The help command can be used in three ways. Typing help at the MATLAB prompt will generate a list of all available help topics. Try it out.

>> help

Near the top of this list, is a topic called matlab\specfun. To see the commands included in a given topic the general format is to type >> help topic. To see the commands included in specfun, type:

>> help specfun

This will generate a list of commands under that topic heading. To get help on a specific command, type >> help command. Let's see how the MATLAB command factor is used:

>> help factor

The following statement should appear:

> FACTOR Prime factors.
> FACTOR(N) returns a vector containing the prime factors of N.
> See also PRIMES, ISPRIME.

Even though MATLAB displays the description in capital letters. MATLAB is case sensitive and will only recognize MATLAB commands in lower case.

**Method 3**—MATLAB has a means of looking for key words in the command descriptions. This enables you to discover command names when you only know what type of operation you wish to perform. For instance, I may wish to learn how to plot. So I would type:

> >> lookfor plot

A list of all the available commands that have the word plot in their descriptions will be generated. How many different commands have the word plot in their description?

# Example 17.2

MATLAB is designed to work with matrices. Therefore, by default, matrix math is performed. Given the following two matrices:

$$\mathbf{A} = \begin{bmatrix} 1 & 0 & 4 \\ 7 & 9 & 3 \\ 3 & 6 & 9 \end{bmatrix} \quad \mathbf{B} = \begin{bmatrix} 9 & 0 & 1 \\ 3 & 4 & 5 \\ 5 & 9 & 7 \end{bmatrix}$$

**Problem 1**—Find the product of A and B (Section 19.5):

> % Assign each matrix to a variable name
> A_matrix = [1 0 4; 7 9 3; 3 6 9]
> B_matrix = [9 0 1; 3 4 5; 5 9 7]
> % Perform matrix multiplication to find C
> C_matrix = A_matrix * B_matrix

The result in MATLAB will be:

```
C_matrix =
    29      36      29
   105      63      73
    90     105      96
```

**Problem 2**—Find the element-by-element product of **A** and **B** and set it equal to **D** (e.g., a11 * b11 = d11):

% Perform element-by-element multiplication to find D
D_matrix = A_matrix. * B_matrix

Note that the notation for element-by-element multiplication is . * . This same notation is used to do any element-by-element operations (e.g., . * , ./, .\, .^). The resulting matrix **D** is:

```
D_matrix  =
   9      0      4
  21     36     15
  15     54     63
```

# Example 17.3

MATLAB can be used to plot data points or functions. We will use both methods to plot a circle of unit radius.

**Method 1**—Plotting Data Points
First generate a vector of 100 equally spaced points between 0 and 2 * π using linspace. Note that the semicolon is used after each command line. This suppresses the echo feature on MATLAB so the contents of the vectors are not displayed on the screen:

>> angle=linspace(0,2 * pi,100);

Define vector of *x* and *y* to plot later:

>> x=sin(angle);
>> y=cos(angle);

Use the plot function to generate the plot:

>> plot(x,y);

Use the axis command to set the scales equal on the *x*- and *y*-axes. Otherwise you might produce an oval.

>> axis('equal');

Label the *x*-axis and *y*-axis and title, respectively.

>> xlabel('x-axis');
>> ylabel('y-axis');
>> title('Circle of Radius One');

**Method 2**—Plotting With a Function

First define the function. The name for this function is func_circle. You may choose any name, but be sure to avoid names that MATLAB already uses. Also note that the .^ is used for the exponent. Remember that MATLAB assumes everything is a matrix. The .^ function indicates that each element in the data should be raised to an exponent individually rather that using a matrix operation:

>> func_circle=('sqrt(1-x.^2)')

Plot the function using the command *fplot* instead of *plot* over the range −1 to 1:

>> fplot(func_circle,[−1,1])

This function is only positive and gives a semi-circle. To get the bottom half of the circle, define a second function that is negative:

>> func_neg=('-sqrt(1-x.^2)')

Activate the hold feature to plot the second function on the same plot:

>> hold on

Plot the negative function over the same range:

>> fplot(func_neg,[-1,1])

Make the axes equal in length to get a circle:

>> axis square

Label the x-axis and y-axis and title, respectively:

>> xlabel('x-axis');
>> ylabel('y-axis');
>> title('Circle of Radius One');

# Example 17.4

MATLAB can be used to perform functional analysis. For examples, find the two local minimums of the function $y = x\cos(x + 0.3)$ between $x = -4$ and $x = 4$.

**Method**—First, you need to create a function assignment. Then it is best to plot the function over the specified range because MATLAB's command for finding minimums (fmin) requires that you provide a range over which to look for the minimums:

% Assign the function to a variable name as a string
func_y = ëx. * cos(x + 0.3)í

% Plot the function over the entire range
fplot(func_y,[-4 4])

% Find each local minimum over the specified sub-ranges
x_min1 = fmin(func_y,-2,0)
x_min2 = fmin(func_y,2,4)
% Find the y values at the minimum locations. Note that MATLAB can only evaluate functions in terms
% of x.
x = [x_min1 x_min2]
y_mins = eval(func_y)

The MATLAB solution will yield:

x_min1 =
−1.0575
x_min2 =
3.1491
y_mins =
    −0.7683   −3.0014

# Example 17.5

MATLAB can also be used to find the roots or zeroes of a function. As with locating the minimums in the previous example, it is best to plot the function first.

Define the function of interest. In this case, we will find the zero of sin(x) near 3:

>> func_zero=('sin(x)')

Plot the function over the interval −5 to 5:

>> fplot(func_zero,[−5,5])

Use the *fzero* command to find the exact location of the zero. An initial guess must be provided (in this case it is 3.0):

>> fzero(func_zero,3)

MATLAB returns the answer:

ans =
3.1416

Note that *fzero* finds the location where the function crosses the *x*-axis. If you have a function that has a minimum or maximum at zero but does not cross the axis, it will not find it.

## Example 17.6

MATLAB can be used to write scripts to be used later. On a PC or Mac, select "New M-File" from the File menu. A new edit window should appear. On a Unix workstation, open an editor. In either case, type the following commands:

```
disp('This program calculates the square root of a number')
x=input('Please enter a number: ');
y=sqrt(x);
disp('The square root of your number is')
disp(y)
```

Save this file under the name "calc_sqrt". Note that on a PC or Mac, a ".m" extension will be added to your file name. Using Unix, you will need to add a ".m" to the end of your file name so MATLAB will recognize it as a script file. In MATLAB, type:

```
>> calc_sqrt
```

MATLAB should respond with the first two lines below. Assuming that "3" is entered and the Return key is pressed, MATLAB will display the last two lines:

```
This program calculates the square root of a number
Please enter a number: 3
The square root of your number is
    1.7321
```

If MATLAB did not recognize "calc_sqrt", check to see that it has a ".m" extension attached to the end of the file name. Alternatively, the file may not be in a directory or folder in which MATLAB is searching.

## PRESENTATION SOFTWARE

At regular intervals during the design process, presentations must be made to research groups and to management. Presentation programs allow a user to communicate ideas and facts through visuals. Presentation programs provide layout and template designs that help the user create slides.

### PowerPoint

Microsoft PowerPoint is a popular electronic presentation package. Figure 17.11 shows a slide view window of PowerPoint. As you have seen in other Microsoft

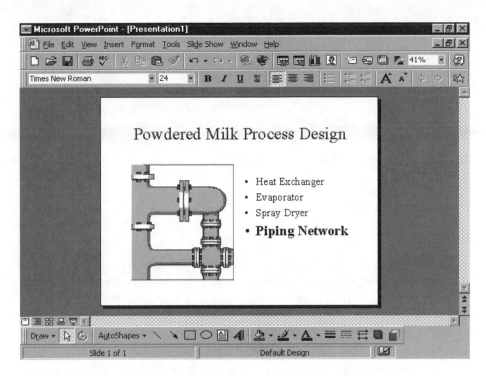

**Figure 17.11 Example of a PowerPoint "slide view" window.**

programs, the GUI for this package has a title bar, a menu bar, a toolbar, and a status bar. The unique feature of this tool is the drawing toolbar.

PowerPoint has five different viewing windows. The "slide view" window displays one slide at a time. Typically, this is the window through which slide editing is done. The "outline view" window lists the text for each slide in a presentation. The "slide sorter view" window graphically displays all the slides. As the name implies, slide sorting is done through this window. The "notes page view" allows the user to attach text with each slide. Since bullets are used on most slides, this feature comes in very handy. It allows the user to come back to a presentation file at a future date, and still remember what the presentation covered. The slides can be viewed sequentially using the "slide show view." This is the view used during a presentation.

*With the proper computer hardware, sound and video clips also can be incorporated into a presentation.*

Presentation packages have a number of features that help the user put together very professional presentations. PowerPoint has a gallery of templates. Templates specify color scheme and arrangement of elements (e.g., graphics and text) on the slide. Figure 17.11 shows a "slide view" with Clip Art. PowerPoint also has animation capabilities. With the proper computer hardware, sound and video clips also can be incorporated into a presentation.

## OPERATING SYSTEMS

An *operating system* is a group of programs that manage the operation of a computer. It is the interface between the computer hardware, the software, and the user. Printing, memory management, and file access are examples of processes that operating systems control. The most recognized operating systems are UNIX, originally from AT&T Bell Laboratories, and MS-DOS, from Microsoft Corporation. Microsoft's Windows family (95, 98, NT, etc.) are also operating systems, as is the Apple OS for Macintosh Computers.

## PROGRAMMING LANGUAGES

### Introduction

*Programming languages* enable users to communicate with the computer. A computer operates using machine language, which is binary code—a series of 0's and 1's (e.g., 01110101 01010101 01101001). Machine language was used primarily before 1953. Assembly language was developed to overcome the difficulties of writing binary code. It is a very cryptic symbolic code which corresponds one-to-one with machine language. FORTRAN, a higher-level language which more closely resembles English, emerged in 1957. Other higher-level languages quickly followed. However, regardless of the language the programmer uses, the computer only understands binary code. A programming language is an artificial language with its own grammar and syntax that allows the user to "talk" to the computer.

A program is written to tell the computer how to complete a specific task. A program consists of lines of code (instructions) written in a specific programming language. The program is then compiled or converted into machine language. If the compiler detects no syntax or flow control problems in the code, the user may then execute the code. Upon execution of the code, the computer performs the specified task. If the code is well written, the computer completes the task correctly. If you are unsatisfied with the computer's execution of the task, or the computer is unsatisfied with the user's code, you must debug (fix) the program and try again.

There are a number of programming languages in use. The following sections will highlight the most popular languages in the context of their classification.

### Procedural Languages

Procedural languages use a fetch-decode-execute paradigm (Impagliazzo and Nagin, 1995). Consider the following two lines of programming code:

$$value = 5$$
$$new\_value = log(value)$$

In the second line of code, the computer "fetches" the variable named value from memory, the log function is learned or "decoded", and then the function log(value) is "executed." FORTRAN, BASIC, Pascal, and C are all procedural languages.

The development of FORTRAN (Formula Translating System) was led by IBM employee John Backus. It quickly became popular with engineers, scientists, and mathematicians due to its execution efficiency. Three versions are in use today: FORTRAN 77, FORTRAN 90, and FORTRAN 95. Each new version has added functionality. The American National Standards Institute (ANSI), which facilitates the establishment of standards for programming languages, recognizes FOR-TRAN 77 and FORTRAN 90. Thousands of FORTRAN programs and routines written by experts are gathered in standard libraries such as IMSL (International Mathematics and Statistics Library) and NAG (Numerical Algorithm Group).

BASIC (Beginner's All-Purpose Symbolic Instruction Code) was created in 1964 at Dartmouth College by John Kemeny and Thomas Kurtz. Its purpose was to encourage more students to use computers by offering the user-friendly, easy-to-learn computing language.

Pascal is a general-purpose language developed for educational use in 1971. The developer was Prof. Nicklaus Wirth (Deitel and Deitel, 1998). Pascal, named for 17th century mathematician and philosopher Blaise Pascal, was designed for teaching structured programming. This language is not widely used in commercial, industrial, or government applications because it lacks features that would make it useful in those sectors.

Dennis Ritchie of AT&T Bell Laboratories developed the C programming language in 1972. It was an outgrowth of the BCPL (Basic Combined Programming Language) and B programming languages. Martin Richards developed BCPL in 1967 as a language for writing operating system software and compilers (Deitel and Deitel, 1998). Ken Thomas of AT&T Bell Laboratories modeled the B programming language after BCPL. B was used to create the early versions of UNIX. ANSI C programming language is now a common industrial programming language and is the developmental language used in creating operating systems and software such as word processors, database management tools, and graphics packages.

## Object-Oriented Languages

*Object-oriented languages* treat data, and routines that act on data, as single objects or data structures. Referring back to the programming code example, value and log are treated as one single object that, when called, automatically takes the log of value. Objects are essentially reusable software components. Objects allow developers to take a modular approach to software development. This means software can be built quickly, correctly, and economically (Deitel and Deitel, 1998).

C++ (C plus plus) is an object-oriented programming language developed by B. Stroustup at AT&T Bell Laboratories in 1985. It is an evolution of the C programming language. C++ added additional operators and functions that support the creation and use of data abstractions as well as object-oriented design and programming (Etter, 1997).

*There are many software tools designed to handle specific types of engineering problems.*

Java's principle designer was James Gosling of Sun Microsystems. The creators of Java were assigned the task of developing a language that would enable the creation of very small, fast, reliable, and transportable programs. The need for such a language arose from a need for software for consumer electronics such as TVs, VCRs, telephones, and pagers (Bell & Par, 1999). Such a language also makes it possible to write programs for the Internet and the Web. Two types of programs can be written in Java—applications and applets. Examples of application programs are word processors and spreadsheets (Deitel and Deitel, 1998). An *applet* is a program that is typically stored on a remote computer that users can connect to through the Web. When a browser sees that a Java applet is referred to in the HTML code for a Webpage, the applet is loaded, interpreted, and executed.

## ADVANCED ENGINEERING PACKAGES

The computer tools introduced so far are useful to all engineers. At one time or another, all engineers need to write papers, work with tabular data, and develop mathematical models. An engineer also needs to be able to solve particular classes of problems. For instance, a food process engineer may need to analyze and optimize processes. A mechanical engineer may need to perform finite element analysis. There are many software tools designed to handle specific types of engineering problems. Described here are just a few key packages that an engineer might encounter.

### Aspen Plus

Aspen Plus is software for the study of processes with continuous flow material and energy. These types of processes are encountered in the chemical, petroleum, food, pulp and paper, pharmaceutical, and biotechnology industries. Chemical engineers, food process engineers, and bioprocess engineers are typical users of Aspen Plus. This tool enables modeling of a process system through the development of a conceptual flowsheet (process diagram). The usual sequence of steps to design and analyze a process in Aspen Plus is:

- Define the problem
- Select a units system (SI or English)
- Specify the stream components (nitrogen, methane, steam, etc.)
- Specify the physical property models (Aspen Plus has an extensive property models library that is user-extendible)
- Select unit operation blocks (heat exchanger, distillation column, etc.)
- Define the composition and flow rates of feed streams
- Specify operating conditions (e.g., temperatures and pressures)
- Impose design specifications

- Perform a case study or sensitivity analysis on the entire process (i.e., see how varying the operating conditions affects the design)

  Software tools allow a single computer to take on the function of dozens of pieces of instrumentation with relative ease.

The output of Aspen Plus is a report that describes the entire process flowsheet. Overall material and energy balances, steam consumption, and unit operation design specifications are just a sampling of what might be included in the final output report.

## GIS Products

Civil engineers, agricultural engineers, and environmental and natural resource engineers work with geographic-referenced data. The Environmental Systems Research Institute, Inc. (ESRI) is the leader in software tool development for *geographical information systems* (GIS). The purpose of these tools is to link geographic information with spatial data. For example, GIS products can be used to study water and waste water management of a particular landmass (watershed). Specifically, ARC/INFO or ArcView (GIS products) could be used to link data sets concerning drainage and weather with a geographic location as specified by latitude and longitude. GIS products also enable visualization of the data through mapping capabilities.

## Virtual Instrumentation

Engineers in all disciplines of engineering have occasions that require them to acquire and analyze data. Modern software tools make it possible to transform a computer into almost any type of instrumentation required to perform these tasks. One such software tool is LabVIEW, created by National Instruments Corporation. This software allows a user to create a virtual instrument on their personal computer. LabVIEW is a graphical programming development environment based on the C programming language for data acquisition and control, data analysis, and data presentation. The engineer creates programs using a graphical interface. LabVIEW then assembles the computer instructions (program) that accomplishes the tasks. This gives the user the flexibility of a powerful programming language without the associated difficulty and complexity.

The software can be used to control data acquisition boards installed in a computer. These boards are used to collect the data the engineer would need. Other types of boards are used to send signals from the computer to other systems. An example of this would be in a control systems application. LabVIEW allows the engineer to quickly configure the data acquisition and control boards to perform the desired tasks.

The tools also can be used to make the computer function as a piece of instrumentation. The signals from the data acquisition boards can be displayed and analyzed like an oscilloscope or a filter. A sample window of a LabVIEW window

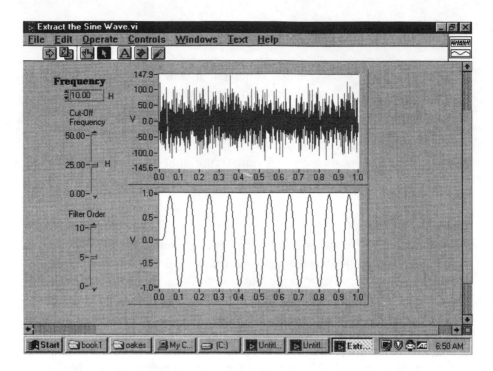

**Figure 17.12  Sample LabVIEW window.**

performing this function is shown in Figure 17.12. In this example, data were collected and processed through a filter to yield a sine wave.

Software tools such as LabVIEW allow a single computer to take on the function of dozens of pieces of instrumentation with relative ease. This reduces cost and improves the speed of data acquisition and analysis.

## Finite Element Tools

Engineers who work with mechanical or structural components include aeronautical, agricultural, biomedical, civil, and mechanical engineers. All must verify the integrity of the systems they design and maintain. Testing these systems on a computer is the most cost-effective and efficient method. Modern finite element tools allow engineers to do just that. One such commonly used tool is ANSYS, created by ANSYS, Inc.

*Finite element tools take the geometry of a system and break it into small elements or pieces.*

Finite element tools take the geometry of a system and break it into small elements or pieces. Each element is given properties and allowed to interact with the other elements in the model. When the elements are made small enough, very accurate simulations of the actual systems can be made.

The first step in using such a tool is to separate the physical component into a mesh. A sample mesh of centrifugal compressor vanes is shown in Figure 17.13.

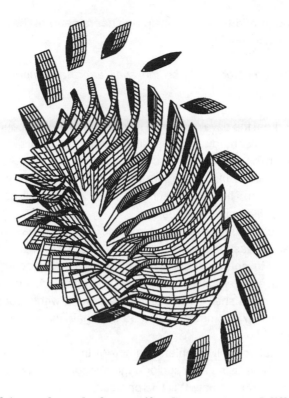

**Figure 17.13 A sample mesh of a centrifugal compressor and diffuser vanes.**

This mesh defines the physical boundaries of the elements. ANSYS contains aids to allow the user to create the mesh from a CAD drawing of the system. Other tools are also available to create the mesh to be used by ANSYS.

After the physical mesh is generated, the properties of the elements must be defined. The properties of the elements represent the material that the engineer is modeling. The elements can be solids or fluids and can change throughout the model. Part of the system may be one material, and another part may be different. Initial conditions such as temperatures and loads also must be defined. The engineer can use tools such as ANSYS to perform steady-state as well as transient analyses.

A final useful aspect of these tools is the post-processing capability. Tools such as ANSYS allow the engineer to examine the results graphically. The finite element model can be colored to show temperature, stress, or displacements. The engineer can then visually interpret the results, rather than searching through lists of numbers, which can highlight problems with a design.

## Other Tools

We would need an entire book, or more likely a series of books, to touch on all the computer tools engineers use today. There are different sets of computer tools within each discipline and each subspecialty of each discipline. Every day more tools are produced, and those existing are improved. New versions are constantly being developed to add features to help engineers perform their jobs more efficiently and more accurately.

> *The 1972 computer would be the snail and the computer of 2002 would be the jet.*

To a large degree, the development of new software tools is capitalizing on rapidly changing computer hardware capabilities. Microprocessors increase in speed according to Moore's Law, which states that microprocessor speed doubles every 18 months. That means that in the four and a half years from your senior year in high school to your first engineering job, processors will go through three doublings—they will be eight times faster.

This may not sound impressive, but consider what this has meant since the first microprocessor was developed in 1972. In 2002, thirty years after the first microprocessor was invented, 20 doublings will have taken place. The processor speed has increased over a million times. An analogy of how much faster this is would be to compare a snail inching along the ground with a commercial jet flying overhead. The 1972 computer would be the snail and the computer of 2002 would be the jet.

Considering that you will be entering a career that will last over 30 years, the computers you will be using as a senior engineer could be a million times faster than the machines you use now. Will microprocessors continue to advance at this pace? We don't know. We do know, however, that the computers of tomorrow will be much faster than the snails you are using now.

What this means in terms of computer tools is that they will continue to change. Experiment with as many tools as you can. Just as mechanics learn how to use an assortment of tools, so should engineers learn to use an assortment of computer tools. They also should always be ready to learn how to use new tools, because there always will be new tools to learn.

## REFERENCES

Andersen, P.K., et. al., *Essential C: an Introduction for Scientists & Engineers*, Saunders College Publishing, Fort Worth, 1995.

Bell, D. and M. Parr, *Java for Students*, 2nd Ed., Prentice-Hall Europe, London, 1998.

Chapman, S.J., *Introduction to FORTRAN 90/95*, WCB/McGraw-Hill, Boston, 1998.

Dean, C.L., *Teaching Materials to accompany PowerPoint 4.0 for Windows,"* McGraw-Hill, New York, 1995.

Deiter, H.M., and P.J. Deiter, *Java: How to Program*, 2nd Ed., Prentice-Hall, Upper Saddle River, 1998.

Etter, D.M., *Microsoft Excel for Engineers*, Addison-Wesley, Menlo Park, CA, 1995.

Etter, D.M., *Introduction to C++ for Engineers and Scientists*, Prentice-Hall, Upper Saddle River, 1997.

Glass, G. and K. Ables, *UNIX for Programmers and Users*, 2nd Ed., Prentice-Hall, Upper Saddle River, 1999.

Gottfried, B.S., *Spreadsheet Tools for Engineers: Excel 97 Version*, WCB/McGraw-Hill, Boston, 1998.

Grauer, R.T. and M. Barber, *Exploring Microsoft PowerPoint 97,* Prentice-Hall, Upper Saddle River, 1998.

Greenlaw, R. and E. Hepp, *Introduction to the Internet for Engineers*, WCB McGraw-Hill, Boston, 1999.

Hanselman, D. and B. Littlefield, *Mastering MATLAB 5: A Comprehensive Tutorial and Reference*, Prentice-Hall, Upper Saddle River, 1998.

Heal, K.M, Hansen, et. al., *Maple V Learning Guide*, Springer-Verlag, New York, 1998.

Impagliazzo, J. and P. Nagin, *Computer Science: A Breadth First Approach with C*, John Wiley & Sons, New York, 1995.

James, S.D., *Introduction to the Internet,* 2nd Ed., Prentice-Hall, Upper Saddle River, 1999.

Kincicky, D.C., *Introduction to Excel*, Prentice-Hall, Upper Saddle River, 1999.

Kincicky, D.C., *Introduction to Word*, Prentice-Hall, Upper Saddle River, 1999

King, J., *MathCad for Engineers*, Addison-Wesley, Menlo Park, 1995.

MathWorks, Inc., *The Student Edition of MATLAB, Version 4*, Prentice-Hall, Upper Saddle River, 1995.

Nyhoff, L.R. and S.C. Leestma, *FORTRAN 90 for Engineers and Scientists*, Prentice-Hall, Upper Saddle River, 1997.

Nyhoff, L.R. and S.C. Leestma, *Introduction to FORTRAN 90 for Engineers and Scientists,* Prentice-Hall, Upper Saddle River, 1997.

Palm, William J., *MATLAB for Engineering Applications,* WCB/McGraw-Hill, Boston, 1999.

Powell, T.A., *HTML: The Complete Reference*, Osbourne/McGraw-Hill, Berkeley, 1998.

Pritchard, P.J., *Mathcad: A Tool for Engineering Problem Solving*, WCB/McGraw-Hill, Boston, 1998.

Rathswohl, E.J., et. al., *Microcomputer Applications—Selected Edition—WordPerfect 6.1,* The Benjamin/Cummings Publishing Company, Redwood City, 1996.

Sorby, S.A., *Microsoft Word for Engineers*, Addison-Wesley, Menlo Park, 1996.

Wright, P.H., *Introduction to Engineering,* 2nd Ed., John Wiley, 1989.

## EXERCISES AND ACTIVITIES

1. Access your school's home page; surf the site and locate information about your department.

2. Perform a keyword search on an engineering major using two different web search engines.

3.  Determine if your school has on-line library services. If so, explore the service options.

4.  Search for the e-mail address of an engineering professional society.

5.  Find the web site of an engineering society. Prepare a one-page report on that organization based on the information found at its web site.

6.  Select the web site of the publisher of this text (see the copyright page). Prepare a critique of the website itself. What could be done to improve it? How does it compare with other web sites you have accessed? If you wish, you may send your suggestions to the publisher via e-mail.

7.  Prepare an oral presentation on the reliability or unreliability of information obtained from the Internet. Include techniques to give yourself confidence in the information.

8.  Prepare a written report on the reliability or unreliability of information obtained from the Internet. Include techniques to give yourself confidence in the information.

9.  Create a table using a word processor.

10.  Prepare a résumé using a word processor.

11.  Prepare two customized résumés using a word processor for two different companies you might be interested in working for someday.

12.  Use a spreadsheet to record your expenses for one month.

13.  Most spreadsheets have financial functions. Use these functions to answer the following question: If you now started saving $200 per month until you were 65, how much money would you have saved? Assume an annual interest rate of a) 4%, b) 6%, and c) 8%.

14.  Assume that you wish to accumulate the same amount of money on your 65th birthday as you had in the previous problem, but you do not start saving until your 40th birthday. How much would you need to save each month?

15.  A car you wish to purchase costs $25,000. Assume that you make a 10% down payment. Prepare a graph of monthly payments versus interest rate for a 48-month loan. Use a) 6%, b) 8%, c) 10%, and d) 12%.

16.  Repeat the previous problem, but include plots for a 36-month and 60-month loan.

17. Prepare a brief report on the software tools used in an engineering discipline.

18. Interview a practicing engineer and prepare an oral presentation or written report on the computer tools he or she uses to solve problems.

19. Select an engineering problem and prepare a brief report on which computer tools could be used in solving the problem. Which are the optimum tools and why? What other tools could be used?

20. Prepare a brief report on the kinds of situations for which engineers find the need to write original programs to solve problems in an engineering discipline. Include the languages most commonly used.

21. Prepare a presentation that speculates what computer tools you will have available to you in 5, 10, and 20 years.

# Chapter 18

# Electrical Circuits and Engineering Economics

## Introduction

We have selected several subjects to be reviewed in this chapter that you have undoubtedly studied in your Physics courses, along with a short presentation of Engineering Economics. They are the introductory topics in several engineering programs. You should be somewhat familiar with these subjects (with the possible exception of Engineering Economics), but there may be some material here that was either omitted in your Physics courses, or that was not included in the texts that you used. It is hoped that this short review may better prepare you for your engineering courses.

## ELECTRICAL CIRCUITS

### Circuits

Electric circuits are an interconnection of electrical components for the purpose of either generating and distributing electrical power; converting electrical power to some other useful form such as light, heat, or mechanical torque; or processing information contained in an electrical form (i.e., electrical signals). Most electrical circuits contain a source (or sources) of electrical power, passive components which store or dissipate energy, and possibly active components which change the electrical form of the energy or information being processed by the circuit.

Circuits may be classified as *Direct Current* (DC) circuits when the currents and voltages do not vary with time or as *Alternating Current* (AC) circuits when the currents and voltage vary sinusoidally with time. Both DC and AC circuits are said to be operating in the *steady state* when their current/voltage time variation is purely constant or purely sinusoidal with time. A *transient circuit* condition occurs when a switch is thrown that turns a source either on or off. This review will cover the DC steady-state. The primary quantities of interest in making circuit calculations are presented in Table 18.1.

**TABLE 18.1 Quantities Used in Electric Circuits**

| Quantity | Symbol | Unit | Defining Equation | Definition |
|----------|--------|------|-------------------|------------|
| Charge | $Q$ | coulomb | $Q = \int I dt$ | |
| Current | $I$ | ampere | $I = \dfrac{dQ}{dt}$ | Time rate of flow of charge past a point in the circuit. |
| Voltage | $V$ | volt | $V = \dfrac{dW}{dQ}$ | Energy per unit charge either gained or lost through a circuit element. |
| Energy | $W$ | joule | $W = \int V dQ = \int P dt$ | |
| Power | $P$ | watt | $P = \dfrac{dW}{dt} = IV$ | Power is the time rate of energy flow. |

## Circuit Components

The circuits reviewed in this section will contain one or more sources interconnected with passive components. These passive circuit components include resistors, inductors and capacitors.

   a. *Resistors* absorb energy and have a resistance value $R$ measured in ohms

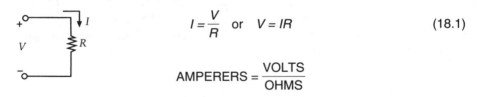

$$I = \frac{V}{R} \quad \text{or} \quad V = IR \qquad (18.1)$$

$$\text{AMPERERS} = \frac{\text{VOLTS}}{\text{OHMS}}$$

   b. *Inductors* store energy and have an inductance value $L$ measured in henries

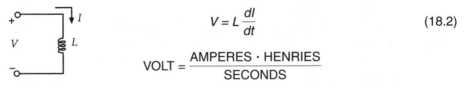

$$V = L\frac{dI}{dt} \qquad (18.2)$$

$$\text{VOLT} = \frac{\text{AMPERES} \cdot \text{HENRIES}}{\text{SECONDS}}$$

   c. *Capacitors* store energy and have a capacitance value $C$ measured in farads

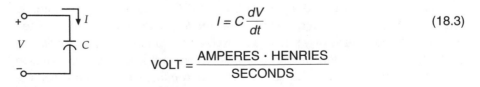

$$I = C\frac{dV}{dt} \qquad (18.3)$$

$$\text{VOLT} = \frac{\text{AMPERES} \cdot \text{HENRIES}}{\text{SECONDS}}$$

## Sources of Electrical Energy

Sources in electric circuits can be either independent of current and/or voltage values elsewhere in the circuit, or they can be dependent upon them. In this sec-

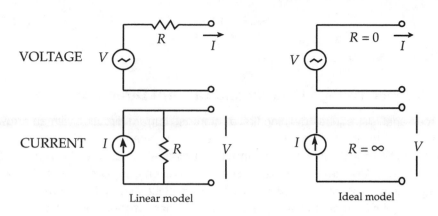

VOLTAGE     CURRENT

Linear model                    Ideal model

**Figure 18.1 Ideal and linear models of current and voltage sources.**

tion only independent sources will be considered. Fig. 18.1 shows both ideal and linear models for current and voltage sources.

## Kirchhoff's Laws

Two laws of conservation govern the behavior of all electrical circuits:

a. *Kirchhoff's Voltage Law* (KVL), for the conservation of energy, states that the sum of voltage rises or drops around any closed path in an electrical circuit must be zero:

$$\Sigma V_{DROPS} = 0 \quad \Sigma V_{RISES} = 0 \text{ (around close path)} \qquad (18.4)$$

b. *Kirchhoff's Current Law* (KCL), for the conservation of charge, states that the flow of charges either into (positive) or out of (negative) any node in a circuit must add to zero:

$$\Sigma I_{IN} = 0 \quad \Sigma I_{OUT} = 0 \text{ (at node)} \qquad (18.5)$$

## Ohm's Law

*Ohm's Law* is a statement of the relationship between the voltage across an electrical component and the current through the component. For DC circuits, where the components are resistors, Ohm's law is

$$V = IR \quad \text{or} \quad I = V/R \qquad (18.6)$$

For AC circuits, with resistors, capacitors and inductors, Ohm's law, stated in terms of the component impedance $Z$ (see Table 18.3), is

$$V = IZ \quad \text{or} \quad I = V/Z \tag{18.7}$$

where all variables can be complex.

## Reference Voltage Polarity and Current Direction

Circuit analysis requires defining first a reference current direction with an arrow placed next to the circuit component. For each of the components a reference current direction is arbitrarily defined. Once the current reference direction is defined, the voltage reference polarity marks can be placed on each component. The polarity marks on passive components are always placed so that the current flows from the plus (+) mark to the minus (-) mark, the passive sign convention.

Current values can be either positive or negative. A positive current value shows that the current does in fact flow in the reference direction. A negative current value shows that the current flows opposite to the reference direction. Voltage values can be either positive or negative. A positive voltage value indicates a loss of energy or reduction in voltage when moving through the circuit from the plus polarity mark to the minus polarity mark. A negative voltage value indicates a gain of energy when moving through the circuit from the plus polarity mark to the minus polarity mark.

It is proper electrical terminology to talk about voltage drops and voltage rises. A voltage drop is experienced when moving through the circuit from the plus (+) polarity mark to the minus (-) polarity mark. A voltage rise is experienced when moving through the circuit from the minus (-) polarity mark to the plus (+) polarity mark.

## Circuit Equations

When writing circuit equations, the current is assumed to have a positive value in the reference direction and the voltage is assumed to have a positive value as indicated by the polarity marks. To write the KVL circuit equation one must move around a closed path in the circuit and sum all the voltage rises or all the voltage drops. For example, for the circuit in Fig. 18.2 begin at point $a$ and move to $b$, then $c$, then $d$, and back to $a$. For $\sum V_{RISES} = 0$ obtain

$$V_s - IR_1 - IR_2 - IR_3 = 0 \tag{18.8}$$

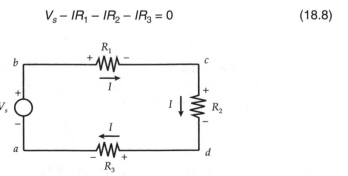

**Figure 18.2 A simple circuit.**

For $\sum V_{DROPS} = 0$ obtain

$$-V_s + IR_1 + IR_2 + IR_3 = 0 \tag{18.9}$$

Either of these equations can now be solved for the one unknown current

$$I = \frac{V_s}{R_1 + R_2 + R_3} = \frac{V_s}{R_{eq}} \tag{18.10}$$

where $R_{eq}$ is the equivalent resistance for the circuit.

## Circuit Equations Using Branch Currents

The circuit in Fig. 18.3 has meshes around which two voltage equations (KVL) can be written. There are, however, three branches in the circuit. An unknown current with a reference direction is assumed in each branch. The polarity marks are indicated for each resistor so the KVL can be written. Write two KVL equations, one around each mesh. Using  obtain:

$$\begin{aligned} -V_s + I_1R_1 + I_3R_2 + I_1R_3 &= 0 \\ -I_3R_2 + I_2R_4 + I_2R_5 + I_2R_6 &= 0 \end{aligned} \tag{18.11}$$

Write one KCL equation at circuit node $a$. This additional equation is necessary since there are three unknown currents:

$$I_1 - I_2 - I_3 = 0 \tag{18.12}$$

These three equations can be solved for $I_1$, $I_2$ and $I_3$. The current $I_1$ is

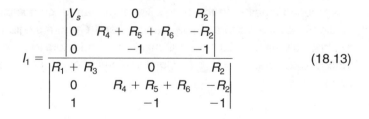

$$I_1 = \frac{\begin{vmatrix} V_s & 0 & R_2 \\ 0 & R_4 + R_5 + R_6 & -R_2 \\ 0 & -1 & -1 \end{vmatrix}}{\begin{vmatrix} R_1 + R_3 & 0 & R_2 \\ 0 & R_4 + R_5 + R_6 & -R_2 \\ 1 & -1 & -1 \end{vmatrix}} \tag{18.13}$$

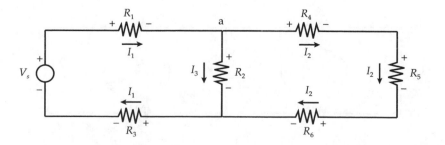

**Figure 18.3 A circuit with two meshes and three branches.**

## Circuit Equations Using Mesh Currents

A simplification in writing the circuit equations for Fig. 17.8 occurs if mesh currents are used. Note that

$$I_3 = I_1 - I_2 \tag{18.14}$$

Redefine the reference currents in the network of Fig. 18.3 as shown in Fig. 18.4. Now there are only two unknown currents to solve for instead of the three in Fig. 18.3. The current through $R_1$ and $R_3$ is $I_1$. The current through $R_4$, $R_5$ and $R_6$ is $I_2$. The current through $R_2$ is $I_1 - I_2$ which is consistent with the KCL equation written for the network of Fig. 18.4. Write two KVL equations, one around each mesh:

$$-V_s + I_1(R_1 + R_2 + R_3) - I_2R_2 = 0$$
$$-I_1R_2 + I_2(R_2 + R_4 + R_5 + R_6) = 0 \tag{18.15}$$

These two equations can be solved for $I_1$ and $I_2$. The current $I_1$ is equivalent to that of Eq. 18.13:

$$I_1 = \frac{\begin{vmatrix} V_s & -R_2 \\ 0 & R_2 + R_4 + R_5 + R_6 \end{vmatrix}}{\begin{vmatrix} R_1 + R_2 + R_3 & -R_2 \\ -R_2 & R_2 + R_4 + R_5 + R_6 \end{vmatrix}} \tag{18.16}$$

## Circuit Simplification

It is possible to simplify a circuit by combining components of the same kind that are grouped together in the circuit. The formulas for combining several $R$'s to a single $R$, several $L$'s to a single $L$ and several $C$'s to a single $C$ are found using Kirchhoff's laws (see Eq.18.10). Consider two inductors in series as shown in Fig. 18.5. We see that $L_{eq} = L_1 + L_2$. Combinations of circuit components are summarized in Table 18.2.

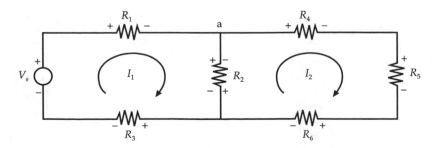

**Figure 18.4 A circuit with two meshes.**

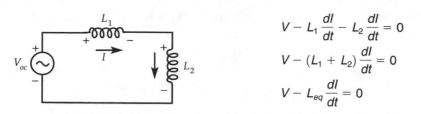

$$V - L_1 \frac{dI}{dt} - L_2 \frac{dI}{dt} = 0$$

$$V - (L_1 + L_2) \frac{dI}{dt} = 0$$

$$V - L_{eq} \frac{dI}{dt} = 0$$

**Figure 18.5 Circuit Simplification.**

## TABLE 18.2 Circuit Components in Series and Parallel

| Component | Series | Parallel |
|---|---|---|
| R | $R_{eq} = R_1 + R_2 + \cdots + R_N$ | $\frac{1}{R_{eq}} = \frac{1}{R_1} + \frac{1}{R_2} + \cdots + \frac{1}{R_N}$ |
| L | $L_{eq} = L_1 + L_2 + \cdots + L_N$ | $\frac{1}{L_{eq}} = \frac{1}{L_1} + \frac{1}{L_2} + \cdots + \frac{1}{L_N}$ |
| C | $\frac{1}{C_{eq}} = \frac{1}{C_1} + \frac{1}{C_2} + \cdots + \frac{1}{C_N}$ | $C_{eq} = C_1 + C_2 + \cdots + C_N$ |

## DC Circuits

In a DC circuit the only crucial components are resistors. Another component, the inductor, appears as a zero resistance connection (or short circuit) and a third component, a capacitor, appears as an infinite resistance, or open circuit. The three circuit components are summarized in Table 18.3.

## *Example 18.1*

Compute the current in the 10Ω resistor.

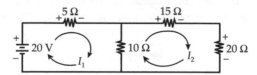

## TABLE 18.3 DC Circuit Components

| Component | | Impedance | Current | Power | Energy Stored |
|---|---|---|---|---|---|
| Resistor | | $R$ | $I = V/R$ | $P = I^2R = V^2/R$ | *None stored* |
| Inductor | | Zero (Short Circuit) | Unconstrained | None dissipated | $W_L = \frac{1}{2} LI^2$ |
| Capacitor | | Infinite (Open Circuit) | Zero | None dissipated | $W_C = \frac{1}{2} CV^2$ |

**Solution:** Assume loop currents $I_1$ and $I_2$. Write KVL around both meshes:

$$\Sigma V_{DROPS} = 0$$

Mesh 1:  $\quad -20 + 5I_1 + 10I_1 - 10I_2 = 0$

Mesh 2:  $\quad -10I_1 + 10I_2 + 15I_2 + 20I_2 = 0$

These are arranged as

$$15I_1 - 10I_2 = 20$$
$$-10I_1 + 45I_2 = 0$$

The solution is

$$I_1 = 1.57 \text{ A}, \qquad I_2 = 0.35 \text{ A}$$

The current in the $10\Omega$ resistor is

$$I = I_1 - I_2$$
$$= 1.57 - 0.35 = 1.22 \text{ A}$$

## Example 18.2

Compute the power delivered to the $6\Omega$ resistor.

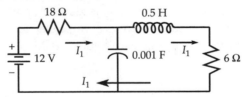

**Solution:** Only one current path exists since no DC current flows through a capacitor; the voltage drop across the inductor is zero. Write KVL:

$$\Sigma V_{DROPS} = 0$$
$$-12 + 18I_1 + 6I_1 = 0$$
$$I_1 = 0.5 \text{ A}$$

The power is then

$$P = I^2 R$$
$$= 0.5^2 \times 6 = 1.5 \text{ W}$$

## EXERCISES—ELECTRICAL CIRCUITS

1. For the circuit below, with voltages' polarities as shown, KVL in equation form is

a. $v_1 + v_2 + v_3 - v_4 + v_5 = 0$
b. $-v_1 + v_2 + v_3 - v_4 + v_5 = 0$
c. $v_1 + v_2 - v_3 - v_4 + v_5 = 0$
d. $-v_1 - v_2 - v_3 + v_4 + v_5 = 0$

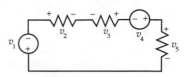

2. Find $I_1$ in amps.
   a. 12
   b. 15
   c. 18
   d. 21

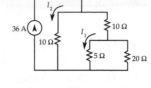

3. Find the magnitude and sign of the power, in watts, absorbed by the circuit element in the box.
   a. -20
   b. -8
   c. 8
   d. 12

4. For the circuit shown, the voltage across the 4 ohm resistor is, with $v = 1$ V
   a. 1/4
   b. 1/2
   c. 2/3
   d. 2

5. The total conductance, in mhos, in the circuit shown below is
   a. 1/5
   b. 1/2
   c. 2
   d. 5

6. The voltage across the 5 ohm resistor in the circuit shown is
   a. 1.0
   b. 2.5
   c. 3.0
   d. 5.83

7. The power delivered to the 5 ohm resistor is
   a. 1.5
   b. 2.15
   c. 2.85
   d. 3.2

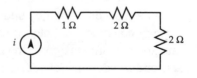

8. The power, in watts, absorbed by the 6 mho conductance in the circuit below is
   a. -0.24
   b. 0.2

c. 0.24

d. 0.48

9. The equivalent resistance, in ohms, between points *a* and *b* in the circuit below is

a) 3

b. 5

c. 7

d. 8

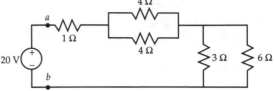

10. The voltage $V_2$ is

a. 6.4

b. 4.0

c. 2.0

d. 5.6

11. Find $I_1$ in amperes.

a. 4.0

b. 2.0

c. 4.11

d. 2.11

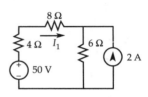

# ENGINEERING ECONOMICS

Engineering designs are intended to produce good results. In general, the good results are accompanied by undesirable effects including the costs of manufacturing or construction. Selecting the best design requires the engineer to anticipate and compare the good and bad outcomes. If outcomes are evaluated in dollars, and if "good" is defined as positive monetary value, then design decisions may be guided by the techniques known as *engineering economy*.

## Value and Interest

"Value" is not synonymous with "amount." The value of an amount of money depends on when the amount is received or spent. For example, the promise that you will be given a dollar one year from now is of less value to you than a dollar received today. The difference between the anticipated amount and its current value is called *interest* and is frequently expressed as a time rate. If an interest rate of 10% per year is used, the expectation of receiving $1.00 one year hence has a value now of about $0.91. In engineering economy, interest usually is stated in percent per year. If no time unit is given, "per year" is assumed.

## Example 18.3

What amount must be paid in two years to settle a current debt of $1,000 if the interest rate is 6%?

**Solution:** Value after one year = 1000 + 1000 × 0.06
$$= 1000(1 + 0.06)$$
$$= \$1060$$
Value after two years = 1060 + 1060 × 0.06
$$= 1000(1 + 0.06)^2$$
$$= \$1124$$

Hence, $1,124 must be paid in two years to settle the debt.

## Cash Flow Diagrams

As an aid to analysis and communication, an engineering economy problem may be represented graphically by a horizontal time axis and vertical vectors representing dollar amounts. The cash flow diagram for Example 18.3 is sketched in Fig. 18.6. Income is up and expenditures are down. It is important to pick a point of view and stick with it. For example, the vectors in Fig. 18.6 would have been reversed if the point of view of the lender had been adopted. It is a good idea to draw a cash flow diagram for every engineering economy problem that involves amounts occurring at different times.

In engineering economy, amounts are almost always assumed to occur at the ends of years. Consider, for example, the value today of the future operating expenses of a truck. The costs probably will be paid in varied amounts scattered throughout each year of operation, but for computational ease the expenses in each year are represented by their sum (computed without consideration of interest) occurring at the end of the year. The error introduced by neglecting interest for partial years is usually insignificant compared to uncertainties in the estimates of future amounts.

## Cash Flow Patterns

Engineering economy problems involve the following four patterns of cash flow, both separately and in combination. They are illustrated in Fig. 18.7.

*P*-pattern:  A single amount *P* occurring at the beginning of *n* years. *P* frequently represents "present" amounts.

*F*-pattern:  A single amount *F* occurring at the end of *n* years. *F* frequently represents "future" amounts.

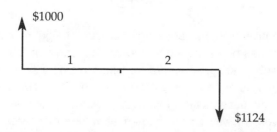

**Figure 18.6 Cash flow diagram for Example 18.3.**

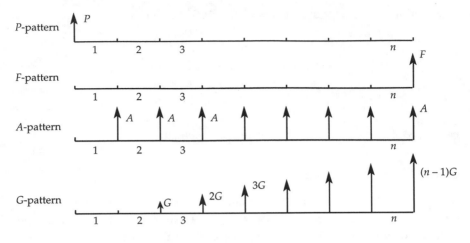

**Figure 18.7 Four cash flow patterns.**

*A*-pattern:   Equal amounts *A* occurring at the ends of each of *n* years. The *A*-pattern frequently is used to represent "annual" amounts.

*G*-pattern:   End-of-year amounts increasing by an equal annual gradient *G*. Note that the first amount occurs at the end of the second year. *G* is the abbreviation of "gradient."

## Equivalence of Cash Flow Patterns

Two cash flow patterns are said to be equivalent if they have the same value. Most of the computational effort in engineering economy problems is directed at finding a cash flow pattern that is equivalent to a combination of other patterns. Example 18.1 can be thought of as finding the amount in an *F*-pattern that is equivalent to $1,000 in a *P*-pattern. The two amounts are proportional, and the factor of proportionality is a function of interest rate *i* and number of periods *n*. There is a different factor of proportionality for each possible pair of the cash flow patterns defined in the next section. To minimize the possibility of selecting the wrong factor, mnemonic symbols are assigned to the factors. For Example 18.4, the proportionality factor is written  and solution is achieved by evaluating

$$F = (F/P)_n^i P.$$

To analysts familiar with the canceling operation of algebra, it is apparent that the correct factor has been chosen. However, the letters in the parentheses together with the sub- and super-scripts constitute a single symbol; therefore, the canceling operation is not actually performed. Table 18.4 lists symbols and formulas for commonly used factors. Table 18.5, located at the end of this chapter, presents a convenient way to find numerical values of interest factors. Those values are tabulated for selected interest rates *i* and number of interest periods

## TABLE 18.4 Formulas for Interest Factors

| Symbol | To Find | Given | Formula |
|--------|---------|-------|---------|
| $(F/P)_n^i$ | $F$ | $P$ | $(1 + i)^n$ |
| $(P/F)_n^i$ | $P$ | $F$ | $\dfrac{1}{(1 + i)^n}$ |
| $(A/P)_n^i$ | $A$ | $P$ | $\dfrac{i(1 + i)^n}{(1 + i)^n - 1}$ |
| $(P/A)_n^i$ | $P$ | $A$ | $\dfrac{(1 + i)^n - 1}{i(1 + i)^n}$ |
| $(A/F)_n^i$ | $A$ | $F$ | $\dfrac{i}{(1 + i)^n - 1}$ |
| $(F/A)_n^i$ | $F$ | $A$ | $\dfrac{(1 + i)^n - 1}{i}$ |
| $(A/G)_n^i$ | $A$ | $G$ | $\dfrac{1}{i} - \dfrac{n}{(1 + i)^n - 1}$ |
| $(F/G)_n^i$ | $F$ | $G$ | $\dfrac{1}{i}\left[\dfrac{(1 + i)^n - 1}{i} - n\right]$ |
| $(P/G)_n^i$ | $P$ | $G$ | $\dfrac{1}{i}\left[\dfrac{(1 + i)^n - 1}{i(1 + i)^n} - \dfrac{n}{(1 + i)^n}\right]$ |

$n$; linear interpolation for intermediate values of $i$ and $n$ is acceptable for most situations.

## Example 18.4

A new widget twister, with a life of six years, would save $2,000 in production costs each year. Using a 12% interest rate, determine the highest price that could be justified for the machine. Although the savings occur continuously throughout each year, follow the usual practice of lumping all amounts at the ends of years.

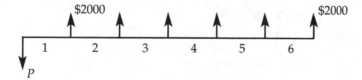

**Solution:** First, sketch the cash flow diagram.

The cash flow diagram indicates that an amount in a *P*-pattern must be found that is equivalent to $2,000 in an *A*-pattern. The corresponding equation is

$$P = (P/A)_n^i\, A$$
$$= (P/A)_6^{12\%} 2000$$

Table 18.5 is used to evaluate the interest factor for $i = 12\%$ and $n = 6$:

## Example 18.5

How soon does money double if it is invested at 8% interest?
   **Solution:** Obviously, this is stated as

$$F = 2P.$$

Therefore,

$$(F/P)_n^{8\%} = 2$$

In the 8% interest table, the tabulated value for ($F/P$) that is closest to 2 corresponds to $n = 9$ years.

## Example 18.6

Find the value in 1987 of a bond described as "Acme 8% of 2000" if the rate of return set by the market for similar bonds is 10%.
   **Solution:** The bond description means that the Acme company has an outstanding debt that it will repay in the year 2000. Until then, the company will pay out interest on that debt at the 8% rate. Unless otherwise stated, the principal amount of a single bond is $1,000. If it is assumed that the debt is due December 31, 2000, interest is paid every December 31, and the bond is purchased January 1, 1987, then the cash flow diagram, with unknown purchase price $P$, is:

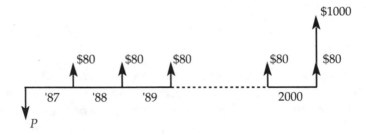

The corresponding equation is

$$P = (P/A)_{14}^{10\%} 80 (P/F)_{14}^{10\%} 1000$$
$$= 7.3667 \times 80 + 0.2633 \times 1000$$
$$= \$853$$

That is, to earn 10% the investor must buy the 8% bond for $853, a "discount" of $147. Conversely, if the market interest rate is less than the nominal rate of the bond, the buyer will pay a "premium" over $1,000.
   The solution is approximate because bonds usually pay interest semiannually, and $80 at the end of the year is not equivalent to $40 at the end of each half year. But the error is small and is neglected.

## Example 18.7

You are buying a new appliance. From past experience you estimate future repair costs as:

|  |  |
|---|---|
| First Year | $ 5 |
| Second Year | 15 |
| Third Year | 25 |
| Fourth Year | 35 |

The dealer offers to sell you a four-year warranty for $60. You require at least a 6% interest rate on your investments. Should you invest in the warranty?

**Solution:** Sketch the cash flow diagram.

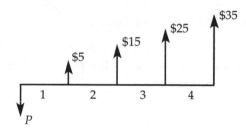

The known cash flows can be represented by superposition of a $5 *A*-pattern and a $10 *G*-pattern. Verify that statement by drawing the two patterns. Now it is clear why the standard *G*-pattern is defined to have the first cash flow at the end of the second year. Next, the equivalent amount *P* is computed:

$$P = (P/A)_4^{6\%} A + (P/G)_4^{6\%} G$$
$$= 3.4651 \times 5 + 4.9455 \times 10$$
$$= \$67$$

Since the warranty can be purchased for less then $67, the investment will earn a rate of return greater than the required 6%. Therefore, you should purchase the warranty .

If the required interest rate had been 12%, the decision would be reversed. This demonstrates the effect of a required interest rate on decision-making. Increasing the required rate reduces the number of acceptable investments.

## Example 18.8

Compute the annual equivalent maintenance costs over a 5-year life of a laser printer that is warranted for two years and has estimated maintenance costs of $100 annually. Use *i* = 10%.

**Solution:** The cash flow diagram appears as:

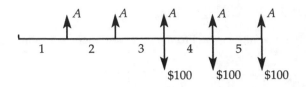

There are several ways to find the 5-year *A*-pattern equivalent to the given cash flow. One of the more efficient methods is to convert the given 3-year *A*-pattern to an *F*-pattern, and then find the 5-year *A*-pattern that is equivalent to that *F*-pattern. That is,

$$A = (A/F)_5^{10\%}(F/A)_3^{10\%}100$$
$$= \$54$$

## Unusual Cash Flows and Interest Periods

Occasionally an engineering economy problem will deviate from the year-end cash flow and annual compounding norm. The examples in this section demonstrate how to handle these situations.

### *Example 18.9*

PAYMENTS AT BEGINNINGS OF YEARS
    Using a 10% interest rate, find the future equivalent of:

**Solution:** Shift each payment forward one year. Therefore,

$$A = (F/P)_1^{10\%}100 = \$110$$

This converts the series to the equivalent *A*-pattern:

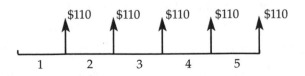

and the future equivalent is found to be

$$F = (F/A)_5^{10\%}110 = \$672$$

**Alternative Solution:** Convert to a six-year series:

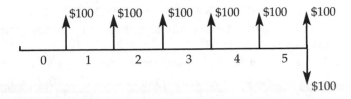

The future equivalent is

$$F = (F/A)_6^{10\%}100 - 100 = \$672$$

## Example 18.10

### ANNUAL PAYMENTS WITH INTEREST COMPOUNDED *m* TIMES PER YEAR

Compute the effective annual interest rate equivalent to 5% nominal annual interest compounded daily. (There are 365 days in a year.)

**Solution:** The legal definition of nominal annual interest is

$$i_n = mi$$

where $i$ is the interest rate per compounding period. For the example,

$$i = i_n/m$$
$$= 0.05/365 = 0.000137 \quad \text{or} \quad 0.0137\% \text{ per day}$$

Because of compounding, the effective annual rate is greater than the nominal rate. By equating $(F/P)$-factors for one year and $m$ periods, the effective annual rate may be computed as follows:

$$(1 + i_e)^1 = (1 + i)^m$$
$$i_e = (1 + i)^m - 1$$
$$= (1.000137)^{365} - 1 = 0.05127 \quad \text{or} \quad 5.127\%$$

## Example 18.11

### CONTINUOUS COMPOUNDING

Compute the effective annual interest rate $i_e$ equivalent to 5% nominal annual interest compounded continuously.

**Solution:** As $m$ approaches infinity, the value for $i_e$ is found as follows:

$$i_e = e^i - 1$$
$$= e^{0.05} - 1$$
$$= 0.051271 \quad \text{or} \quad 5.1271\%$$

## Example 18.12

### ANNUAL COMPOUNDING WITH *m* PAYMENTS PER YEAR

Compute the year-end amount equivalent to twelve end-of-month payments of $10 each. Annual interest rate is 6%.

**Solution:** The usual simplification in engineering economy is to assume that all payments occur at the end of the year, giving an answer of $120. This approximation may not be acceptable for a precise analysis of a financial agreement. In such cases, the agreement's policy on interest for partial periods must be investigated.

## Example 18.13

### ANNUAL COMPOUNDING WITH PAYMENT EVERY *m* YEARS

With interest at 10% compute the present equivalent of

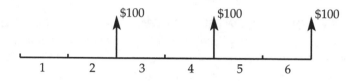

**Solution:** First convert each payment to an *A*-pattern for the *m* preceding years. That is,

$$A = (A/F)_2^{10\%}100$$
$$= \$47.62$$

Then, convert the *A*-pattern to a *P*-pattern:

$$P = (P/A)_6^{10\%}47.62$$
$$= \$207$$

## EXERCISES—ECONOMICS

1. Which of the following would be most difficult to monetize?
   a. maintenance cost
   b. selling price
   c. fuel cost
   d. prestige

2. If $1,000 is deposited in a savings account that pays 6% annual interest and all the interest is left in the account, what is the account balance after three years?
   a. $840
   b. $1,000
   c. $1,180
   d. $1,191

3. Your perfectly reliable friend, Frank, asks for a loan and promises to pay back $150 two years from now. If the minimum interest rate you will accept is 8%, what is the maximum amount you will loan him?
   a. $119
   b. $126
   c. $129
   d. $139

4. $12,000 is borrowed now at 12% interest. The first payment is $4000 and is made 3 years from now. The balance of the debt immediately after the payment is
   a. $4000
   b. $8000
   c. $12,000
   d. $12,860

5. An alumnus establishes a perpetual endowment fund to help Saint Louis University. What amount must be invested now to produce income of $100,000 one year from now and at one-year intervals forever? Interest rate is 8%.
   a. $8000
   b. $100,000
   c. $1,250,000
   d. $10,000,000

6. The annual amount of a series of payments to be made at the end of each of the next twelve years is $500. What is the present worth of the payments at 8% interest compounded annually?
   a. $500
   b. $3,768
   c. $6,000
   d. $6,480

7. Consider a prospective investment in a project having a first cost of $300,000, operating and maintenance costs of $35,000 per year, and an estimated net disposal value of $50,000 at the end of thirty years. Assume an interest rate of 8%.
   
   What is the present equivalent cost of the investment if the planning horizon is thirty years?
   a. $670,000
   b. $689,000
   c. $720,000
   d. $791,000
   
   If the project replacement will have the same first cost, life, salvage value, and operating and maintenance costs as the original, what is the capitalized cost of perpetual service?

a. $670,000
b. $689,000
c. $720,000
d. $765,000

8. Maintenance expenditures for a structure with a twenty-year life will come as periodic outlays of $1,000 at the end of the fifth year, $2,000 at the end of the tenth year, and $3,500 at the end of the fifteenth year. With interest at 10%, what is the equivalent uniform annual cost of maintenance for the twenty-year period?
   a. $200
   b. $262
   c. $300
   d. $325

9. An alumnus has given Michigan State University ten million dollars to build and operate a laboratory. Annual operating cost is estimated to be one hundred thousand dollars. The endowment will earn 6% interest. Assume an infinite life for the laboratory and determine how much money may be used for its construction.
   a. $5.00 × 10^6
   b. $8.33 × 10^6
   c. $8.72 × 10^6
   d. $9.90 × 10^6

10. An investment pays $6000 at the end of the first year, $4000 at the end of the second year, and $2000 at the end of the third year. Compute the present value of the investment if a 10% rate-of-return is required.
    a. $8333
    b. $9667
    c. $10,300
    d. $12,000

11. An amount $F$ is accumulated by investing a single amount $P$ for $n$ compounding periods with interest rate of $i$. Select the formula that relates $P$ to $F$.
    a. $P = F(1 + i)^{-n}$
    b. $P = F(1 + i)^{n}$
    c. $P = F(1 + n)^{-i}$
    d. $P = F(1 + ni)^{-1}$

12. At the end of each of the next ten years, a payment of $200 is due. At an interest rate of 6%, what is the present worth of the payments?
    a. $27
    b. $200
    c. $1472
    d. $2000

13. The purchase price of an instrument is $12,000 and its estimated mainte-
    nance costs are $500 for the first year, $1500 for the second and $2500 for
    the third year. After three years of use the instrument is replaced; it has no
    salvage value. Compute the present equivalent cost of the instrument using
    10% interest.
    a. $14,070
    b. $15,570
    c. $15,730
    d. $16,500

14. If an amount invested five years ago has doubled, what is the annual inter-
    est rate?
    a. 15%
    b. 12%
    c. 10%
    d. 6%

15. After a factory has been built near a stream, it is learned that the stream oc-
    casionally overflows its banks. A hydrologic study indicates that the proba-
    bility of flooding is about 1 in 8 in any one year. A flood would cause about
    $20,000 in damage to the factory. A levee can be constructed to prevent
    flood damage. Its cost will be $54,000 and its useful life is thirty years.
    Money can be borrowed at 8% interest. If the annual equivalent cost of the
    levee is less than the annual expectation of flood damage, the levee should
    be built. The annual expectation of flood damage is (1/8)20,000 = $2,500.
    Compute the annual equivalent cost of the levee.
    a. $1,261
    b. $1,800
    c. $4,320
    d. $4,800

16. If $10,000 is borrowed now at 6% interest, how much will remain to be paid
    after a $3,000 payment is made four years from now?
    a. $7,000
    b. $9,400
    c. $9,625
    d. $9,725

17. The maintenance costs associated with a machine are $2,000 per year for
    the first ten years, and $1,000 per year thereafter. The machine has an
    infinite life. If interest is 10%, what is the present worth of the annual
    disbursements?
    a. $16,145
    b. $19,678
    c. $21,300
    d. $92,136

18. A bank currently charges 10% interest compounded annually on business loans. If the bank were to change to continuous compounding, what would be the effective annual interest rate?

a. 10%

b. 10.517%

c. 12.5%

d. 12.649%

# TABLE 18.5 Interest Tables (Compound Interest Factors)

i = 6.00%

| n | (P/F) | (P/A) | (P/G) | (F/P) | (F/A) | (A/P) | (A/F) | (A/G) | n |
|---|-------|-------|-------|-------|-------|-------|-------|-------|---|
| 1 | .9434 | 0.9434 | −0.0000 | 1.0600 | 1.0000 | 1.0600 | 1.0000 | −0.0000 | 1 |
| 2 | .8900 | 1.8334 | 0.8900 | 1.1236 | 2.0600 | 0.5454 | 0.4854 | 0.4854 | 2 |
| 3 | .8396 | 2.6730 | 2.5692 | 1.1910 | 3.1836 | 0.3741 | 0.3141 | 0.9612 | 3 |
| 4 | .7921 | 3.4651 | 4.9455 | 1.2625 | 4.3746 | 0.2886 | 0.2286 | 1.4272 | 4 |
| 5 | .7473 | 4.2124 | 7.9345 | 1.3382 | 5.6371 | 0.2374 | 0.1774 | 1.8836 | 5 |
| 6 | .7050 | 4.9173 | 11.4594 | 1.4185 | 6.9753 | 0.2034 | 0.1434 | 2.3304 | 6 |
| 7 | .6651 | 5.5824 | 15.4497 | 1.5036 | 8.3938 | 0.1791 | 0.1191 | 2.7676 | 7 |
| 8 | .6274 | 6.2098 | 19.8416 | 1.5938 | 9.8975 | 0.1610 | 0.1010 | 3.1952 | 8 |
| 9 | .5919 | 6.8017 | 24.5768 | 1.6895 | 11.4913 | 0.1470 | 0.0870 | 3.6133 | 9 |
| 10 | .5584 | 7.3601 | 29.6023 | 1.7908 | 13.1808 | 0.1359 | 0.0759 | 4.0220 | 10 |
| 11 | .5268 | 7.8869 | 34.8702 | 1.8983 | 14.9716 | 0.1268 | 0.0668 | 4.4213 | 11 |
| 12 | .4970 | 8.3838 | 40.3369 | 2.0122 | 16.8699 | 0.1193 | 0.0593 | 4.8113 | 12 |
| 13 | .4688 | 8.8527 | 45.9629 | 2.1329 | 18.8821 | 0.1130 | 0.0530 | 5.1920 | 13 |
| 14 | .4423 | 9.2950 | 51.7128 | 2.2609 | 21.0151 | 0.1076 | 0.0476 | 5.5635 | 14 |
| 15 | .4173 | 9.7122 | 57.5546 | 2.3966 | 23.2760 | 0.1030 | 0.0430 | 5.9260 | 15 |
| 16 | .3936 | 10.1059 | 63.4592 | 2.5404 | 25.6725 | 0.0990 | 0.0390 | 6.2794 | 16 |
| 17 | .3714 | 10.4773 | 69.4011 | 2.6928 | 28.2129 | 0.0954 | 0.0354 | 6.6240 | 17 |
| 18 | .3503 | 10.8276 | 75.3569 | 2.8543 | 30.9057 | 0.0924 | 0.0324 | 6.9597 | 18 |
| 19 | .3305 | 11.1581 | 81.3062 | 3.0256 | 33.7600 | 0.0896 | 0.0296 | 7.2867 | 19 |
| 20 | .3118 | 11.4699 | 87.2304 | 3.2071 | 36.7856 | 0.0872 | 0.0272 | 7.6051 | 20 |
| 21 | .2942 | 11.7641 | 93.1136 | 3.3996 | 39.9927 | 0.0850 | 0.0250 | 7.9151 | 21 |
| 22 | .2775 | 12.0416 | 98.9412 | 3.6035 | 43.3923 | 0.0830 | 0.0230 | 8.2166 | 22 |
| 23 | .2618 | 12.3034 | 104.7007 | 3.8197 | 46.9958 | 0.0813 | 0.0213 | 8.5099 | 23 |
| 24 | .2470 | 12.5504 | 110.3812 | 4.0489 | 50.8156 | 0.0797 | 0.0197 | 1.87951 | 24 |
| 25 | .2330 | 12.7834 | 115.9732 | 4.2919 | 54.8645 | 0.0782 | 0.0182 | 9.0722 | 25 |
| 26 | .2198 | 13.0032 | 121.4684 | 4.5494 | 59.1564 | 0.0769 | 0.0169 | 9.3414 | 26 |
| 28 | .1956 | 13.4062 | 132.1420 | 5.1117 | 68.5281 | 0.0746 | 0.0146 | 9.8568 | 28 |
| 30 | .1741 | 13.7648 | 142.3588 | 5.7435 | 79.0582 | 0.0726 | 0.0126 | 10.3422 | 30 |
| ∞ | .0000 | 16.6667 | 277.7778 | ∞ | ∞ | 0.0600 | 0.0000 | 16.667 | ∞ |

i = 8.00%

| n | (P/F) | (P/A) | (P/G) | (F/P) | (F/A) | (A/P) | (A/F) | (A/G) | n |
|---|-------|-------|-------|-------|-------|-------|-------|-------|---|
| 1 | .9259 | 0.9259 | −0.0000 | 1.0800 | 1.0000 | 1.0800 | 1.0000 | −0.0000 | 1 |
| 2 | .8573 | 1.7833 | 0.8573 | 1.1664 | 2.0800 | 0.5608 | 0.4808 | 0.4808 | 2 |
| 3 | .7938 | 2.5771 | 2.4450 | 1.2597 | 3.2464 | 0.3880 | 0.3080 | 0.9487 | 3 |
| 4 | .7350 | 3.3121 | 4.6501 | 1.3605 | 4.5061 | 0.3019 | 0.2219 | 1.4040 | 4 |
| 5 | .6806 | 3.9927 | 7.3724 | 1.4693 | 5.8666 | 0.2505 | 0.1705 | 1.8465 | 5 |
| 6 | .6302 | 4.6229 | 10.5233 | 1.5869 | 7.3359 | 0.2163 | 0.1363 | 2.2763 | 6 |
| 7 | .5835 | 5.2064 | 14.0242 | 1.7138 | 8.9228 | 0.1921 | 0.1121 | 2.6937 | 7 |
| 8 | .5403 | 5.7466 | 17.8061 | 1.8509 | 10.6366 | 0.1740 | 0.0940 | 3.0985 | 8 |
| 9 | .5002 | 6.2469 | 21.8081 | 1.9990 | 12.4876 | 0.1601 | 0.0801 | 3.4910 | 9 |
| 10 | .4632 | 6.7101 | 25.9768 | 2.1589 | 14.4866 | 0.1490 | 0.0690 | 3.8713 | 10 |
| 11 | .4289 | 7.1390 | 30.2657 | 2.3316 | 16.6455 | 0.1401 | 0.0601 | 4.2395 | 11 |
| 12 | .3971 | 7.5361 | 34.6339 | 2.5182 | 18.9771 | 0.1327 | 0.0527 | 4.5957 | 12 |
| 13 | .3677 | 7.9038 | 39.0463 | 2.7196 | 21.4953 | 0.1265 | 0.0465 | 4.9402 | 13 |
| 14 | .3405 | 8.2442 | 43.4723 | 2.9372 | 24.2149 | 0.1213 | 0.0413 | 5.2731 | 14 |
| 15 | .3152 | 8.5595 | 47.8857 | 3.1722 | 27.1521 | 0.1168 | 0.0368 | 5.5945 | 15 |
| 16 | .2919 | 8.8514 | 52.2640 | 3.4259 | 30.3243 | 0.1130 | 0.0330 | 5.9046 | 16 |
| 17 | .2703 | 9.1216 | 56.5883 | 3.7000 | 33.7502 | 0.1096 | 0.0296 | 6.2037 | 17 |
| 18 | .2502 | 9.3719 | 60.8426 | 3.9960 | 37.4502 | 0.1067 | 0.0267 | 6.4920 | 18 |
| 19 | .2317 | 9.6036 | 65.0134 | 4.3157 | 41.4463 | 0.1041 | 0.0241 | 6.7697 | 19 |
| 20 | .2145 | 9.8181 | 69.0898 | 4.6610 | 45.7620 | 0.1019 | 0.0219 | 7.0369 | 20 |
| 21 | .1987 | 10.0168 | 73.0629 | 5.0338 | 50.4229 | 0.0998 | 0.0198 | 7.2940 | 21 |
| 22 | .1839 | 10.2007 | 76.9257 | 5.4365 | 55.4568 | 0.0980 | 0.0180 | 7.5412 | 22 |
| 23 | .1703 | 10.3711 | 80.6726 | 5.8715 | 60.8933 | 0.0964 | 0.0164 | 7.7786 | 23 |
| 24 | .1577 | 10.5288 | 84.2997 | 6.3412 | 66.7648 | 0.0950 | 0.0150 | 8.0066 | 24 |
| 25 | .1460 | 10.6748 | 87.8041 | 6.8485 | 73.1059 | 0.0937 | 0.0137 | 8.2254 | 25 |
| 26 | .1352 | 10.8100 | 91.1842 | 7.3964 | 79.9544 | 0.0925 | 0.0125 | 8.4352 | 26 |
| 28 | .1159 | 11.0511 | 97.5687 | 8.6271 | 95.3388 | 0.0905 | 0.0105 | 8.8289 | 28 |
| 30 | .0994 | 11.2578 | 103.4558 | 10.0627 | 113.2832 | 0.0888 | 0.0088 | 9.1897 | 30 |
| ∞ | .0000 | 12.500 | 156.2500 | ∞ | ∞ | 0.0800 | 0.0000 | 12.5000 | ∞ |

**TABLE 18.5 (*Continued*)**

*i* = 10.00%

| n | (P/F) | (P/A) | (P/G) | (F/P) | (F/A) | (A/P) | (A/F) | (A/G) | n |
|---|-------|-------|-------|-------|-------|-------|-------|-------|---|
| 1 | .9091 | 0.9091 | –0.0000 | 1.1000 | 1.0000 | 1.1000 | 1.0000 | –0.0000 | 1 |
| 2 | .8264 | 1.7355 | 0.8264 | 1.2100 | 2.1000 | 0.5762 | 0.4762 | 0.4762 | 2 |
| 3 | .7513 | 2.4869 | 2.3291 | 1.3310 | 3.3100 | 0.4021 | 0.3021 | 0.9366 | 3 |
| 4 | .6830 | 3.1699 | 4.3781 | 1.4641 | 4.6410 | 0.3155 | 0.2155 | 1.3812 | 4 |
| 5 | .6209 | 3.7908 | 6.8618 | 1.6105 | 6.1051 | 0.2638 | 0.1638 | 1.8101 | 5 |
| 6 | .5645 | 4.3553 | 9.6842 | 1.7716 | 7.7156 | 0.2296 | 0.1296 | 2.2236 | 6 |
| 7 | .5132 | 4.8684 | 12.7631 | 1.9487 | 9.4872 | 0.2054 | 0.1054 | 2.6216 | 7 |
| 8 | .4665 | 5.3349 | 16.0287 | 2.1436 | 11.4359 | 0.1874 | 0.0874 | 3.0045 | 8 |
| 9 | .4241 | 5.7590 | 19.4215 | 2.3579 | 13.5795 | 0.1736 | 0.0736 | 3.3724 | 9 |
| 10 | .3855 | 6.1446 | 22.8913 | 2.5937 | 15.9374 | 0.1627 | 0.0627 | 3.7255 | 10 |
| 11 | .3505 | 6.4951 | 26.3963 | 2.8531 | 18.5312 | 0.1540 | 0.0540 | 4.0641 | 11 |
| I2 | .3186 | 6.8137 | 29.9012 | 3.1384 | 21.3843 | 0.1468 | 0.0468 | 4.3884 | 12 |
| 13 | .2897 | 7.1034 | 33.3772 | 3.4523 | 24.5227 | 0.1408 | 0.0408 | 4.6988 | 13 |
| 14 | .2633 | 7.3667 | 36.8005 | 3.7975 | 27.9750 | 0.1357 | 0.0357 | 4.9955 | 14 |
| 15 | .2394 | 7.6061 | 40.1520 | 4.1772 | 31.7725 | 0.1315 | 0.0315 | 5.2789 | 15 |
| 16 | .2176 | 7.8237 | 43.4164 | 4.5950 | 35.9497 | 0.1278 | 0.0278 | 5.5493 | 16 |
| 17 | .1978 | 8.0216 | 46.5819 | 5.0545 | 40.5447 | 0.1247 | 0.0247 | 5.8071 | 17 |
| 18 | .1799 | 8.2014 | 49.6395 | 5.5599 | 45.5992 | 0.1219 | 0.0219 | 6.0526 | 18 |
| 19 | .1635 | 8.3649 | 52.5827 | 6.1159 | 51.1591 | 0.1195 | 0.0195 | 6.2861 | 19 |
| 20 | .1486 | 8.5136 | 55.4069 | 6.7275 | 57.2750 | 0.1175 | 0.0175 | 6.5081 | 20 |
| 21 | .1351 | 8.6487 | 58.1095 | 7.4002 | 64.0025 | 0.1156 | 0.0156 | 6.7189 | 21 |
| 22 | .1228 | 8.7715 | 60.6893 | 8.1403 | 71.4027 | 0.1140 | 0.0140 | 6.9189 | 22 |
| 23 | .1117 | 8.8832 | 63.1462 | 8.9543 | 79.5430 | 0.1126 | 0.0126 | 7.1085 | 23 |
| 24 | .1015 | 8.9847 | 65.4813 | 9.8497 | 88.4973 | 0.1113 | 0.0113 | 7.2881 | 24 |
| 25 | .0923 | 9.0770 | 67.6964 | 10.8347 | 98.3471 | 0.1102 | 0.0102 | 7.4580 | 25 |
| 26 | .0839 | 9.1609 | 69.7940 | 11.9182 | 109.1818 | 0.1092 | 0.0092 | 7.6186 | 26 |
| 28 | .0693 | 9.3066 | 73.6495 | 14.4210 | 134.2099 | 0.1075 | 0.0075 | 7.9137 | 28 |
| 30 | .0573 | 9.4269 | 77.0766 | 17.4494 | 164.4940 | 0.1061 | 0.0061 | 8.1762 | 30 |
| ∞ | .0000 | 10.0000 | 100.0000 | ∞ | ∞ | 0.1000 | 0.0000 | 10.0000 | ∞ |

*i* = 12.00%

| n | (P/F) | (P/A) | (P/G) | (F/P) | (F/A) | (A/P) | (A/F) | (A/G) | n |
|---|-------|-------|-------|-------|-------|-------|-------|-------|---|
| 1 | .8929 | 0.8929 | –0.0000 | 1.1200 | 1.0000 | 1.1200 | 1.0000 | –0.0000 | 1 |
| 2 | .7972 | 1.6901 | 0.7972 | 1.2544 | 2.1200 | 0.5917 | 0.4717 | 0.4717 | 2 |
| 3 | .7118 | 2.4018 | 2.2208 | 1.4049 | 3.3744 | 0.4163 | 0.2963 | 0.9246 | 3 |
| 4 | .6355 | 3.073 | 4.1273 | 1.5735 | 4.7793 | 0.3292 | 0.2092 | 1.3589 | 4 |
| 5 | .5674 | 3.6048 | 6.3970 | 1.7623 | 6.3528 | 0.2774 | 0.1574 | 1.7746 | 5 |
| 6 | .5066 | 4.1114 | 8.9302 | 1.9738 | 8.1152 | 0.2432 | 0.1232 | 2.1720 | 6 |
| 7 | .4523 | 4.5638 | 11.6443 | 2.2107 | 10.0890 | 0.2191 | 0.0991 | 2.5515 | 7 |
| 8 | .4039 | 4.9676 | 14.4714 | 2.4760 | 12.2997 | 0.2013 | 0.0813 | 2.9131 | 8 |
| 9 | .3606 | 5.3282 | 17.3563 | 2.7731 | 14.7757 | 0.1877 | 0.0677 | 3.2574 | 9 |
| 10 | .3220 | 5.6502 | 20.2541 | 3.1058 | 17.5487 | 0.1770 | 0.0570 | 3.5847 | 10 |
| 11 | .2875 | 5.9377 | 23.1288 | 3.4785 | 20.6546 | 0.1684 | 0.0484 | 3.8953 | 11 |
| 12 | .2567 | 6.1944 | 25.9523 | 3.8960 | 24.1331 | 0.1614 | 0.0414 | 4.1897 | 12 |
| 13 | .2292 | 6.4235 | 28.7024 | 4.3635 | 28.0291 | 0.1557 | 0.0357 | 4.4683 | 13 |
| 14 | .2046 | 6.6282 | 31.3624 | 4.8871 | 32.3926 | 0.1509 | 0.0309 | 4.7317 | 14 |
| 15 | .1827 | 6.8109 | 33.9202 | 5.4736 | 37.2797 | 0.1468 | 0.0268 | 4.9803 | 15 |
| 16 | .1631 | 6.9740 | 36.3670 | 6.1304 | 42.7533 | 0.1434 | 0.0234 | 5.2147 | 16 |
| 17 | .1456 | 7.1196 | 38.6973 | 6.8660 | 48.8837 | 0.1405 | 0.0205 | 5.4353 | 17 |
| 18 | .1300 | 7.2497 | 40.9080 | 7.6900 | 55.7497 | 0.1379 | 0.0179 | 5.6427 | 18 |
| 19 | .1161 | 7.3658 | 42.9979 | 8.6128 | 63.4397 | 0.1358 | 0.0158 | 5.8375 | 19 |
| 20 | .1037 | 7.4694 | 44.9676 | 9.6463 | 72.0524 | 0.1339 | 0.0139 | 6.0202 | 20 |
| 21 | .0926 | 7.5620 | 46.8188 | 10.8038 | 81.6987 | 0.1322 | 0.0122 | 6.1913 | 21 |
| 22 | .0826 | 7.6446 | 48.5543 | 12.1003 | 92.5026 | 0.1308 | 0.0108 | 6.3514 | 22 |
| 23 | .0738 | 7.7184 | 50.1776 | 13.5523 | 104.6029 | 0.1296 | 0.0096 | 6.5010 | 23 |
| 24 | .0659 | 7.7843 | 51.6929 | 15.1786 | 118.1552 | 0.1285 | 0.0085 | 6.6406 | 24 |
| 25 | .0588 | 7.8431 | 53.1046 | 17.0001 | 133.3339 | 0.1275 | 0.0075 | 6.7708 | 25 |
| 26 | .0525 | 7.8957 | 54.4177 | 19.0401 | 150.3339 | 0.1267 | 0.0067 | 6.8921 | 26 |
| 28 | .0419 | 7.9844 | 56.7674 | 23.8839 | 190.6989 | 0.1252 | 0.0052 | 7.1098 | 28 |
| 30 | .0334 | 8.0552 | 58.7821 | 29.9599 | 241.3327 | 0.1241 | 0.0041 | 7.2974 | 30 |
| ∞ | .0000 | 8.333 | 69.4444 | ∞ | ∞ | 0.1200 | 0.0000 | 8.3333 | ∞ |

*Chapter 19*

# Engineering Work Experience

## A Job and Experience

How do you get experience without a job, and how do you get a job without experience?" This is the job seeker's first, natural question. It was used in a national publicity campaign by the National Commission for Cooperative Education a number of years ago. Career-related experience, commonly called experiential learning, has become an important supplement to a student's engineering education. Throughout the years, and despite the changing times and levels of job markets, employers have continued to stress that they seek to fill their annual openings by selecting those candidates who have done outstanding things during their school years to distinguish themselves from their peers. In most cases, those students who have participated in some form of experiential learning have gained the types of skills and competencies that prospective employers and graduate schools are seeking.

An education has evolved over the past century to be quite a different experience from that of our predecessors. Since studying engineering has always been a rigorous experience, most students chose to devote their time and efforts to experiences in the classroom, library, or laboratory. Those who chose to work while going to school were viewed by the conventional student population as "non-traditional," representing the "working class" who were forced to work for pay solely as a means of affording their schooling. Now the rule, rather than the exception, is that most students work for pay in order to keep up with steadily costs.

For many students, the idea of combining their academic studies with a career-related employment experience is something they had not previously considered.

Consider the change in employment trends. The current generation of students was born, for the most part, in the mid- to late1980's. During their lifetime, this group has witnessed several major shifts firsthand within corporate America. During the years when many of these students were born, a major recession hit some of the largest manufacturing sectors. Especially hard hit was the automobile industry and the surrounding Midwest region of the country. Dubbed the Rust Belt, this area saw a significant reduction in the workforce, especially in the traditional blue collar jobs. Jobs that had traditionally been viewed as secure for a worker's lifetime had eroded and disappeared.

From 1983 to 1986, the U.S. economy revived and this was reflected by an increase in hiring of new engineers, with salaries well above the rate of inflation.

From 1988 to 1994, major restructuring within corporate America occurred with many major firms instituting a series of layoffs and downsizing. This period was notable for how the white collar workforce was affected. Professionals were either being laid off or were provided with early retirement incentives. The net effect was a significant restructuring, reflected by one of the weakest labor markets ever for engineers.

From 1994 to March of 2000, the economy rebounded. Increases in productivity, the general vitality of the global marketplace, mergers and acquisitions, and a lower birth rate which has produced one of the lowest supplies of graduating engineers in decades contributed to one of the strongest job markets.

From 2000 to the present, the economy experienced a bear market and a recession. A further collapse of the economy was attributed to the catastrophic terrorist attacks on U.S. soil on of September 11, 2001. Insult was added to injury with the war with Afghanistan, and then an immediate war with Iraq. All of these factors in an already poor economic environment did not help. The war with Iraq was short and the coalition suffered little loss of life. As a result, the market started rebounding as of the end of the official war with Iraq.

As for the future, we cannot control events, such as terrorist attacks and wars. A prudent individual follows a strategy for investing and planning. Human nature is to consume and as the world becomes more of a global economy things will improve drastically.

The current generation of college students has witnessed a roller-coaster economy. Many realize that the rules of the game have changed and that lifelong job security with a single employer is gone.

However, one fact remains clear from employers in good times and bad: The need for college graduates with experience is a constant. Many employers expect today's college graduates to have significant work experience. Why? Because engineering graduates with career-related work experience require less training and can produce results more quickly than the typical college graduate. To an employer, this is sound economic policy.

Many forms of experiential learning are available to engineering students. This chapter will discuss several of the most popular options, including summer work, volunteer experiences, academic internships, research assistantships, and perhaps the most popular and beneficial program for students, cooperative education. Each of these has its own advantages and disadvantages, depending on a student's particular situation and background. These issues will be examined as part of this chapter.

## Summer Jobs and Other Work Experience

Engineering employers prefer career-related work experience. Most students have some type of work experience. Some have mowed lawns, done baby-sitting, or worked in fast food franchises, while others may have had the opportunity to work in a business or industry that was involved in some type of engineering activity. Whatever the experience, it can be helpful in preparing you for your future. You have to start someplace in developing a work history.

Working in a printing company, delivering pizzas, or helping out at the local hospital has value. These experiences build a foundation, which your classroom learning and other career-related experiences will supplement. These jobs and summer work experiences may not seem relevant, but they can help develop important skills. Through such experiences, any students strengthen their communication skills, learn to work as part of a team, and develop problem-solving abilities and an organizational style that can be applied to engineering courses and future employment positions. Your student jobs will influence your perceptions about the things you like to do, the conditions and environments in which you like to do them, and the types of people and things with whom and which you like to work. These jobs can be useful in helping you identify your personal strengths and weaknesses, interests, personal priorities, and the degree of challenge you are willing to accept in a future position.

Real-world experience can come in many forms. While career-related experience is important, understand the importance of learning some basic on-the-job skills. If you are seeking to develop some basic job skills or merely earning some extra money to assist with your expenses, you should explore the many employment options available during the school year and in the summer.

## Volunteer or Community Service Experience

Another popular form of experiential learning is volunteer or community service experiences. While these programs are available to engineering students of all class levels, they can be particularly beneficial to young students who want to gain some practical experience. Generally, volunteer or community service projects can be short-term or long-term arrangements. Typically, these positions are with nonprofit organizations such as human service groups, educational institutions, local, state and federal government agencies, or small businesses. These groups are particularly interested in engineering students for their strong backgrounds in math, science, and computer skills, which can be applied to projects and activities.

Through these typically unpaid positions, students can gain significant experience and develop important skills. Students have found that working with different groups can enhance their communication abilities, as they are often interacting with people different from those in their school environment. Through the various projects and activities, students can develop organizational skills, initiative, and the desire for independent learning that can be useful in many settings including the classroom and other employment situations. Students involved in volunteer and community service experiences find these opportunities helpful with career decisions and build confidence with the satisfaction of helping others while providing needed services. For those students interested in developing a basic set of skills which can be useful throughout their professional engineering career, volunteer and community service work is worth exploring.

## Internships

What is an internship? Of all forms of experiential learning, this is the most difficult to describe as it can have different meanings, uses, and applications. Internships can be paid or unpaid work experience arranged for a set period. The internship usually occurs during the summer, which disrupts a student's academic schedule the least. For most students and employers, an internship is a one-time arrangement with no obligations by either party for future employment. Objectives and practices will vary depending on the employer and the student's school level. Some internships involve observing practicing engineers and professionals in the field. The intern's actual work often involves menial tasks designed to support the office. However, other internships can be structured as capstone experiences, which permit students to apply the principles and theory taught in school to some highly technical and challenging real-world engineering problems.

At most schools, internships are treated as an informal arrangement between the student and the employer. Therefore, an experienced professional usually supervises the student in a job situation, which places much responsibility for success or achievement. While many placement centers provide facilities and assist with arrangements for internship interviews, the instructor does not usually evaluate the internship job description. As a result, students have trouble understanding how, and if, this internship will be related to their engineering study. Since the typical engineering internship does not involve instructor input, school officials usually do not evaluate or monitor the experience. For these reasons, credit is usually not awarded for these experiences.

Many employers will use internship programs to screen candidates for possible full-time employment.

For many students, however, an internship can be a valuable part of their education. Under the best circumstances, the employer will structure the internship to be directly related to the student's study. The student should obtain, in advance, a description of the position's duties and responsibilities. The student may wish to review this material with an academic advisor or faculty member to gain an opinion on the relevance to the his or her field. These advisors may make suggestions and comments on how students can maximize the benefits of the experience. Students gain some practical real-world experience, providing them with a better perspective of the engineering field. For employers, internships provide a mechanism to supplement their workforce in order to complete many short-term engineering projects. It also enables employers to evaluate students in work-related situations for possible full-time employment.

For students with time constraints, curriculum inflexibility, scheduling difficulties, location or geographical preferences, the one-time internship program can supplement their engineering education. While internships do not provide the depth of experience found in other experiential learning, they can help students gain an appreciation for the engineering profession and perhaps provide additional opportunities for full-time employment.

## Cooperative Education

At most schools, cooperative education programs are viewed and promoted as the preferred form of experiential learning. The Cooperative Education Division of the American Society of Engineering Education states "Engineering cooperative education (co-op) programs integrate theory and practice by combining academic study with work experiences related to the student's academic program. Employing organizations are invited to participate as partners in the learning process and should provide experiences that are an extension of and complement to, classroom learning." This background statement describes the basic foundation of cooperative education and some of the critical components of a quality engineering co-op program.

As an academic program, a co-op is structured so student assignments directly relate to their major field of study. Typically, participating employers, who have been previously approved by the program administrators will forward detailed job descriptions, including salary information, to the co-op faculty and staff. These descriptions are evaluated for their correlation to the academic content and objectives. Employers also provide candidate qualifications to match the students with the available positions.

One component of the co-op programs is the integration of school and work. This integration is achieved by alternating periods of complex co-op assignments with periods of school study, usually with the same employer. Upon returning to school, the students then relate the classroom learning to the job experience. After another school semester, students return to the employers with a greater knowledge of engineering and scientific principles, which can be used for more complex workplace assignments. As students mature academically and professionally, they become stronger students and more valuable employees.

## Advantages of Cooperative Education Programs

### Advantages for Students

Co-op programs provide students with an employment advantage since many employers fill permanent positions with students with significant work experience. These employers use the co-op program to evaluate candidates on the job. For many students, this program can lead to an offer of permanent employment upon graduation. Students are under no obligation to accept full-time employment from their co-op employer, but studies show that many will accept these offers. Students who choose to work for their co-op employer begin with higher seniority and benefits, including larger starting salaries, increased vacation time, and other fringe benefits. Those students who elect to interview with other firms find that their co-op experience places them ahead of their peers.

The nature of co-op assignments allows students to use their classroom skills and apply them to engineering problems. In the classroom, student's assignments and laboratory experiments are graded based on how closely they come to achieving the correct answer. On the job, they learn the real-life assignments and projects are often not directed toward a specific answer but to potential solutions. This is where the student's technical background is developed and applied. Much of the actual technical learning occurs when students must apply principles to these situations. The co-op learning provides students with knowledge that will better prepare them for classroom challenges.

Many students who participate in cooperative education programs also find that the experiences help solidify career decisions. Many students are not fully aware of the available options and opportunities. By engaging in the daily engineering responsibilities, interacting with the technical and non-technical individuals who work with engineers, and then applying their academic background, students gain a better understanding of the engineering profession. As a result, many find that they confirm and strengthen their job selection. Others enjoy the exposure to these new fields and technologies of which they were not aware. Some students change their minds about specific disciplines to reflect their new interests, and others modify their course schedule to include courses to reflect their engineering exposure.

## *Advantages for Employers*

Many students wonder why employers become involved in co-op programs. Here are some of the most common reasons:

1. The cost of recruiting co-op students is less than recruiting new employees.
2. Almost 50% of co-op students accept permanent positions with their co-op employers.
3. Typically, co-op students receive lower salaries and fewer fringe benefits than permanent employees. Total wages average 40% less for co-op students. In addition, employers are not required to pay unemployment compensation taxes on wages of co-op students if they are enrolled in a qualified program.
4. Co-op programs provide an opportunity to evaluate employees prior to offering them full-time employment.
5. The co-op graduate's work performance is often superior to that of an employee without co-op experience. Co-op students are more flexible and more easily adapt to a professional environment.
6. Regular staff members are freed up from more rudimentary aspects of their jobs to focus on more complex or profitable assignments.
7. Co-op programs often supply students who have fresh ideas and approaches, and who bring expert technical knowledge to their work assignments.
8. Co-op graduates can be promoted sooner (and farther) than other graduates.
9. Co-op programs build positive relationships between businesses and schools, which in turn helps employers with their recruiting.

## *Advantages for Schools*

Here are many benefits to educational institutions which offer cooperative education programs to their students:

1. Cooperative work experiences provide for an extension of classroom experience, thus integrating theory and practice.
2. Cooperative education keeps instructors better informed of current business and industry trends.
3. Co-op programs build positive relationships between schools and businesses and provide instructors with knowledgeable people.
4. Co-op programs enhance the institution's reputation and attract students interested in the co-op plan.
5. Cooperative education provides schools with additional business and industry training facilities that would have been unaffordable.

## More Benefits of Co-op:

Co-op students have found that many professional development skills so critical for success in today's workplace can be enhanced through their co-op experiences.

1. **Written and oral communication skills**—The ability to communicate ideas, in writing and orally, is a critical skill. Most co-op assignments require students to document their findings and report them. This may be an e-mail to a supervisor, a technical memo, a presentation to engineers and technicians, or even a formal presentation to directors and executives. While communications and presentations may not be a favorite student activity, co-ops provide the opportunity to improve and develop these critical skills. For those who find this communication a liability, co-op assignments can help turn it into an asset.
2. **Networking**—Most co-op students find the opportunity to develop new contacts and work a great asset to their short-term and long-term career objectives. Co-op students meet many who can provide advice, lend project support, and help work more efficiently. These same people can recommend new assignments to co-op students and even write a letter of recommendation.
3. **Self-discipline**—Co-op students discover that the transition from school to work requires that they develop an organizational style superior to the one used in school. Within the co-op structure, they are given much responsibility and freedom to succeed with their projects. They learn time management, punctuality, adequate preparation, organization, and protocol and procedures. As they acquire and develop these skills, they will apply them in school, making them more productive and successful students.
4. **Interactions with a variety of people and groups**—Co-op students deal with many individuals as they complete their tasks, e.g., engineers, technicians, scientists, production workers, labor union representatives, clerical

staff, managers, directors, and CEOs. Each of these will contribute to the students' learning as they experience the successes, challenges, frustrations, and friendships that will evolve from these interactions. These individuals can help the co-op student evolve from a student to a successful engineer.

5. **Supervisory and management experience**—Many cooperative education assignments require that students take responsibility for other individuals or groups to accomplish their tasks. These opportunities to manage others' work can provide invaluable experience, especially for those students who aspire to management or leadership positions.

## A Note of Caution

As discussed above, co-op programs provide many positive benefits to students, employers, and schools. However, a co-op program may not be the appropriate form of experiential education all. Many aspects of the co-op structure may not be consistent with a student's overall goals and objectives.

The biggest drawback to co-op participation is the intermingling of work and study. Balance between work and school is critical when a student takes on this responsibility. Overextending oneself in any situation can lead to high stress, a lack of sleep, and a grade drop if too much time is spent on a co-op job. The overwhelming majority of co-op students, schools, and employers feel that this sacrifice is minor compared to the outstanding benefits received from this type of education.

Other factors should be considered. The relocation between school and the work is a minor inconvenience but one which takes time and effort if the two sites are far apart.

The engineering curricula also can present a difficulty due to the nature of the sequential offering of classes and their prerequisites. Students should work out a long-term schedule with their academic advisor and co-op coordinator to handle any class scheduling problems.

To find out more about co-op programs at your school, contact your academic advisor or a faculty member. If your school doesn't have a co-op program, you may want to write or talk with a corporation of interest to you.

## Which Is Best for You?

We have discussed many forms of experiential learning. Each of these opportunities has advantages and disadvantages. Only you can determine which program is best for you. Begin by asking yourself these questions:

1. Is obtaining career-related work experience a high priority in my educational planning?
2. Am I willing to sacrifice personal convenience to gain the best possible experience?

3. Will I be flexible in considering all available work opportunities, or do I have special personal circumstances (class schedules, time constraints, geographic or location restrictions) which limit my choices?

4. Do I have the drive and commitment necessary to succeed in a co-op program?

Your answers to such questions will be important as you move forward in your engineering career. Work with your academic advisor or an instructor to get as much information as possible to make an informed choice. You must take advantage of the available resources.

## Exercises and Activities

1. Develop a list of your skills and abilities. Determine which could be improved with work experience and which will be developed in the classroom.

2. Talk with other students who have done volunteer and community service projects. List the benefits they have gained which have helped them in their engineering classes.

3. Attend a career/job fair. Talk with at least three recruiters from different firms. Write a report comparing their views of co-op experiences, internships, and other experiential learning.

4. Visit with students who have been co-ops or interns. Write an essay that compares and contrasts their experiences. Which students do you feel gained the most from their experience and why?

5. Meet with your academic advisor and develop a long-range course schedule. Modify this plan so it shows how a co-op and an internship would affect this schedule. Write a short paper discussing the advantages and disadvantages of extending your education to include extra time for work experience.

6. Develop a list of your personal career priorities. How can you enhance this list with experiential learning?

7. Develop a list of your personal and educational abilities that need to be strengthened. Write an essay that discusses how work experience could improve your skills.